VOLUME ONE

METHODS IN CHEMICAL PROCESS SAFETY

Edited by

FAISAL KHAN
Centre for Risk, Integrity and Safety Engineering (C-RISE)
Faculty of Engineering & Applied Science
Memorial University
St John's, NL, A1B 3X5
Canada

ELSEVIER

ACADEMIC PRESS
An imprint of Elsevier

Academic Press is an imprint of Elsevier
50 Hampshire Street, 5th Floor, Cambridge, MA 02139, United States
525 B Street, Suite 1800, San Diego, CA 92101–4495, United States
The Boulevard, Langford Lane, Kidlington, Oxford OX5 1GB, United Kingdom
125 London Wall, London, EC2Y 5AS, United Kingdom

First edition 2017

ISBN: 978-0-12-811547-3
ISSN: 2468-6514

For information on all Academic Press publications
visit our website at https://www.elsevier.com/books-and-journals

Publisher: Zoe Kruze
Acquisition Editor: Poppy Garraway
Editorial Project Manager: Shellie Bryant
Production Project Manager: Surya Narayanan Jayachandran
Senior Cover Designer: Miles Hitchen

VOLUME ONE

METHODS IN CHEMICAL PROCESS SAFETY

CONTENTS

CONTRIBUTORS

Paul R. Amyotte
Dalhousie University, Halifax, NS, Canada

Jeffrey E. Arbogast
American Air Liquide, Newark, DE, United States

Daryl Attwood
Lloyd's Register EMEA, London, UK

Seyed J. Hashemi
Centre for Risk, Integrity and Safety Engineering (C-RISE), Faculty of Engineering & Applied Science, Memorial University of Newfoundland, St. John's, NL, Canada

Faisal Khan
Centre for Risk, Integrity and Safety Engineering (C-RISE), Faculty of Engineering & Applied Science, Memorial University of Newfoundland, St. John's, NL, Canada

Cathleen S. Lupien
Dalhousie University, Halifax, NS, Canada

Kathryn Mearns
Human Factors Consultant

Ian Moskowitz
American Air Liquide, Newark, DE, United States

Ulku G. Oktem
University of Pennsylvania; Near-Miss Management LLC, Philadelphia, PA, United States

Ankur Pariyani
Near-Miss Management LLC, Philadelphia, PA, United States

Howard Pike
Centre for Risk, Integrity and Safety Engineering (C-RISE), Faculty of Engineering, Memorial University, St. John's, NL, Canada

Warren D. Seider
University of Pennsylvania, Philadelphia, PA, United States

Masoud Soroush
Drexel University, Philadelphia, PA, United States

PREFACE

Chemical Process Safety is an area of growing importance within the chemical, oil and gas, and allied industries. Worldwide, the chemical sector alone represents a $5 trillion industry[a] and directly employs millions. The chemical sector is a fast-developing industry due to the growing dependence of society on energy resources and the rapid development of exploration and production technologies. The need for safe, well-managed processes within the industry has been highlighted by several recent disasters including the Deepwater Horizon oil spill in 2010 (caused by poor safety systems and cost cutting measures, according to a White House report) and the Tianjin disaster in China in 2015. These disasters not only cost human life but had a huge impact on the local environment and represent major financial losses for the companies involved.

The chemical, oil and gas, and allied industries are inherently risk-laden sectors. The continued occurrence of major process incidents has increased the awareness within the process industry about the importance of making development and operational decisions based on a thorough assessment of the associated risks to identify measures that can be taken to prevent potential losses. This increased awareness has shaped and influenced process safety science. Chemical process safety is a rapidly evolving area and is moving to more dynamic and adaptive methods of design and management to improve health and safety across the industry.

This book is the first volume of the *Methods in Chemical Process Safety* book series. This book series intends to be a one-stop resource for both academic researchers and professional practitioners. It aims to publish fundamentals of process safety science leading state-of-the-art advances occurring in the field while maintaining a practical approach for their application to the industries. An international editorial board and authorship ensures that this book series depicts the latest research developments from around the globe. Each volume will cover fully commissioned methods across the field of process safety, risk assessment and management, and loss prevention. This first volume discusses the Fundamentals of Process Safety from a practical perspective to make the book applicable for practitioners working within the industry.

[a] http://www.statista.com/statistics/302081/revenue-of-global-chemical-industry/.

This volume presents six chapters. Chapter 1 provides an overview of process safety regulatory and technological evolutions. It also provides examples of different Methods in Chemical Process Safety, including methods to identify process hazards and to implement, measure, evaluate, monitor, and manage safety of hazardous processes. Chapter 2 reviews the process safety incidents in the last decades and highlights the importance of learning from both major incidents and near misses. Main elements and foundational blocks of process safety are discussed in Chapter 3. This chapter helps safety practitioners to design and implement elements, which are required to optimize process safety management performance, efficiency, and effectiveness. Human error is frequently used to describe a cause of losses. Chapter 4 discusses the role of human factor in process safety and the importance of improving safety system designs rather than focusing on human failure as the root cause of accidents. Chapter 5 introduces the concept of risk-based process safety and the importance of considering both probability and consequences of process safety incidents in decision making. Finally, the process safety regulatory context is discussed in Chapter 6.

I am indebted to all members of the editorial and the contributory authors; this book could not have been published without their dedication, time, and commitment. On behalf of everyone who contributed in this volume, I hope that this book contributes to a safer future by serving as a source of knowledge in the field of chemical process safety. It is personally a great pleasure for me to bring together experts and compiling their contribution. I am imperfect and still learning and improving, I sincerely apology in advance for potential errors and misses in this volume. I encourage readers to share them with me for my self-learning and also to serve the community better. I look forward to learning from your feedback.

Faisal Khan
Memorial University of Newfoundland, St. John's, NL, Canada

CHAPTER ONE

Introduction

Faisal Khan[1], Seyed J. Hashemi
Centre for Risk, Integrity and Safety Engineering (C-RISE), Faculty of Engineering & Applied Science, Memorial University of Newfoundland, St. John's, NL, Canada
[1]Corresponding author: e-mail address: fikhan@mun.ca

Contents

1. BACKGROUND

1.1 Major Accidents Are Still Occurring

The growing dependence of society on energy resources has resulted in extensive exploration of hydrocarbon resources and the rapid development of the process industry. But, has process safety technology developed proportionally to the growth of the process industry? This needs to be investigated but what is clear is that major accidents are still occurring. Does this mean that we do not know what is the right thing to do? Or, do we know

Methods in Chemical Process Safety, Volume 1
ISSN 2468-6514
http://dx.doi.org/10.1016/bs.mcps.2017.01.001

but have failed to act? Considering the alarming rate of the continued occurrence of major accidents in the oil and gas industry, the answer to both questions is "Yes," at least partially. Until we can answer "No" to these questions with confidence, if the future is similar to the past, unfortunately we should expect more accidents. For those who have been in the industry for very long, this is a given assumption.

1.2 People or Systems? Where Does the Blame Lie?

The answer to this question seems obvious. Crowl and Louvar reviewed the causes of the largest hydrocarbon and chemical plant accidents from Marsh's 100 largest losses report in the period from 1972 to 2001 and came to this conclusion: "Human error is frequently used to describe a cause of losses. Almost all accidents, except those caused by natural hazards, can be attributed to human error. For instance, mechanical failures could all be due to human error as a result of improper maintenance or inspection" (Crowl & Louvar, 2011). This conclusion is also aligned with Kletz's statement in his work titled "Still Going Wrong!" that: "Missing from this book is a chapter on human error. This is because all accidents are due to human error" (Kletz, 2004).

In 2005, an explosion at BP's Texas City refinery claimed 15 lives and caused much more injury and destruction. The company's vice president of North American refining testified in 2007 that: "Our people did not follow their start-up procedures [...] If they'd followed the start-up procedures, we wouldn't have had this accident" (Calkins & Fisk, 2007). Later, when it was found that the equipment was substandard, the company questioned managerial decisions to use it. Examples such as these are familiar in the field of professional safety, where expert investigators, managers, and the public respond to accidents by pointing to a "Bad Apple" tendency, and focusing on human failure as the root cause of accidents (Holden, 2009). Accordingly, the oil and gas industry has focused on making progress in process safety by protecting the system from unreliable employees/workers through selection, procedures, automation, training, and discipline.

However, it appears that outside the oil and gas industry, the situation is different, for example, in the aviation and nuclear energy sectors. Studies such as those by Dekker (2001) and Holden (2009) have referred to the "Bad Apple" approach of safety management (in industries such as oil and gas) as the "old view," according to which one "identifies bad apples (unreliable human components) somewhere in an organization, and gets rid of them or somehow constrains their activities" (Dekker, 2001). In contrast to the "old view" that posits human error as the cause of many accidents, the "new view" considers human error as a symptom of more complicated systematic issues (Holden, 2009). Perhaps this focus on flawed systems rather than human is one of the reasons behind the significantly fewer major accidents in the aviation and nuclear industries. Of course, humans, the creators of the systems, are involved in accidents, but they are not necessarily the sole or primary cause of losses. There is an emerging need for modern safety professionals to "reinvent" the understanding of human error more holistically by tracing the connections between human error and the tools through system-centered solutions. More discussions on the role of human factor in process safety are provided in the chapter "Role of Human Factor in Process Safety" by Mearns.

1.3 Learning From the Experience

Unfortunately, there is no shortage of new accident reports. Accident investigation reports are influential documents in the growth of process safety science. They are very beneficial in the light of what the industry currently knows—or assumes—about the nature of accidents (Lundberg, Rollenhagen, & Hollnagel, 2009). However, time, distance, and cultural challenges such as litigation, fear of adverse publicity, internal procedure, and disclosure of confidential information may influence how well the industry learns from its previous mistakes (Kletz, 2004). Usually, only those incidents that have had catastrophic consequences are publicized and used for developing new safety barriers. Moreover, some of the incident reports still describe only the immediate technical causes, failing to investigate the accident from different aspects.

The scope of improving process safety should also include learning from near misses. Near misses are symptoms of underlying process issues and provide valuable information to understand how systems work (Dahle et al., 2012). Learning from accidents and near misses requires a system-thinking approach to assess the interdependence of people, technology, and organizations rather than considering these elements in isolation (Dahle et al., 2012; Tjorhom & Aase, 2010; Wiig & Aase, 2007). The main assumption in system-thinking approach is that accidents do not occur solely because of incompetent operators, wrong procedures, poor techniques, faulty processes, or organizational failures. Instead, it is the combination of correlated, multicausal variables that interact to create the conditions in which accidents may occur. Failure to consider all aspects of organizational, cultural, technological, and human factors in accident investigation has usually resulted in the identification of lack of competence, experience, and risk awareness of operators as the primary cause of accidents (Dahle et al., 2012).

According to Lundberg et al. (2009), the causes found during investigations reflect the assumptions in the accident model following the principle of "What you look for is what you find." Moreover, the identified causes typically become specific problems to be solved during implementation of remedial actions, which follows the principle of "What you find is what you fix" (Lundberg, Rollenhagen, & Hollnagel, 2010). Therefore, the learning process from these experiences should focus on a bigger picture of causes of the wide range of accidents, using a holistic approach that includes all factors involved in accidents.

Another important issue relates to our ability—or lack of ability—to learn from positive results. There are several good examples of where the industry does appear to have learnt from incidents and has made improvements on a global basis (Marsh, 2016). However, in addition to focusing on "what went wrong," the oil and gas industry can benefit from asking "what went well," perhaps by examining the aviation, nuclear, and healthcare sectors. The chapter "Learning From the Experience" by Mannan is devoted to learning from the past success and failure experiences.

1.4 Are Major Accidents Black Swans?

"Black Swan" events are extreme and rare events, and in practice impossible to anticipate. Understanding and evaluating the potential exposure to "black swan" events has been a topic of discussion in recent years, particularly in the actuarial industry (Taleb, 2007). As expressed in Marsh's 100 Largest Losses report: "none of the losses listed in this document should be considered black swan events" (Marsh, 2016). Blowouts in drilling and well operations, flooding in distillation processes, runaway conditions in reactor systems, and other extremely dangerous process conditions are all inherent hazards of the process industry and are foreseeable. The disastrous Deepwater Horizon incident in the Gulf of Mexico in 2010, the largest oil spill in oil and gas history, was also not a "black swan" event. "On the day of the April 20 tragedy, no effective safeguards were in place to eliminate or minimize the consequences of a process safety incident" (CSB, 2014b).

Effective process safety management (PSM) should be able to identify safety and security issues that threaten process safety and include appropriate safety barriers and mitigation measures to prevent serious losses. For example, an attack on security (such as a cyber-attack) along with failure of all protection layers in place should not be considered as a "black swan" event. The identification of new and developing threats and forming strategies to prevent and mitigate their impact are integral parts of a continuously improved PSM system.

1.5 Low Oil Prices and Process Safety

There is a concern, from the process safety point of view, that the oil and gas industry will implement cost-cutting strategies to counteract low oil prices, disregarding the fact that lowering safety standards increases not only the possibility of human and environmental harm but also financial loss. Other than the cancellation of new projects and hiring freezes, these strategies unfortunately may include risk control measures, reductions, fewer investments in inspection and maintenance activities, and training cuts. According to research by Marsh, different periods of price falls have been followed by a significant increase in the total values of upstream losses over the past 40 years (Marsh, 2016). Although this correlation between oil price and losses does not imply their causal relationship (Marsh, 2016), it should serve as a reminder that chemical companies and refineries need to continue to invest in process safety and preventive maintenance, even as the economic downturn cuts into sales and profits (Bresland, 2008).

It is understood that the aging infrastructures, limited resources, and increasing production demands, along with fluctuating oil prices, have made investment decision making a challenge for energy companies. Nevertheless, decision makers, if they value safety, sustainability, and profit, should holistically look into this challenge and seek for a balance among cost, risk, and value of cost-saving initiatives over the asset's entire life cycle (ISO, 2014). As famously declared by Dr. Trevor Kletz: "downturns and recessions can actually be a good time to take care of deferred maintenance." This is because of the reduced financial impact from temporarily shutting down a process during periods when sales are depressed (Bresland, 2008). Cost saving decisions should be assessed carefully to ensure that all risks introduced to the system are measured and mitigated effectively and also to evaluate their long-term value and impact.

2. OVERVIEW OF PROCESS SAFETY

2.1 What Is Process Safety?

The oil and gas industry is an inherently risk-laden sector. The extraction, transportation, and processing of hydrocarbons involve complex operations with high pressure, reactive chemicals, and complex chemistry. Increasing energy demand and related business opportunities have been the main drivers to push the operators to take higher risks by exploring deeper waters and more remote locations with extreme climates. The results have been larger facilities, more complex processes, and compact facility designs.

The continued occurrence of major process incidents has increased the awareness within the process industry about the importance of making development and operational decisions based on a thorough assessment of the associated risks to identify measures that can be taken to prevent potential losses. This increased awareness has shaped and influenced process safety science. Accordingly, API 754 defines process safety as: "a disciplined framework for managing the integrity of hazardous operating systems and processes by applying good design principles, engineering, and operating and maintenance practices" (API, 2010).

Unlike the occupational safety approach, which focuses on hazards that could result in health issues (e.g., slips, trips, and falls), process safety focuses on the identification, prevention, and mitigation of process hazards and near misses that may result in the release of chemicals or energy (HSE, 2015a). Such hazards could ultimately result in serious impacts including human health loss, environmental damage, asset loss, and loss of production. Because of this difference in the scope, the blend of engineering and management skills required to manage process safety exceeds those required for managing occupational safety.

2.2 The Evolution of Process Safety Regulations

"The driving force for process safety has been primarily based on catastrophic events" (Mannan, Reyes-Valdez, Jain, Tamim, & Ahammad, 2016). Lessons learnt from events such as Flixborough (1974, United Kingdom), Seveso (1976, Italy), Bhopal (1984, India), Piper Alpha (1988, United Kingdom), Phillips 66 (1989, United States), BP Texas City (2005, United States), and Deepwater Horizon (2010, United States) have improved safety management systems and technological solutions. These disasters have also framed the global regulatory environment.

Fig. 1 illustrates a few examples of significant industry and process-related incidents and the legislative responses. It was mostly after highly publicized disasters such as those in Flixborough (United Kingdom, 1974), Seveso (Italy, 1976), and Bhopal (India, 1984) that governments and regulatory agencies began to establish what is now called PSM (Khan et al., 2016).

2.2.1 European Union

In Europe, a serious explosion at a chemical manufacturing plant in northern Italy in 1976 spurred the European Union (EU) to adopt, on June 24, 1982, Directive 82/501/EEC on the major-accident hazards of certain industrial activities, also known as the Seveso Directive (EC, 1982). Subsequent incidents led to amendments to the Seveso Directive, and Seveso II (Directive 96/82/EC) was eventually adopted on December 9, 1996 (EC, 1997). In response to incidents such as explosions in the Sandoz chemical plant near Basel, Switzerland in 1986 and the AZF fertilizer factory in Toulouse, France in 2001, the Seveso III (Directive 2012/18/EU) was adopted in 2012 to:

- integrate new categories of fertilizers;
- accommodate additional requirements for preventing environmental damage;
- align national Seveso legislation with the new classification scheme for chemical substances; and
- acknowledge the community's right to know provisions.

The Seveso III Directive was amended in 2013 to include safety of offshore oil and gas operations (Macza, 2008; Sreenevasan, 2015).

2.2.2 United Kingdom

Following the Flixborough (1974) and Piper Alpha (1988) accidents, the Control of Major Accident Hazards (COMAH) Regulations came into force in the United Kingdom on April 1, 1999. The COMAH regulations were further amended in 2005. The purpose of these amendments was to take all necessary measures to prevent major accidents involving dangerous substances. Major onshore hazard regulation in the United Kingdom now derives primarily from the EU's Seveso Directives and is largely implemented by the COMAH Regulations, including Safety Case Regulations (HSE, 2005).

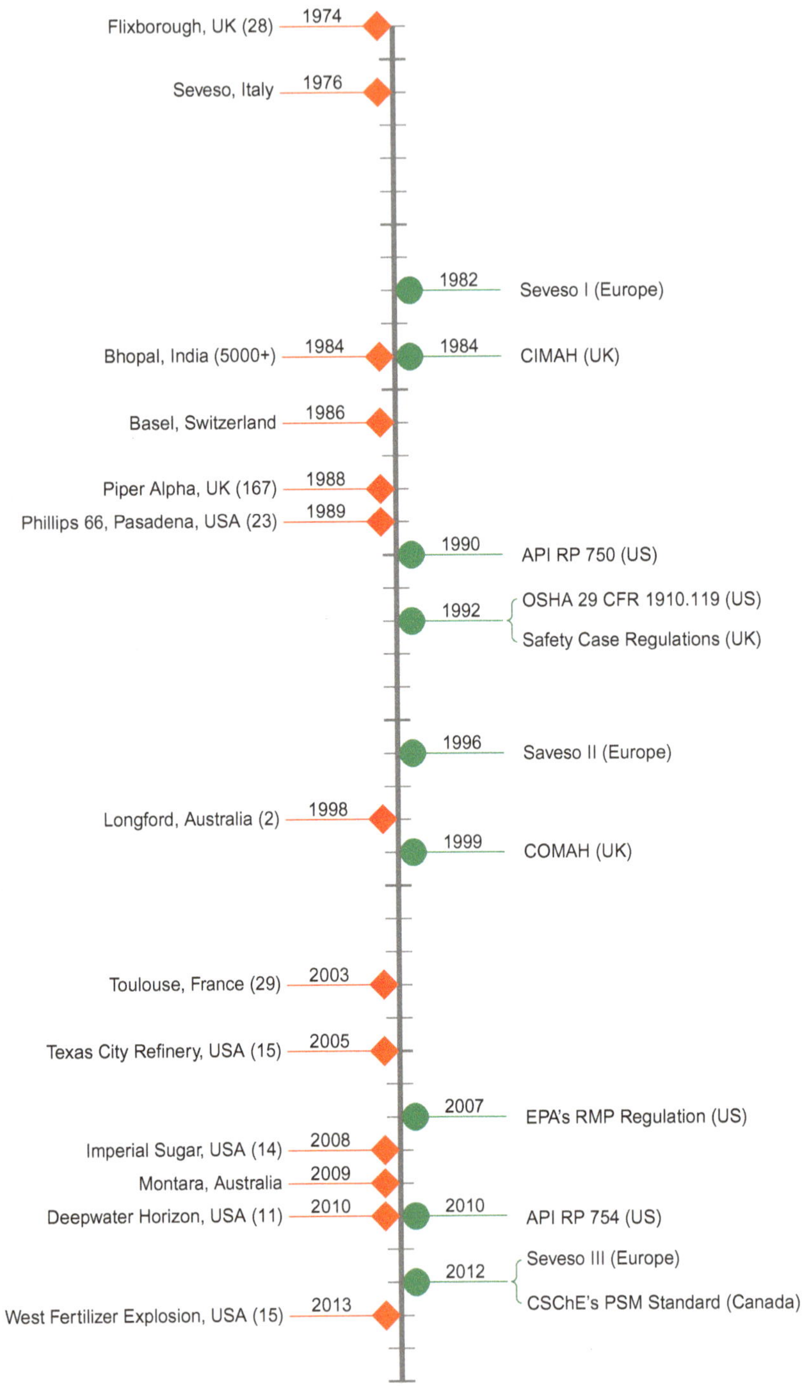

Fig. 1 Landmark disasters and main regulatory legislations and process safety standards. *Numbers in parentheses* in the *left panel* show the number of deaths in each accident.

A Safety Case is a structured argument, supported by evidence, intended to provide criteria to judge when a system is acceptably safe for a specific application in a specific operating environment. The Safety Case regulations have implemented the main recommendations of Lord Cullen's Report, derived from the Public Inquiry into the Piper Alpha Disaster (Macza, 2008). The latest revised version of this agreement was slated to be enacted in 2015 (HSE, 2015b). This common-sense approach is a recognized practice in Europe prior to an operating permit being issued and is likely to be adopted internationally by other legislative bodies.

2.2.3 United States

Despite major accidents involving highly hazardous substances, the United States did not have a process safety standard until the early 1990s. In response, Congress required action by both the Occupational Safety and Health Administration (OSHA) and the Environmental Protection Agency (EPA), through amendments to the 1990 Clean Air Act. In 1990, the American Petroleum Institute (API) published API 750, Management of Process Hazards voluntary guidelines. The API standard was a basis for the proposed PSM standard by OSHA, which was published in 1992. One year later, the US EPA released its Risk Management Program (RMP) Rule (McAteer & Whiteman, 1993). Based on this regulation, RMPs must be revised and resubmitted to the EPA every 5 years.

2.2.4 Australia/New Zealand

The Longford explosion in 1998 and the Montara oil spill in 2009 are landmark events that have influenced Australian process safety regulations. These events resulted in the development of the Major Hazard Facilities Regulations in 2004 (Macza, 2008) and the establishment of the National Offshore Petroleum Safety and Environmental Management Authority (NOPSEMA) in January 2012. Now, NOPSEMA operates as a single regulatory body for all offshore operations including major accidents (Sreenevasan, 2015). Australia also follows the Safety Case regime developed in the United Kingdom following Piper Alpha (CAPP, 2014). Unlike Australia, New Zealand does not rely only on regulations. A Business Leaders' Health & Safety Forum provides guidance and training to Chief Executives in New Zealand. No such initiative or program is available in Australia (Sreenevasan, 2015).

2.2.5 Canada

The 1982 sinking of the oil rig Ocean Ranger, which took the lives of the entire crew of 84 men on the Grand Banks of Newfoundland, is a major disaster in Canadian oil and gas history that has influenced Canadian offshore petroleum safety regulations. As an example, the Newfoundland Offshore Petroleum Drilling and Production Regulations has the requirements for a "Safety Plan," which requires the identification of hazards and a plan to manage them (C-NLOPB, 2010).

Unlike the offshore industry, PSM in the Canadian chemical industry follows a voluntary path. Canada responded to Bhopal by creating the Major Industrial Accident Council of Canada (MIACC) in 1987 (CAPP, 2014). From 1987 until its dissolution in 1999, this partnership included the federal, provincial and municipal governments, industry, labor, emergency response groups, public interest groups, and academia. This collaboration resulted in the development of a series of industrial guidelines. The application of these guidelines is not mandatory, and the industrial organizations and companies voluntarily adopted them.

The MIACC dissolved in 1999 due to the decline in funding and participation, and the developed intellectual property was transferred in 2000 to the newly created Process Safety Management Division (PSMD) of the Canadian Society for Chemical Engineering (CSChE). CSChE published its PSM standard in 2012 based on the original version of the United States' Center for Chemical Process Safety (CCPS) PSM guidelines (CSChE, 2012). In 2012, PSDM and the Canadian Chemical Producers Association (CCPA) jointly initiated the Process Safety Network (PSN) and are likely to work more closely together to form a larger PSM body in Canada (Macza, 2008; Sreenevasan, 2015).

2.2.6 Future of Regulatory Focus

A review of the state of current regulations shows an evolution in regulatory thinking that provides a path for advancing process safety within the process industry. Three important shifts are taking place and gaining momentum (CAPP, 2014):

(i) a shift from rule-based to performance-based regulations,
(ii) a shift from personal safety to process safety, and
(iii) increased attention to corporate safety culture.

The regulatory environment will continue to evolve, at least until the process industry achieves a "zero harm" milestone. Hopefully, the global PSM initiatives will not wait for more process-related disasters to ensue. The future evolution of process safety regulations can be obtained through a higher degree of involvement and collaboration among regulators, regulated organizations, and employees. The regulatory context of process safety is discussed in more details in the chapter "Regulatory Context" by Attwood.

2.3 The Evolution of Process Safety Technology

The body of knowledge of process safety in the 1970s was mainly focused on engineering principles. However, process accidents in the early 1980s, such as the Bhopal disaster, brought process safety knowledge into mainstream consideration within the industry. Consequently, the industry started to recognize the importance of engaging management and incorporating social concerns in PSM. In the 1990s, risk-based approaches were developed to include safety in design as well as economic considerations (Khan et al., 2016). Recently acknowledged is the important analysis of the "human factor" to reduce and prevent process safety incidents. Moreover, safety management systems began to be incorporated in regulations and procedural systems around the world (Dahle et al., 2012; Mannan et al., 2016). In the 2000s, the inherent safety approach gained recognition and process safety engineering made significant progress. More recently, the concept of dynamic risk management has been emphasized in process safety literature due to the limitations of traditional risk analysis (Amyotte et al., 2016; Khan et al., 2016). Table 1 demonstrates the industrial initiatives and recommendations that resulted from the investigation of the causes of major accidents since the Flixborough disaster in 1974.

Other than the regulatory and sociotechnical developments identified in Table 1, another major change in process safety programs has been the introduction of risk-based approaches in the late 1980s. Although different organizations may choose to implement multiple strategies at the same time (CCPS, 2014), in general, there has been an increased focus on risk-based strategies in recent years. The main objective of the Risk-Based Process Safety (RBPS) approach is to help process industries build and operate a more effective PSM system (CCPS, 2007). The chapter "Risk-Based Process Safety" by Seider is devoted to a risk-based approach to process safety management.

In Europe, the Canvey Island (HSE, 1978, 1981) and Rijnmond area (Rijnmond Public Authority, 1982) pilot studies are considered as the precursors of RBPS in the chemical sector. In 1988, a project aiming at risk-based decision making was developed in the corporate chemical

Table 1 Process Safety Initiatives and Technology Development Influenced by Major Incidents

Year	Accident	Description of Technology Development
1974	Flixborough	• Increased attention to risk assessment • Introduction of Management of Change (MOC) • Introduction of Hazard and Operability (HAZOP) study
1976	Seveso	• Enhanced inherently safer design • Improved emergency planning and response
1984	Bhopal	• Improved release modeling and understanding of runaway reactions • Inherently safer design • Release modeling
1988	Piper Alpha	• Stronger implementation of permits to work systems and incident reporting systems • Enhanced emergency response systems • Introduction of "Step Change in Safety" • Enhanced safety training • Changes to platform design
2005	BP Texas City	• Improved risk assessment and management of change • Improved process knowledge among senior/corporate management and board members • Increased liability for senior/corporate management and board members • Adequate physical devices and technology (barriers) • Occupational vs process safety indicators
2009	Montara	• New requirements for safety barriers risk assessment • Lower threshold for conducting risk assessments and independent reviewing • Operators emergency assistance • Communication and information sharing between stakeholders • Broader context when conducting decisions and risk assessments
2010	Deepwater Horizon	• Proposed introduction of a Safety Excellence Institute (Graham et al., 2011) • Stricter operating permit conditions (well integrity and oil-spill response) • Enhanced safety culture • Increased system, operation and risk understanding • Enhanced blowout prevention (BOP) and oil-spill recovery technologies
2013	West Fertilizer explosion	• Enhanced emergency response procedures • Establishing safety procedures • Enhanced safety culture

Adapted from Dahle, I. B., Dybvig, G., Ersdal, G., Guldbrandsen, T., Hanson, B. A., Tharaldsen, J. E., & Wiig, A. (2012). Major accidents and their consequences for risk regulation. In *Advances in Safety, Reliability and Risk Management: ERSEL 2011* (pp. 33–41). France: Taylor & Francis Group; Mannan, M. S., Reyes-Valdez, O., Jain, P., Tamim, N., & Ahammad, M. (2016). The evolution of process safety: Current status and future directions. *Annual Review of Chemical and Biomolecular Engineering*, 7, 135–162. https://doi.org/10.1146/annurev-chembioeng-080615-033640.

cluster of Ravenna (Egidi, Foraboschi, Spadoni, & Amendola, 1995). In the European framework, RBPS has been applied to manage, control, and reduce the risk in single facilities and also in large chemical clusters (Rijnmond Public Authority, 1982) and for land-use planning around chemical sites to reduce risk to the population (Christou, Amendola, & Smeder, 1999; Cozzani, Bandini, Basta, & Christou, 2006; Spadoni, Egidi, & Contini, 2000). Publications such as the Dutch "Purple Book" (Uijt de Haag & Ale, 1999) provide guidelines, practical procedures, and data to support such practices. In the United Kingdom and in the Netherlands, such methods are now required to support the implementation of Seveso Directives (HSE, 2015a; Uijt de Haag & Ale, 1999). In the United States, however, risk-based approaches to PSM were introduced much later. It was in 2007 that the CCPS proposed the introduction of risk-based process safety (CCPS, 2007).

The conditions in process facilities are dynamic, with changes in operating parameters often being reflected in changed operating procedures and equipment (NOPSEMA, 2012c). Raw materials' quality and availability, services' quality and availability, product quality and throughput, plant equipment availability and environmental conditions, links with other plants, drifting and decaying factors, process materials behavior, plant equipment malfunction, and control system malfunction (Mannan, 2012) along with causes such as mechanical integrity degradation, improper methods, and human factors can cause abnormal situations that can eventually result in severe consequences (Hashemi, Ahmed, & Khan, 2014). However, due to their static structure, conventional risk assessment approaches fail to capture the variation of risks as deviations or changes in the process and plant take place. Dynamic risk assessment has gained increased attention in academia as the next generation of risk and management approaches that help to enable safer complex process systems operating in extreme environments (Amyotte et al., 2016; Khan et al., 2016). However, there is still a compelling need to conduct research on dynamic risk assessment techniques in order to apply them effectively for practical process safety.

3. METHODS IN CHEMICAL PROCESS SAFETY

Methods in Chemical Process Safety (MCPS) are defined here as systematic procedures to identify process hazards and to implement, measure, evaluate, monitor, and manage the safety of hazardous processes. Table 2 shows different categories of process safety methods and provides examples

Table 2 Steps and Methods in Chemical Process Safety

Safety Management System Steps	Examples of Related Methods
Identification	• Hazard and Operability (HAZOP) Study • Checklist • Failure Modes, Effects and Criticality Analysis (FMECA) • What If
Implementation	• Inherent safety design • Inspection programs, repair and replacement, metallurgy changes • Safety and control systems to reduce probability of failure: – Basic process control system (BPCS) – Alarms and human intervention – Safety instrumented system (SIS) – Emergency shutdown devices – Pressure relief devices • Mitigations systems to reduce consequences of failure: – Physical devices for containment (dykes, blast walls, firefighting, etc.) – Plant and community emergency response
Measurement (risk analysis)	• Consequence assessment • Likelihood assessment • Risk assessment
Evaluation	• As-Low-As-Reasonably-Practicable (ALARP) • Layers of Protection Analysis (LOPA) • Safety Integrity Levels (SIL)
Monitoring	• Lagging indicators • Leading indicators
Management	• Performance standards • Audit procedures • Safety culture

for each category. As shown in Table 2, process safety engineering requires a broad understanding of interactions among process design, integrity management methods, process monitoring and control, safety barriers, operating procedures, and PSM systems (Khan et al., 2015; Mannan et al., 2009). The following sections describe different steps and methods identified in Table 2.

3.1 Methods to Identify Process Hazards

A hazard is defined as a situation with a potential for causing harm. Hazards are diverse in nature, but they are all potential sources of harm. In the context of the process industry, the potential harm may relate to human injury, damage to the environment, damage to property, or a combination of these. Minimally, a hazard identification (HAZID) process provides the following outcomes (NOPSEMA, 2012b; WorkSafe, 2011):

- identifies all major incidents that could occur at the facility (irrespective of existing control measures);
- shows clear links between hazards, causes, and the potential events;
- provides a systematic record of all identified hazards and major incidents along with related assumptions; and
- provides a basis for identifying, evaluating, defining, and justifying the selection (and rejection) of control measures for eliminating or reducing risk.

The HAZID step is inherent in all risk assessment approaches. There are several methods for performing HAZID, not all of which serve necessarily as the best technique for a particular application. Most companies use their experience to choose or adapt a HAZID technique for their operations (Crowl & Louvar, 2011). In selecting the appropriate HAZID technique, the nature and scale of the installation, the stage in the process life cycle, and experience of similar installations should all be considered. The level of effort devoted to HAZID should be based on the anticipated level of risk and any limitations in knowledge (ISO, 2002). In broad terms, the HAZID technique selection can be quite separate from the subsequent risk assessment approach. Thus, an initial HAZID may support both qualitative and semiquantitative risk assessments, whereas a more detailed HAZID can support any level of risk assessment. Table 3 provides a list of frequently used HAZID methods along with their benefits and disadvantages.

3.2 Methods to Implement Process Safety

3.2.1 Diversity Is a Key to Implement Process Safety

Different strategies can be applied to implement process safety. Crowl and Louvar classified process safety into four different strategies (Crowl & Louvar, 2011), as shown in Table 4. Inherent safety design (ISD) is the most preferred strategy, focusing on the elimination of hazards. If the elimination of a hazard is not possible, ISD focuses on minimizing the hazard. Passivity is

Table 3 Hazard Identification Popular Methods (CCPS, 2008a; Crowl & Louvar, 2011; ISO, 2002; NOPSEMA, 2012b)

HAZID Method	Advantages	Disadvantages
HAZOP	• A highly structured technique • Provides a detailed understanding of the possible "deviations from design intent"	• Less suitable for identification of hazardous scenarios associated with mechanical integrity failures and external events (such as collision) • Since the HAZOP analysis uses a "section-by-section" approach, it may not identify hazards associated with the interactions between different nodes
Safety Review	• An effective but less formal type of the HAZOP study	• The results are highly dependent on the experience and integrity of the Safety Review team
Checklists	• An effective way of capturing and passing on the experience of others • Easy to implement using a set of prewritten questions developed by experts to stimulate discussion • Can be used effectively to demonstrate compliance with an engineering standard	• "Closed" in nature, can only observe what is stated in the checklist • Checklists should only be used as a final check that nothing has been neglected or missed by other studies • Checklists should not be used as the sole tool in a hazard identification process, since they may not cover all types of hazard, particularly facility-specific hazards
What-if	• Easy to implement using a set of preprepared and customized "What-if" questions on potential deviations and upsets at the facility (Example: what if the reboiler control system fails?)	• The questions are often based on the experience of others and hence this technique has some of the same limitations as a checklist approach

Table 3 Hazard Identification Popular Methods (CCPS, 2008a; Crowl & Louvar, 2011; ISO, 2002; NOPSEMA, 2012b)—cont'd

HAZID Method	Advantages	Disadvantages
FMECA and FMEA	• Failure Modes and Effects Analysis (FMEA) and Failure Modes, Effects, and Criticality Analysis (FMECA) are highly structured techniques, usually used for hazard identification of complex equipment by breaking down the analysis to the component level • A comprehensive method to identify potential causes and effects of each component failure	• Could be cumbersome to implement as they require a thorough knowledge of the system and failures • A brainstorming session with several team members involved from conception to operation is required to ensure the effectiveness

Table 4 Process Safety Strategies (Crowl & Louvar, 2011; IAEA, 2009)

Strategy	Description	Examples
Inherent	Methods to eliminate or significantly reduce hazards, rather than to develop add-on protective systems and procedures	• Use smaller quantities of hazardous materials • Replace a substance with a less hazardous material
Passive	Safety features that do not require action by any device or human intervention	• Dikes • Passive flame arrestors • Elevated gravity drain tanks
Active	Safety devices and systems to prevent accidents	• Process control systems • Automatic shutdown systems • Pressure relief devices
Procedural	Rules, procedure, and techniques that limit or prevent people's exposure to hazards	• Standard operating procedures • Safety rules • Operator training • Emergency response procedures

the next preferred strategy in process safety implementation since the operation of passive safety components (PSCs) is connected only with the triggering event. The passivity concept, which is a highly recognized approach in nuclear engineering (IAEA, 2009), eliminates the dependency of safety components on the operation of power sources and other active components such as a control system.

Inherently safer designs and PSCs are the most reliable and robust process safety strategies. However, achieving process safety solely through ISD and PSCs is difficult in practice and elements of active and procedural safety strategies are also required to minimize process safety issues. This diversification is key to ensure the reduction of failure probabilities while mitigating the consequences of potential failures. While active safety barriers such as control systems reduce the probability of failures, passive and procedural strategies primarily help to mitigate the severity of losses (Crowl & Louvar, 2011). ISD, which is the most preferred approach to implement process safety, helps to reduce both the probability and consequences of failures (Amyotte, Goraya, Hendershot, & Khan, 2007). These different strategies supplement each other and, generally, none of them can be substituted for other methods.

3.2.2 Layers of Protection

To implement process safety strategies, an array of protective layers are required to reduce the risk of an operation due to identified hazards. Fig. 2 demonstrates the concept of layers of protection using the "onion model." Layers of protection include the physical features of a facility and elements of human intervention which eliminate, prevent, reduce, or mitigate the risk of hazardous events. They can take many forms, including physical equipment, process control systems, and safety devices as well as operating and management processes (NOPSEMA, 2012a).

The principles of inherent safety should be taken into consideration when designing layers of protection to ensure their reliability and robustness. These protection layers should be independent to avoid potential failure of multiple layers due to a common cause. For example, malfunction of a pressure sensor can cause failure of both control and safety instrumented systems if they share the same sensor. Moreover, a proper maintenance management program is required to avoid the degradation of protection layers' effectiveness.

A review of incident investigation reports indicates that usually a coincidence of several events and failure of multiple safety barriers causes process

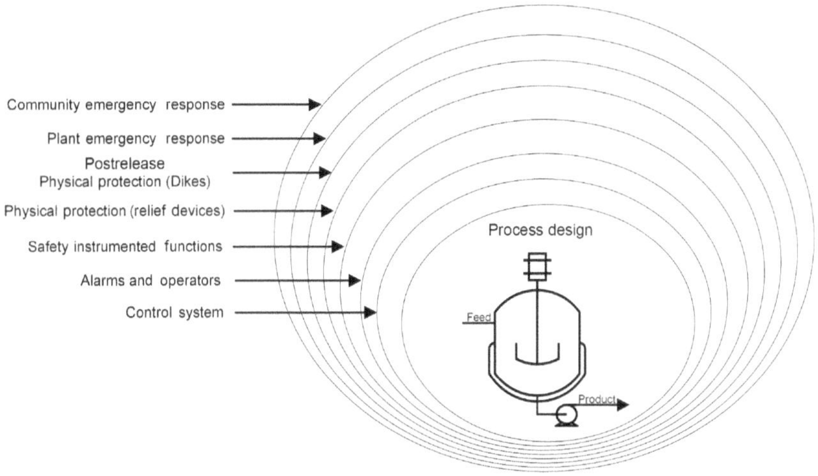

Fig. 2 Layers of protection. *Adapted from Crowl, D. A., & Louvar, J. F. (2011).* Chemical process safety: fundamentals with applications *(3rd ed.). New Jersey: Prentice Hall.*

safety incidents. Fig. 3 shows the "Swiss Cheese Model," originally proposed by Reason (1990), which is frequently used to demonstrate the relationship between hazards and sequential failures of multiple safety barriers that eventually cause an accident. Hart (2003) represented Reason's model as a set of spinning disks with variably sized holes, recognizing the dynamic relationship between process hazards and safety barriers. Based on the "Spinning Disk Model," the type and size of holes (weaknesses) in each protection layer (safety barrier) change over time, and the alignment of the holes constantly shifts. Fig. 3 illustrates both models. In both models, protection layers can be active, passive, or procedural. Holes in each protective layer can be latent, incipient, or actively open (API, 2010).

As discussed earlier, protective layers have their own life cycle and, therefore, their effectiveness degrades over time due to natural and assignable causes. Consequently, putting too much reliance on a single, or a few, protective layer(s) should be avoided. Moreover, the type of required protective layer for a given system could vary during the system's life cycle. Although process design is an important process safety factor for a new system, inspection and maintenance may need to receive more consideration at later stages of the system's life. Finally, methods such as Layers of Protection Analysis (LOPA) should be used to investigate the amount of risk reduction of each protective layer to ensure an adequate level of protection for each hazard (NOPSEMA, 2012a).

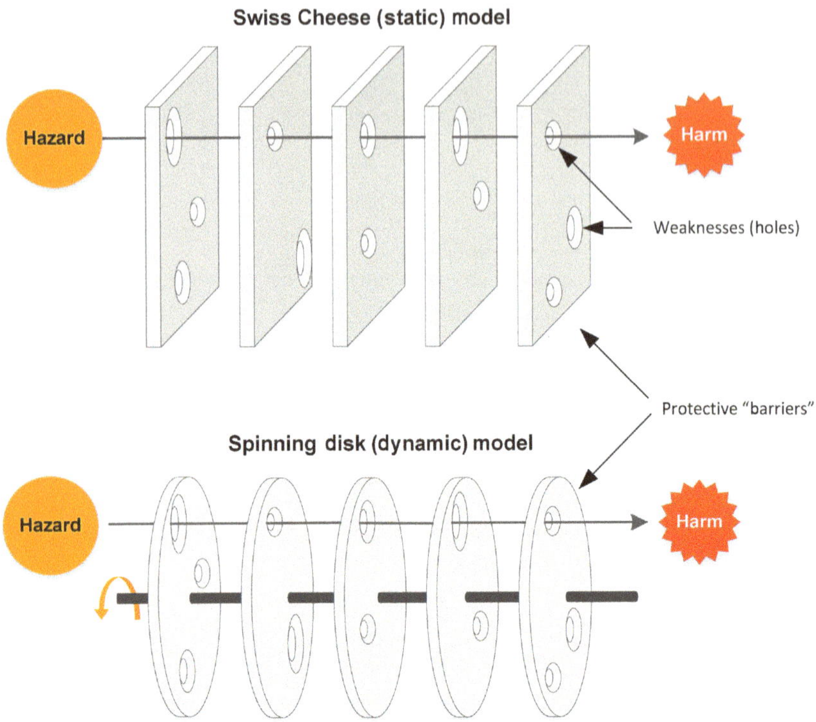

Fig. 3 "Swiss Cheese (static) model" and "spinning disk (dynamic) model."

3.2.3 *Inherent Safety Design*

Explicit incorporation of the principles of inherent safety in safety management can help to enhance the effectiveness of the safety management process and minimize the inherent risk (Amyotte et al., 2007). There is a fundamental difference between the inherent safety approach to loss prevention and engineered and procedural safety. While the latter two approaches accept the hazard, and use physical barriers and administrative controls to mitigate the effects, the inherent safety approach seeks to eliminate the hazard at the source. Considering the criticality of operations, therefore, the "explicit incorporation" of inherent safety principles in marine and offshore engineering is of paramount importance (Amyotte et al., 2007).

Minimization (intensification), substitution, moderation (attenuation), and simplification are four fundamentals of inherently safer design of offshore facilities (Khan et al., 2013), also referred to as ISD guidewords. Table 5 provides the description of these guidewords, along with example checklist questions to implement them.

3.2.4 Automation and Redundancy to Ensure Safety

Process safety system automation is required to ensure timely action when equipment and operators fail. Automated safety systems consist of different components, failure of which can result in failure of the associated

Table 5 Inherent Safety Guidewords and Example Checklist

Guideword	Description
Minimize	Use smaller quantities of hazardous materials when the use of such materials cannot be avoided. Perform a hazardous procedure as few times as possible when the procedure is unavoidable. *Checklist*: • Is the size of storage of all hazardous gases, liquids, and solids minimized? • Are just in time deliveries used when dealing with hazardous materials? • Are all hazardous materials removed or properly disposed of when they are no longer needed or not needed in the next *X* days? • Is shift rotation optimized to avoid fatigue?
Substitute	Replace a substance with a less hazardous material or a processing route with one that does not involve hazardous material. Replace a hazardous procedure with one that is less hazardous. *Checklist*: • Can a less toxic, flammable, or reactive material be substituted for use? • Is there an alternate way of moving the product or equipment to eliminate human strain? • Can a water-based product be used in place of a solvent- or an oil-based product? • Are all allergenic materials, products, and equipment replaced with nonallergenic materials, products, and equipment when possible?
Moderate	Use hazardous materials in their least hazardous forms or identify processing options that involve less severe processing conditions. *Checklist*: • Can potential releases be reduced via lower temperatures or pressures, or elimination of equipment? • Are all hazardous gases, liquids, and solids stored as far away as possible to eliminate disruption to people, property, production, and the environment in the event of an incident? • When purchasing new equipment, are acceptable models available that operate at lower speeds, pressures, temperatures, or volumes? • Are workplaces designed such that employee entrapment is minimized?

Continued

Table 5 Inherent Safety Guidewords and Example Checklist—cont'd

Guideword	Description
Simplify	Design processes, processing equipment, and procedures to eliminate opportunities for errors by eliminating excessive use of add-on (engineered) safety features and protective devices. *Checklist*: • Are all manuals, guides, and instructional materials clear and easy to understand, especially those that are used in an emergency situation? • Are equipment and procedures designed so that they cannot be operated incorrectly or carried out incorrectly? • Are machine controls located to prevent unintentional activation while allowing easy access for stopping the machine? • Are all machines, equipment, and electrical installations easily isolated from all sources of power?

Adapted from Amyotte, P. R., Goraya, A. U., Hendershot, D. C., & Khan, F. I. (2007). Incorporation of inherent safety principles in process safety management. *Process Safety Progress, 26*(4), 333–346. https://doi.org/10.1002/prs.10217; Khan, F., & Amyotte, P. R. (2005). I2SI: A comprehensive quantitative tool for inherent safety and cost evaluation. *Journal of Loss Prevention in the Process Industries, 18*(4–6), 310–326. https://doi.org/10.1016/j.jlp.2005.06.022; Khan, F. I., & Amyotte, P. R. (2002). Inherent safety in offshore oil and gas activities: A review of the present status and future directions. *Journal of Loss Prevention in the Process Industries, 15*(4), 279–289. https://doi.org/10.1016/S0950-4230(02)00009-8.

safety system that can eventually result in unsafe process conditions. To tackle this problem, redundancy is used in process safety engineering to ensure that safety devices continue to perform their defined function when a single instrument or control function fails. An example is the inclusion of an additional flow meter in an abrasive process to increase the reliability of a flow control system and protect the system from a potential flow meter failure. The level of redundancy required for a particular safety system depends on the criticality of the operation. Standards such as ANSI/ISA-S84 (ISA, 1996) and methods such as Safety Integrity Level (SIL) analysis (ISA, 2002) can be used to determine the required level of redundancy.

3.3 Methods to Measure Process Safety

3.3.1 Overview

"Risk" is used as a measure of safety and is defined as the combination of three attributes: what can go wrong, how bad could it be, and how often might it happen. The purpose of risk assessment is to help all stakeholders understand the risks to health and safety and address potential major-accident events in a structured manner. Moreover, risk assessment is now a regulatory requirement in different parts of the world. For example, the Safety Case regulations in the United Kingdom, EU, and Australia require the operators of offshore facilities to conduct a detailed and systematic formal safety assessment, which includes the assessment of risk in relation to all potential major-accident events before an operating permit is issued (HSE, 2015b; Macza, 2008; NOPSEMA, 2012c; Sreenevasan, 2015). In the United States, the RMP Rule requires facilities that use extremely hazardous substances to develop a Risk Management Plan and submit revised plans to the EPA every 5 years.

Different industries and regulatory agencies have developed several risk assessment techniques, including qualitative, semiqualitative, and quantitative approaches. Qualitative approaches provide the least degree of insight, but are easiest to apply because they require the fewest resources and fewer skill sets. Quantitative Risk Assessment (QRA) approaches deliver the most detailed understanding and provide the best analysis if supported by adequate resources. Semiquantitative approaches lie in between these extremes.

The aim of all three approaches is to help operators understand the risk levels. Risk assessment also provides a basis for identifying, evaluating, defining, and justifying the selection or rejection of protective layers for eliminating or reducing risk. No matter which risk assessment approach is used, subjective and expert judgments are an essential part of risk assessment. Therefore, it should be noted that the estimated risk using any approach provides a relative criterion, not an absolute measure, to rank assets and prioritize plans.

3.3.2 Selection of Approach

Selecting the best approach to risk assessment could be challenging due to the existence of different options. While there is no single correct approach for a specific activity, there are approaches that are more suitable than others, and a decision framework is helpful in the selection process (DNV, 2001). The operators should therefore choose the right level of risk assessment to help them link the identified hazards, the adopted protective layers, and the demonstration of As-Low-As-Reasonably-Practicable (ALARP) risk within their operations.

The HSE's Guide to the Offshore Installations (Safety Case) Regulations (HSE, 1998) gives a brief indication of the type of risk assessment expected: "The evaluation of risk should involve both a qualitative and quantitative approach. Where relevant good or best practice is clear, the balance should be in favor of qualitative arguments to show that the risks have been properly controlled. Where relevant good or best practice is less clear, appropriate support from quantitative arguments will be necessary" (HSE, 1998).

The type, complexity, and criticality of operations are important decision factors to select an appropriate risk assessment approach. The selected approach should assist the users in understanding and selecting control and risk reduction measures. Furthermore, it should be capable of assessing the potential effects of risk reduction measures (NOPSEMA, 2012c). Overall, the selection of the risk assessment technique should be done based on the level of detail required as well as on the available resources.

Table 6 depicts the most popular risk assessment techniques. The methods identified in Table 6 are selected examples among many to illustrate different approaches. The purpose of Table 6 is to provide a high-level comparison of different techniques. A detailed or comprehensive description of risk assessment techniques is not within the scope of this work and an interested reader may refer to several published works on this topic (API, 2016; Modarres, 2006; Rathnayaka, Khan, & Amyotte, 2011; Rüschendorf, 2013).

3.4 Methods to Evaluate Process Safety

While risk assessment techniques such as Risk Matrix and Bowtie help to measure the safety of a particular process, safety evaluation methods use risk assessment techniques to assess the adequacy of the layers of protection provided for an activity. Table 7 provides a list of frequently used process safety evaluation methods, which are ALARP and LOPA. These are very established techniques in process industries with a well-defined and specific purpose. The main objective of these methods is to ensure an adequate reduction in risk by taking into account the effectiveness of existing control measures. The outcomes of these methods can help decision makers evaluate the necessity of adding alternative control measures to ensure compliance with defined risk thresholds.

Many process safety regulations, such as the Safety Case regulations in the United Kingdom, EU, and Australia, require the operators to demonstrate that they have reduced the risks to a level that is ALARP (HSE, 2015b; Macza, 2008; NOPSEMA, 2015; Sreenevasan, 2015). This means that the operator has to show, through reasoned and supported arguments, that there are no other practical measures that could reasonably be taken to additionally reduce risks (NOPSEMA, 2015). In practice, a combination of

Table 6 Popular Risk Assessment Methods

Method	Advantages	Disadvantages
Risk Matrix	• A simple scoring system to represent the combined estimates of likelihood and consequence • An effective and practical tool to present complex risk data in a concise visual fashion	• Assessment of likelihood and consequence and resulting risk ratings require subjective interpretation and different users may obtain contrary ratings of the same quantitative risks • Oversimplification of the complexity or volatility of dynamic risks • Inability to model complex dependencies and uncertain variables
Fault Tree	• A structured way to estimate and quantify the likelihood of failure occurring • A reproducible and justifiable tool to identify failures' causes and major contributors to the likelihood of the top event • Ability to provide both qualitative and quantitative representation of the modeled accident scenario	• Limited capability to handle uncertainty, multistate variables, and dependent failures due to application of simple Boolean functions in Fault Tree analysis • Application of deterministic probability values • Inability to model complex dependencies among variables
Event Tree	• A structured way to estimate and quantify the possible outcomes of a single accident event • A reproducible and justifiable tool to investigate accident scenarios and to estimate the likelihood of potential outcomes of an initiating event • Ability to provide both qualitative and quantitative representation of the modeled accident scenario	• Similar to Fault Tree
Bowtie	• Combination of the advantages of Fault Tree and Event Tree	• Similar to Fault Tree and Event Tree analyses

Continued

Table 6 Popular Risk Assessment Methods—cont'd

Method	Advantages	Disadvantages
Bayesian Analysis	• An effective graphical tool to provide both quantitative and qualitative representation of causal relationships among risk factors • Capability to update the prior beliefs about the probability of accidents by incorporating new system information • Capability to handle uncertainty, multistate variables, complex causal relationships, and sequentially dependent failures	• High computational burden to construct conditional probability tables • Inability to model nonlinear dependencies among variables • Application of deterministic and/or normally distributed probabilities

Table 7 Popular Methods to Evaluate the Adequacy of Control Measures (Dowell, 1998; NOPSEMA, 2015)

Method	Advantages	Disadvantages/Limitations
ALARP	• Allows operators to set goals for their own safety performance rather than following prescriptive requirements • Provides flexibility for regulators to accept or reject the operator's arrangements for the safety case	• Implementation of ALARP can be challenging because it requires operators to exercise judgment with respect to risk management • For complex processes, it is difficult to decide what is "reasonably practicable"
LOPA	• A structured and reproducible method to evaluate the adequacy of protective layers • A justifiable tool to determine the required SIL of a safety instrumented system • Focuses greater risk reduction efforts on hazardous events with high frequency and high likelihood • Encourages thinking from a system perspective • Gives clarity in reasoning process and it documents everything considered	• Inability to deal with uncertainty connected with input data • LOPA is not intended to be a hazard identification tool. LOPA is usually performed after a HAZOP study

approaches is likely to be employed to provide the required evidence for ALARP in a process plant. A systematic and detailed HAZID and risk assessment are fundamental requirements to demonstrate that the risks are reduced to a level that is ALARP, as they provide the foundation on which to base the control measure selection. The guidance notes provided by the United Kingdom's Health and Safety Executive (HSE) (HSE, 2001) and Australia's National Offshore Petroleum Safety and Environmental Management Authority (NOPSEMA, 2015) are good ALARP guidelines for operators.

The SIL is a target level of risk reduction provided by a safety function and is defined in terms of Probability of Failure on Demand (PFD) (ISA, 1996). In plain words, SIL is a measurement of performance required for a Safety Instrumented Function (SIF). SILs are usually determined using the LOPA method. LOPA is usually performed after the HAZID step (such as a HAZOP study). The first step in LOPA involves the application of risk assessment techniques, such as Fault Tree and Event Tree analyses, to identify causes of potential consequences and estimate their likelihood. Then, a PFD will be assigned to each Independent Protection Layer (IPL), including the control system, alarms, and emergency shutdown devices. These PFDs will be multiplied by the probability of an initiating event to estimate the probability of potential outcome events. The mitigated event probabilities will be compared with the corporation's threshold risk values to make a decision about the adequacy of IPLs (Dowell, 1998; ISA, 1996).

3.5 Methods to Monitor Process Safety Performance

3.5.1 Overview

Although development of Safety Performance Indicators (SPIs) is a relatively new concept in the field of process engineering, the process industry appears to be a role model in terms of standard harmonization. Following several major process safety-related incidents, organizations such as the United Kingdom's Health & Safety Executive (HSE, 2006) and the United States' Center for Chemical Process Safety (CCPS, 2008b) recognized the emerging need for improved SPIs through publication of recommended guidelines to develop and implement SPIs. This resulted in a significant effort by the American Petroleum Industry (API) to develop and publish the ANSI/API Recommended Practice (RP) 754 on Process SPIs in 2010 (API, 2010). CCPS elected to update the original document (CCPS, 2008b) using the CCPS metric recommendations with minor revisions with the intent to align the CCPS and API RP 754 documents. Later, in 2011, the International Association of Oil & Gas Producers (OGP) adopted the ANSI/API

RP 754 standard to build a similar framework for the upstream industry (OGP, 2011). Academia has also responded to the need for improving SPIs. An example is the risk-based process performance indicators developed by Khan, Abunada, John, and Benmosbah (2010).

3.5.2 Lagging vs Leading Indicators

One of the most important and challenging issues for process safety is detecting early signs of deterioration of process safety performance caused by operation, maintenance, management, organization, and safety culture factors before accidents happen. A review of safety management procedures and corporate annual reports shows that the process industry tends to rely on "lagging" indicators such as lost time injuries, OSHA incidence and fatality rates, and lost workdays to assess safety performance. Although these indictors serve a purpose, they monitor events after their occurrence (Khan et al., 2010). As mentioned by the US CCPS, "facilities should monitor the real-time performance of management system activities rather than wait for accidents to happen. Such performance monitoring allows problems to be identified and corrective actions to be taken before a serious incident occurs" (CCPS, 2007). Early warning of dangerous deterioration within critical systems provides an opportunity to avoid major incidents (HSE, 2006).

Recognizing the importance of both reactive and active monitoring of safety management systems, the United Kingdom's HSE introduced the concept of "dual insurance" in 2006. HSE recommended that leading and lagging indicators be set in a structured and systematic way for each critical risk control system within the whole PSM system (HSE, 2006). Later, API, CCPS, and OGP adopted the same concept. Lagging indicators are a form of reactive monitoring (HSE, 2006) and describe events that have already occurred which may indicate potential recurring problems and may include fires, releases, and explosions (API, 2010). Leading indicators are a form of active monitoring (HSE, 2006) and indicate the performance of the key work processes, operating discipline, and protective barriers that prevent incidents.

Lagging indicators tend to be outcome oriented and retrospective, while leading indicators tend to be forward looking (API, 2010). As discussed by API RP 754, the differentiation or classification of indicators as lagging or leading is not important. Instead, the focus should be on capturing and analyzing the information to correct a situation, identify lessons learned, and communicate this knowledge (API, 2010).

3.5.3 Selection of Indicators

API RP 754 presented a process safety pyramid (API RP 754, 1st ed.) with four classifications or tiers (API, 2010). The tiers of the pyramid represent a continuum of leading and lagging process safety indicators. The four tiers are expressed as a triangle to emphasize that statistically larger data sets are available from the indicators at the lower tiers (OGP, 2011). Tiers 1 and 2, at the top of the pyramid, are more lagging and cover less severe incidents. Some examples of Tiers 1 and 2 lagging indicators include metrics to measure injury, fire and explosions, Loss of Primary Containment (LOPC), releases, and pressure relief devices' activation. Tiers 3 and 4 provide more leading measures. Metrics to measure mechanical integrity, follow-up action items, management of change, process safety training and competency, safety culture, operating and maintenance procedures, and fatigue risk management are primary examples of leading indicators.

Selection of indicators is important since some indicators may not provide the needed insights to ensure desired performance. The United Kingdom's HSE provided a six-step approach to identify appropriate barriers and select indicators (HSE, 2006), which was later further developed in a CCPS book on process safety metrics (CCPS, 2011). Fig. 4 depicts a procedure for selection and review of process safety indicators. Companies should choose the actual indicators based on their understanding of the most critical risk control barriers, whether the barriers are facility specific or apply to groups of similar facilities or even apply across the whole company. API RP 754 for the refining and petrochemical industries and the OGP recommendations for drilling and production operations are great resources to help companies select and use their process safety indicators.

3.6 Methods to Manage Process Safety

3.6.1 How Are Different Process Safety Methods Related?

Review of the investigation reports of several major process-related accidents reveals the fact that the main root causes of these accidents were primarily cultural issues, not just technical problems. It has been shown in numerous studies that these accidents might have been prevented if a risk-focused approach was integrated into the management framework (CSB, 2014a; Khan et al., 2016; Paltrinieri, Khan, & Cozzani, 2014). Hendershot states: "We know how to improve process safety performance. Our biggest challenge is not technical, it is cultural. We need to actually do what we already know how to do, we need to do it well, and we need to do it everywhere and all of the time" (Hendershot, 2012). Thus, the

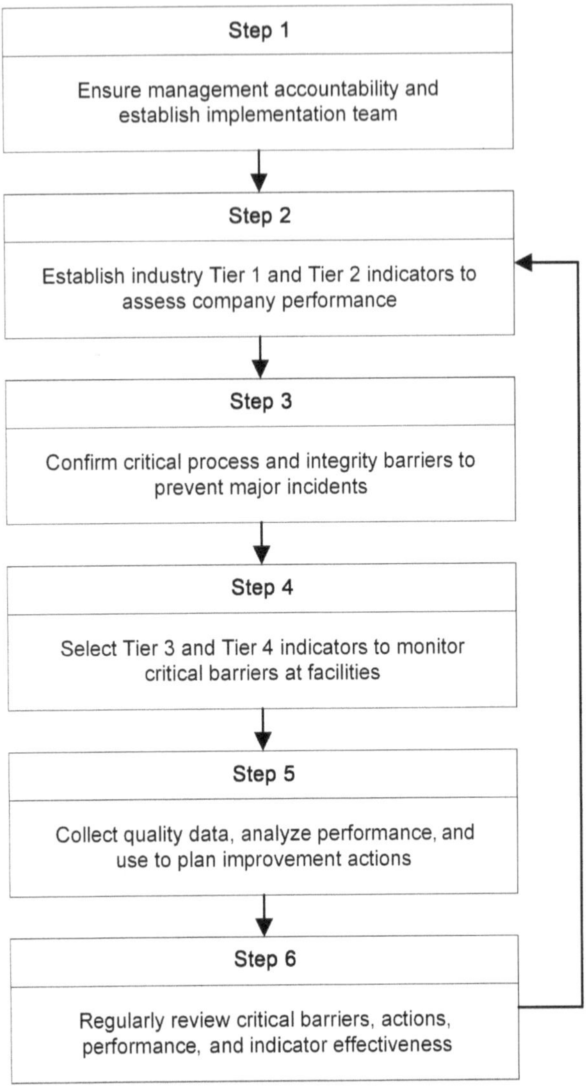

Fig. 4 Six-step approach for selection and review of process safety performance indicators. *Adopted from CCPS. (2008b).* Process safety leading and lagging metrics. *New York: Center for Chemical Process Safety of the American Institute of Chemical Engineers; HSE. (2006).* Developing process safety indicators—A step-by-step guide for chemical and major hazard industries. *London: HSE; OGP. (2011).* Process safety—Recommended practice on key performance indicators. *Retrieved from www.ogp.org.uk.*

development of a strong safety culture is required to continuously monitor process safety performance and record incidents (Paltrinieri et al., 2014). Then, the new evidence from the system should be incorporated in the risk assessment process to revise and update the risk assessment results. Moreover, the integration of the HAZID and risk assessment processes within a management framework is essential to ensure continuous improvement through the application of the revised risk profile in the decision-making process.

Each category of the process safety methods described in this section serves a specific purpose. Different methods complement each other to ensure the development of a robust and effective PSM system. However, the selection and application of proper process safety methods and their integration with a management framework may seem demanding and complicated. To address this potential challenge, the integration of the Plan-Do-Check-Adjust (PDCA) management method (ISO, 2015) with different process safety methods is proposed in Fig. 5.

Fig. 5 also shows the relationship between different categories of process safety methods with an overall safety management system. As highlighted in Fig. 5, a fundamental principle of a successful PSM system is iteration—once triggered, the presented PDCA cycle ensures control and continuous improvement of process safety and provides support to avoid the lack of a reporting and learning culture (Khan et al., 2016). The integration of the PDCA cycle with process safety methods ensures the consistent implementation of these methods on a sustainable basis to achieve a "zero harm" culture for process operations.

3.6.2 PSM Elements

The United States' CCPS published its guidelines for RBPS in 2007 and introduced 20 elements of a successful PSM system. These elements expand on the original CCPS PSM elements to reflect PSM implementation experience, best practices from a variety of industries, and worldwide regulatory requirements. These PSM elements can be designed and implemented at varying levels of enforcement to optimize PSM performance, efficiency, and effectiveness. The new elements also help eliminate gaps and inconsistencies that have contributed to PSM failures (CCPS, 2014).

The RBPS elements are meant to apply to the entire process life cycle. Some elements may not be active in early life cycle stages; but for some elements, the early life cycle stages provide a unique opportunity to minimize risk, for example, by identifying and incorporating inherently safer process

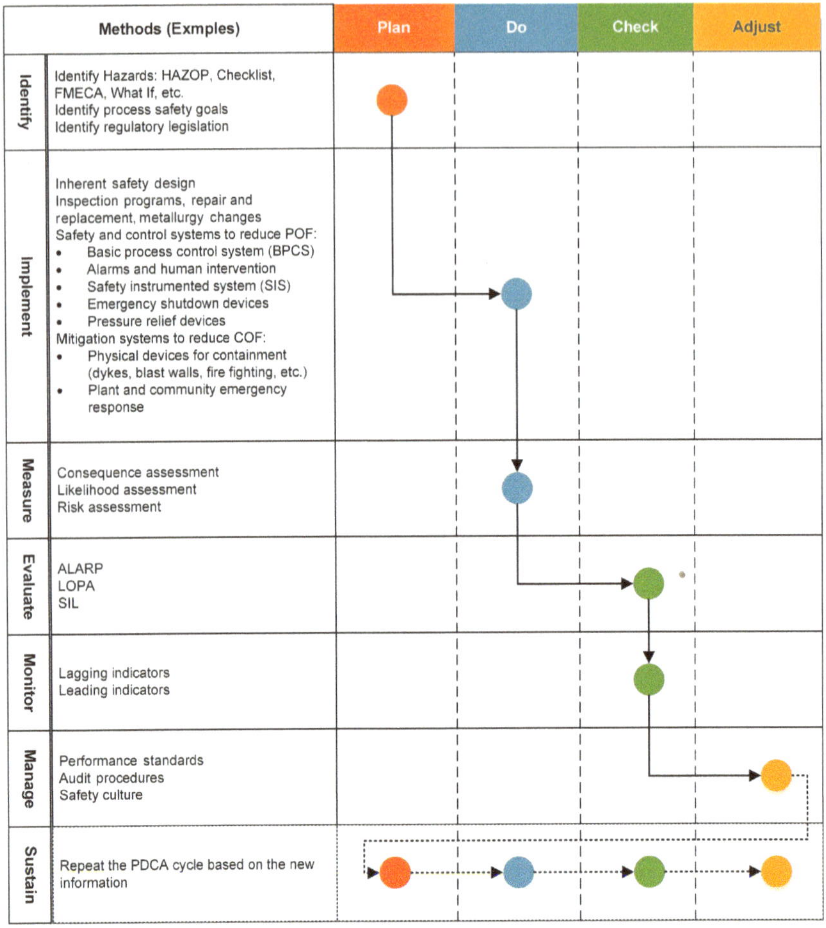

Fig. 5 Plan-Do-Check-Adjust (PDCA) approach to process safety management.

characteristics early in project development. In later stages, such as decommissioning, some work activities may not be as important or may no longer be needed, while others may be simplified (CCPS, 2014). The chapter "Elements of Process Safety" by Amyotte discusses PSM elements in more details.

REFERENCES

Amyotte, P. R., Berger, S., Edwards, D. W., Gupta, J. P., Hendershot, D. C., Khan, F. I., … Willey, R. J. (2016). Why major accidents are still occurring. *Current Opinion in Chemical Engineering*, *14*, 1–8. https://doi.org/10.1016/j.coche.2016.07.003.

Amyotte, P. R., Goraya, A. U., Hendershot, D. C., & Khan, F. I. (2007). Incorporation of inherent safety principles in process safety management. *Process Safety Progress*, *26*(4), 333–346. https://doi.org/10.1002/prs.10217.

API. (2010). *ANSI/API RP 754 process safety performance indicators for the refining and petrochemical industries* (1st ed.). Washington, DC: American Petroleum Institute.
API. (2016). *Recommended practice 581: Risk-based inspection technology* (3rd ed.). Washington, DC: American Petroleum Institute.
Bresland, J. (2008). *Maintain process safety during the recession: Safety messages from the U.S. Chemical Safety Board.* Washington, DC: U.S. Chemical Safety Board (CSB). Retrieved from http://www.csb.gov/videos/.
Calkins, L. B., & Fisk, M. C. (2007). *International Herald Tribune: BP executive blames workers for Texas refinery blast.* London: International Herald Tribune. Retrieved from http://royaldutchshellplc.com/2007/09/17/international-herald-tribune-bp-executive-blames-workers-for-texas-refinery-blast/.
CAPP. (2014). *Process safety management: Regulatory scan.* Calgary: The Canadian Association of Petroleum Producers (CAPP).
CCPS. (2007). *Guidelines for risk based process safety.* Hoboken, NJ: Center for Chemical Process Safety and John Wiley & Sons, Inc.
CCPS. (2008a). *Guidelines for hazard evaluation procedures.* New York: Center for Chemical Process Safety of the American Institute of Chemical Engineers.
CCPS. (2008b). *Process safety leading and lagging metrics.* New York: Center for Chemical Process Safety of the American Institute of Chemical Engineers.
CCPS. (2011). *Process safety leading and lagging metrics.* New York: Center for Chemical Process Safety of the American Institute of Chemical Engineers.
CCPS. (2014). *Risk based process safety overview.* New York: Center for Chemical Process Safety of the American Institute of Chemical Engineers.
Christou, M. D., Amendola, A., & Smeder, M. (1999). The control of major accident hazards: The land-use planning issue. *Journal of Hazardous Materials, 65*(1–2), 151–178. https://doi.org/10.1016/S0304-3894(98)00261-1.
C-NLOPB. (2010). *Newfoundland offshore petroleum drilling and production regulations.* St. John's, NL: Canada-Newfoundland and Labrador Offshore Petroleum Board.
Cozzani, V., Bandini, R., Basta, C., & Christou, M. D. (2006). Application of land-use planning criteria for the control of major accident hazards: A case-study. *Journal of Hazardous Materials, 136*(2), 170–180. https://doi.org/10.1016/j.jhazmat.2005.12.031.
Crowl, D. A., & Louvar, J. F. (2011). *Chemical process safety: Fundamentals with applications* (3rd ed.). New Jersey: Prentice Hall.
CSB. (2014a). *Investigation report overview: Explosion and fire at the Macondo Well.* Washington, DC: The U.S. Chemical Safety Board.
CSB. (2014b). In *Investigation report volume 1: Explosion and fire at the Macondo Well:* Vol. 1. Washington, DC: The U.S. Chemical Safety Board.
CSChE. (2012). *Process safety management standard* (1st ed.). Ottawa, ON: Canadian Society for Chemical Engineering.
Dahle, I. B., Dybvig, G., Ersdal, G., Guldbrandsen, T., Hanson, B. A., Tharaldsen, J. E., & Wiig, A. (2012). Major accidents and their consequences for risk regulation. In A. G. Bérenguer & C. G. Soares (Eds.), *Advances in safety, reliability and risk management: ERSEL 2011* (pp. 33–41). France: Taylor & Francis Group.
Dekker, S. (2001). The reinvention of human error. *Human Factors and Aerospace Safety, 1*(3), 247–265.
DNV. (2001). *Marine risk assessment. Offshore technology report (2001/063).* London: Health and Safety Executive (HSE).
Dowell, A. M., III (1998). Layer of protection analysis for determining safety integrity level. *ISA Transactions, 37*(3), 155–165.
EC. (1982). Council Directive 82/501/EEC of 24 June 1982 on the major-accident hazards of certain industrial activities. *Brussels: Official Journal of the European Communities, L230,* 1–18.

EC. (1997). Council directive 96/82/EC of 9 December 1996 on the control of major-accident hazards involving dangerous substances. *Brussels: Official Journal of the European Communities, 010*, 13–33.

Egidi, D., Foraboschi, F. P., Spadoni, G., & Amendola, A. (1995). The ARIPAR project: Analysis of the major accident risks connected with industrial and transportation activities in the Ravenna area. *Reliability Engineering & System Safety, 49*(1), 75–89. https://doi.org/10.1016/0951-8320(95)00026-X.

Graham, B., Reilly, W. K., Beinecke, F., Boesch, D. F., Garcia, T. D., Murray, C. A., & Ulmer, F. (2011). *Deep water: The gulf oil disaster and the future of offshore drilling—Report to the president*. Washington, DC: National Commission on the BP Deepwater Horizon Oil Spill and Offshore Drilling. https://doi.org/10.3723/ut.30.113.

Hart, C. (2003). Stuck on a plateau: A common problem. In J. R. Phimister, V. M. Bier, & H. C. Kunreuther (Eds.), *Accident precursor analysis and management: Reducing technological risk through diligence* (pp. 147–154). Washington, DC: The National Academic Press.

Hashemi, S. J., Ahmed, S., & Khan, F. (2014). Risk-based operational performance analysis using loss functions. *Chemical Engineering Science, 116*, 99–108. https://doi.org/10.1016/j.ces.2014.04.042.

Hendershot, D. C. (2012). Process safety management—You can't get it right without a good safety culture. *Process Safety Progress, 31*(1), 1–5.

Holden, R. J. (2009). People or systems? To blame is human. The fix is to engineer. *Professional Safety, 54*(12), 34–41. https://doi.org/10.1016/j.str.2010.08.012.Structure.

HSE. (1978). *Canvey: An investigation of potential hazards from operations in the Canvey Island/Thurrock area*. HM Stationery Office, London: Health and Safety Executive (HSE).

HSE. (1981). *Canvey: A second report. A review of the potential hazards from operations in the Canvey Island/Thurrock area three years after publication of the Canvey report*. HM Stationery Office, London: Health and Safety Executive.

HSE. (1998). *A guide to the offshore installations (safety case) regulations 1992*. London: Health and Safety Executive (HSE).

HSE. (2001). *Reducing risks, protecting people—HSE's decision-making process*. London: Health and Safety Executive (HSE). https://doi.org/10.1205/095758203762851994.

HSE. (2005). *The offshore installations (safety case) regulations 2005*. London: Health and Safety Executive (HSE).

HSE. (2006). *Developing process safety indicators—A step-by-step guide for chemical and major hazard industries*. London: HSE.

HSE. (2015a). *The control of major accident hazards (COMAH) regulations* (3rd ed.). London: Health and Safety Executive (HSE).

HSE. (2015b). *The offshore installations (offshore safety directive) (safety case etc.) regulations*. London: Health and Safety Executive (HSE).

IAEA. (2009). *Passive safety systems and natural circulation in water cooler nuclear power plants*. Austria, Vienna: IAEA. Retrieved from http://www-pub.iaea.org/MTCD/publications/PDF/te_1624_web.pdf.

ISA. (1996). *ANSI/ISA-S84.01: Application of safety instrumented systems for the process industries*. North Carolina: ISA—The Instrumentation, Systems, and Automation Society.

ISA. (2002). *ISA-TR84.00.02: Safety instrumented functions (SIF)-safety integrity level (SIL) evaluation techniques Part 2: Determining the SIL of a SIF via simplified equations ISA-TR84.00.02*. North Carolina: ISA—The Instrumentation, Systems, and Automation Society.

ISO. (2002). *BS EN ISO 17776:2002 petroleum and natural gas industries—Offshore production installations—Guidelines on tools and techniques for hazard identification and risk assessment* (1st ed.). Brussels: International Organization for Standardization.

ISO. (2014). *ISO 55000: Asset management—Overview, principles and terminology*. Geneva: International Organization for Standardization.

ISO. (2015). *ISO 9001:2015 quality management systems—Requirements.* Geneva: The International Organization for Standardization.
Khan, F., Abunada, H., John, D., & Benmosbah, T. (2010). Development of risk-based process safety indicators. *Process Safety Progress*, *29*(2), 133–143. https://doi.org/10.1002/prs.10354.
Khan, F., Ahmed, S., Hashemi, S. J., Yang, M., Caines, S., & Oldford, D. (2013). Safety challenges in harsh environments: Lessons learned. In *1st CCPS Asia-Pacific Conference on Process Safety. Qingdao, China.*
Khan, F., Ahmed, S., Hashemi, S. J., Yang, M., Caines, S., Rathnayaka, S., & Oldford, D. (2015). Safety challenges in harsh environments: Lessons learned. *Process Safety Progress*, *34*(2), 191–195. https://doi.org/10.1002/prs.11704.
Khan, F., Hashemi, S. J., Paltrinieri, N., Amyotte, P., Cozzani, V., & Reniers, G. (2016). Dynamic risk management: a contemporary approach to process safety management. *Current Opinion in Chemical Engineering*, *14*, 9–17. https://doi.org/10.1016/j.coche.2016.07.006.
Kletz, T. (2004). *Still going wrong!: Case histories of process plant disasters and how they could have been avoided.* Burlington, MA: Gulf Professional Publishing.
Lundberg, J., Rollenhagen, C., & Hollnagel, E. (2009). What-you-look-for-is-what-you-find—The consequences of underlying accident models in eight accident investigation manuals. *Safety Science*, *47*(10), 1297–1311. https://doi.org/10.1016/j.ssci.2009.01.004.
Lundberg, J., Rollenhagen, C., & Hollnagel, E. (2010). What you find is not always what you fix—How other aspects than causes of accidents decide recommendations for remedial actions. *Accident; Analysis and Prevention*, *42*(6), 2132–2139. https://doi.org/10.1016/j.aap.2010.07.003.
Macza, M. (2008). *A Canadian perspective of the history of process safety management legislation.* Germany: Cologne, pp. 1–22.
Mannan, S. (2012). *Lees' loss prevention in the process industries: Hazard identification, assessment and control, Volume 1* (4th ed.). United States: Elsevier.
Mannan, M. S., Reyes-Valdez, O., Jain, P., Tamim, N., & Ahammad, M. (2016). The evolution of process safety: Current status and future directions. *Annual Review of Chemical and Biomolecular Engineering*, 7, 135–162. https://doi.org/10.1146/annurev-chembioeng-080615-033640.
Mannan, M. S., Sachdeva, S., Chen, H., Reyes-Valdes, O., Liu, Y., & Laboureur, D. M. (2009). Modeling and simulation of the polymeric nanocapsule formation process. *IFAC Proceedings Volumes (IFAC-Papers Online)*, 7(Pt. 1), 405–410. https://doi.org/10.1002/aic.
Marsh. (2016). *The 100 largest losses 1974–2015.* Texas: Marsh's Risk Consulting Practice.
McAteer, J. D., & Whiteman, L. (1993). Learning from hamlet: The case for a national safety and health board. *New Solutions*, *3*(2), 54–59. https://doi.org/10.2190/NS3.2.j.
Modarres, M. (2006). *Risk analysis in engineering: Techniques, tools, and trends.* Boca Raton, FL: CRC Press.
NOPSEMA. (2012a). *Guidance note: Control measures and performance standards.* Perth, Australia: National Offshore Petroleum Safety and Environmental Management Authority.
NOPSEMA. (2012b). *Guidance note N-04300-GN0107—Hazard identification.* Perth, Australia: National Offshore Petroleum Safety and Environmental Management Authority.
NOPSEMA. (2012c). *Guidance note N-04300-GN0165—Risk assessment.* Perth, Australia: National Offshore Petroleum Safety and Environmental Management Authority.
NOPSEMA. (2015). *Guidance note: ALARP.* Perth, Australia: National Offshore Petroleum Safety and Environmental Management Authority.

OGP. (2011). *Process safety—Recommended practice on key performance indicators. Retrieved from* www.ogp.org.uk.

Paltrinieri, N., Khan, F., & Cozzani, V. (2014). Coupling of advanced techniques for dynamic risk management. *Journal of Risk Research, 9877*, 1–21. https://doi.org/10.1080/13669877.2014.919515.

Rathnayaka, S., Khan, F., & Amyotte, P. (2011). SHIPP methodology: Predictive accident modeling approach. Part I: Methodology and model description. *Process Safety and Environmental Protection, 89*(3), 151–164. https://doi.org/10.1016/j.psep.2011.01.002.

Reason, J. T. (1990). The contribution of latent human failures to the breakdown of complex systems. *Philosophical Transactions of the Royal Society B, 327*, 475–484.

Rijnmond Public Authority. (1982). *Risk analysis of six potentially hazardous industrial objects in the Rijnmond area: A pilot study*. Reidel, Dordrecht (NL): Springer Science & Business Media.

Rüschendorf, L. (2013). *Mathematical risk analysis: Dependence, risk bounds, optimal allocations and portfolios*. Heidelberg: Springer. https://doi.org/10.1007/978-3-642-33590-7.

Spadoni, G., Egidi, D., & Contini, S. (2000). Through ARIPAR-GIS the quantified area risk analysis supports land-use planning activities. *Journal of Hazardous Materials, 71*(1–3), 423–437. https://doi.org/10.1016/S0304-3894(99)00091-6.

Sreenevasan, R. (2015). The effect of regulations in improving process safety. In *Risk engineering society WA's technical event. Perth, Australia.*

Taleb, N. N. (2007). Black swans and the domains of statistics. *The American Statistician, 61*(3), 198–200. https://doi.org/10.1198/000313007X219996.

Tjorhom, B. B., & Aase, K. (2010). The role of complexity in accident investigation practice. *International Journal of Emergency Management*, 7(2), 167. https://doi.org/10.1504/IJEM.2010.033655.

Uijt de Haag, P. A. M., & Ale, B. J. M. (1999). *Guidelines for quantitative risk assessment (purple book)*. The Hague (NL): Committee for the Prevention of Disasters.

Wiig, S., & Aase, K. (2007). Fallible humans in infallible systems? Learning from errors in health care. *Safety Science Monitor, 3*, 1–13.

WorkSafe. (2011). *Guidance note: Hazard identification at a major hazard facility*. Melbourne: WorkSafe Victoria Advisory Service.

CHAPTER TWO

Learning From the Experience

Howard Pike[1]
Centre for Risk, Integrity and Safety Engineering (C-RISE), Faculty of Engineering, Memorial University, St. John's, NL, Canada
[1]Corresponding author: e-mail address: p79hlp@mun.ca

Contents

The evolution of the term "process safety management" is closely associated with the major accidents that occurred in the chemical processing industry during the 20th century. Those accidents include: Flixborough, UK (1974), Seveso, Italy (1976), Bhopal, India (1984), and Piper Alpha, UK (1988) (Khan et al., 2015).

The study of case histories of these accidents, and others, provides valuable information and learning opportunities. It also provides for the industry, regulators, and governments, the opportunity to make changes in an effort to reduce and prevent future incidents.

Methods in Chemical Process Safety, Volume 1
ISSN 2468-6514
http://dx.doi.org/10.1016/bs.mcps.2017.02.002

This chapter presents some examples and cases to illustrate the discussion about process safety management, human factors in process safety, and risk-based process safety. The experiences from two major corporations are used: three incidents involve ExxonMobil and two incidents are from BP.

The first example is the *Exxon Valdez* oil spill. It highlights the role of human factors in an incident, both impairment and fatigue. The accident was also the driving force behind ExxonMobil's development of its safety management system. The second is the explosion and fire at ExxonMobil's Esso Longford Gas Plant in Australia. It highlights the difficulty in implementing a safety management system and the value of performing risk studies on an existing plant and, more particularly, on an older existing plant. It also shows the role of human factors in accidents, specifically training and supervision. The third event, Blackbeard, did not result in an incident, but illustrates the use of risk assessment in decision making. When the risk was considered too high, the project was shut down and abandoned.

The fourth case study covers BP's major expansion program and the acquisition of a number of older refineries in the United States, one with a history of fatal accidents. There were 23 fatalities at the Texas City Refinery over 30 years, with 3 deaths in 2004 alone. Despite that history, BP failed to take necessary measures to identify various failings in such broad areas as equipment, risk management, staff management, working culture, maintenance and inspection regimes, and general health and safety assessments. The safety culture and management at BP was the subject of an independent investigation and highlighted the shortcomings of relying on lost time injury rates as a measure of overall safety. The final example is arguably among the most notorious of recent history and is referred to as either the BP Macondo, after the oil well being drilled, or the Deepwater Horizon (DWH), after the name of the drilling unit destroyed. It highlights the complexity of offshore drilling operations in the modern era. In addition to the implications for BP, the corporation, also brought the global offshore industry under intense scrutiny during and after the accident. It resulted in eight separate investigations.

The accident and investigation findings from that incident were a game changer for the offshore drilling industry. It prompted reviews of existing practices around the globe, including the review and amendment of government regulations, and changes in drilling and response procedures by industry associations. As well, the International Association of Oil and Gas Producers (IOGP) developed a coordinated international response capability to be deployed to cap an offshore well. Across the breath of the offshore industry, regulations, procedures, and standards were reviewed and revised in light of the lessons learned from Macondo.

These incidents will be reviewed by providing context, a synopsis of the event, highlight some of the key findings of the investigations, and look at the lessons learned.

1. EXXON VALDEZ (1989)

By March 1989, tankers carrying crude oil from the Alaska North Slope had safely transited Prince William Sound more than 8700 times in the 12 years, since oil began flowing through the trans-Alaska pipeline (Parker, 1990). In that decade-plus there were no major disasters and, in fact, very few serious incidents. This history gave little reason to suspect impending disaster. The system designed to carry 2-million barrels of North Slope oil daily to West Coast and Gulf Coast markets had worked well, perhaps too well. At least partly because of the success of the Valdez tanker trade, a general complacency had come to permeate the operation and oversight of the entire system. That complacency and success were shattered shortly after midnight on March 24, 1989. Industry's insistence on regulating the Valdez tanker trade in its own way, and government's incremental accession to industry pressure, resulted in a disastrous failure of the system.

That night, the US tankship *Exxon Valdez*, loaded with about 1,263,000 barrels of crude oil, ran aground on Bligh Reef in Prince William Sound, near Valdez, Alaska. There were no injuries, but about 258,000 barrels of oil were spilled when eight cargo tanks ruptured, resulting in catastrophic damage to the environment. Damage to the vessel was estimated at $25 million. The cost of the lost cargo was estimated at $3.4 million. The cost of the oil spill cleanup came in at roughly $1.85 billion.

The National Transportation Safety Board (NTSB, 1990) determined that the cause of the grounding of the *Exxon Valdez* was a combination of factors. They included: the failure of the third mate to properly maneuver the vessel because of fatigue and excessive workload; the failure of the master to provide a proper navigation watch because of impairment from alcohol; the failure of Exxon Shipping Company to provide a fit master and a rested and sufficient crew; the lack of an effective Vessel Traffic Service because of inadequate equipment and manning levels, inadequate personnel training, and deficient management oversight; and the lack of effective pilotage services.

1.1 Synopsis of the Event

The *Exxon Valdez* arrived Alyeska Marine Terminal at 11:30 p.m. on March 22, 1989 to take on cargo. It carried a crew of 19 plus the master. The third mate, who became a central figure in the grounding, was relieved of watch duty at 11:50 p.m. Vessel and terminal crews began loading crude oil onto the tanker at 5:05 a.m. on 23 March under the supervision of the chief mate.

The master, chief engineer, and radio officer left the *Exxon Valdez* about 11:00 a.m. on 23 March, and where driven from the Alyeska terminal into the town of Valdez. They expected the *Exxon Valdez*'s sailing time to be 10 p.m. that evening. They left Valdez by taxi at about 7:30 p.m., got through Alyeska terminal gate security at 8:24 p.m. and boarded the vessel. Loading of the *Exxon Valdez* had been completed for an hour by the time the group returned to the vessel and the sailing time had been revised to 9:00 p.m. Both the cab driver and the gate security guard later testified that no one in the party appeared to be intoxicated. A ship's agent who met with the master after he got back on the vessel said it appeared the master may have been drinking because his eyes were watery, but she did not smell alcohol on his breath. However, the marine pilot, assigned to the vessel, indicated that later he did detect the odor of alcohol on the master's breath.

The master's activities in town that day and on the ship that night would become a key focus of accident inquiries, the cause of a state criminal prosecution, and the basis of widespread sensational media stories.

The *Exxon Valdez*'s deck log shows it was clear of the dock at 9:21 p.m. under the direction of the marine pilot and accompanied by a single tug for the passage through Valdez Narrows, the constricted harbor entrance about seven miles from the berth. According to the marine pilot, the master left the bridge at 9:35 p.m. and did not return until about 11:10 p.m., even though company policy requires two ship's officers on the bridge during transit of Valdez Narrows.

The passage through Valdez Narrows proceeded uneventfully. At 10:49 p.m., the ship reported to the Valdez Vessel Traffic Center that it had passed out of the Narrows and was increasing speed. At 11:05 p.m., the marine pilot asked that the master be called to the bridge in anticipation of his disembarking from the ship, and at 11:10 p.m. the master returned. The marine pilot disembarked at 11:24 p.m., with assistance from the third mate. While the third mate was helping the marine pilot and then helping stow the pilot ladder, the master was the only officer on the bridge and according to the NTSB report there was no lookout, even though one was required.

At 11:25 p.m., the master informed the Vessel Traffic Center that the marine pilot had departed and that he was increasing to sea speed. He also reported that they would probably divert from the Traffic Separation Scheme (TSS) and travel in the inbound lane if there was no conflicting traffic. The traffic center indicated concurrence, stating there was no reported traffic in the inbound lane.

The TSS was designed to separate incoming and outgoing tankers in Prince William Sound and keep them in clear, deep waters during their transit. It consists of inbound and outbound lanes, with a half-mile-wide separation zone between them. Small icebergs from nearby Columbia Glacier occasionally enter the traffic lanes. Masters had the choice of slowing down to push through them safely or deviating from their lanes if traffic permitted. The master's report, and the Valdez traffic center's concurrence, meant the ship would change course to leave the western, outbound lane, cross the separation zone and, if necessary, enter the eastern, inbound lane to avoid floating ice. At no time did the *Exxon Valdez* report or seek permission to depart farther east from the inbound traffic lane; but that is exactly what it did.

At 11:30 p.m., the master informed the Valdez traffic center that he was turning the ship toward the east on a heading of 200 degrees and reducing speed to "wind my way through the ice" (engine logs, however, show the vessel's speed continued to increase). At 11:39 p.m., the third mate plotted a fix that showed the ship in the middle of the TSS. The master ordered a further course change to a heading of 180 degrees (due south) and, according to the helmsman, directed that the ship be placed on autopilot. The second course change was not reported to the Valdez traffic center. For 19 or 20 min the ship sailed south—through the inbound traffic lane, then across its easterly boundary and on toward its peril at Bligh Reef. Traveling at approximately 12 knots, the *Exxon Valdez* crossed the traffic lane's easterly boundary at 11:47 p.m.

At 11:52 p.m., the command was given to place the ship's engine on "load program up"—a computer program that, over a span of 43 min, would increase engine speed from 55 RPM to sea speed full ahead at 78 RPM. After conferring with the third mate about where and how to return the ship to its designated traffic lane, the master left the bridge. The time, according to NTSB testimony, was approximately 11:53 p.m.

By this time, the third mate had been on duty for 6 h and was scheduled to be relieved by the second mate. But the third mate, knowing the second mate had worked long hours during loading operations during the day, had told the second mate he could take his time in relieving him. The third mate did not call the second mate to awaken him for the midnight to 4 a.m. watch, instead remaining on duty himself. Testimony before the NTSB suggests that the third mate may have been awake and generally at work for up to 18 h preceding the accident.

The US Coast Guard was responsible for setting minimum crew numbers. It had certified Exxon tankers for a minimum of 15 persons (14 if the radio officer is not required). The president of Exxon Shipping Company, had stated that his company's policy was to reduce its standard crew complement to 16 on fully automated, diesel-powered vessels by 1990. Exxon maintained that modern automated vessel technology permitted the reduction in the number of crew without compromising safety or function.

Sometime during the critical period before the grounding and during the first few minutes of 24 March, the third mate plotted a fix indicating it was time to turn the vessel back toward the traffic lanes. About the same time, the lookout reported that Bligh Reef light appeared broad off the starboard bow—that is, off the bow at an angle of about 45 degrees. The light should have been seen off the port side (the left side of a ship, facing forward). Its position off the starboard side indicated looming and great peril for a supertanker that was out of its lanes and accelerating through close waters. The third mate gave right rudder commands to cause the desired course change and took the ship off autopilot. He also phoned the master in his cabin to inform him the ship was turning back toward the traffic lanes and that, in the process, it would be getting into ice. When the vessel did not turn swiftly enough, the third mate ordered further right rudder with increasing urgency. Finally, realizing the ship was in serious trouble, the third mate phoned the master again to report the danger. At the end of the conversation, he felt an initial shock to the vessel. The grounding, described by helmsman as "a bumpy ride" and by the third mate as six "very sharp jolts," occurred at 12:04 a.m.

After feeling the grounding, the master rushed to the bridge arriving as the vessel came to rest. He immediately gave a series of rudder orders in an attempt to free the vessel, and power to the vessel's engine remained in the "load program up" condition for about 15 min after impact. The chief mate went to the engine control room and determined that eight cargo tanks and two ballast tanks had been ruptured. He concluded the cargo tanks had lost an average of 10 feet of cargo, with approximately 67 feet of cargo remaining in each. He informed the master of his initial damage assessment and was instructed to perform stability and stress analysis. At 12:19 a.m., the master ordered that the vessel's engine be reduced to idle speed.

At 12:26 a.m., the master radioed the Valdez traffic center and reported his predicament.

> *We've fetched up, ah, hard aground, north of Goose Island, off Bligh Reef and, ah, evidently leaking some oil and we're gonna be here for a while and, ah, if you want, ah, so you're notified.*

The master, meanwhile, continued efforts to power the *Exxon Valdez* off the reef. At approximately 12:30 a.m., the chief mate used a computer program to determine that though stress on the vessel exceeded acceptable limits, the ship still had required stability. He went to the bridge to advise the master that the vessel should not go to sea or leave the area. The master directed him to return to the control room to continue assessing the damage and to determine available options. At 12:35 p.m., the master ordered the engine back on—and eventually to "full ahead"—and began another series of rudder commands in an effort to free the vessel. After running his computer program again another way, the chief mate concluded that the ship did not have acceptable stability without being supported by the reef. The chief mate relayed his new analysis to the master at 1:00 a.m. and again recommended that the ship not leave the area. Nonetheless, the master kept the engine running until 1:41 a.m., when he finally abandoned efforts to get the vessel off the reef.

The vessel came to rest roughly facing southwest, perched across its middle on a pinnacle of Bligh Reef. Computations aboard the *Exxon Valdez* showed that 5.8 million gallons had leaked out of the tanker in the first 3¼ h. Weather conditions at the site were reported to be 33°F, slight drizzle with a rain/snow mix, north winds at 10 knots, and visibility 10 miles.

The response capabilities of Alyeska Pipeline Service Company to deal with the spreading oil spill would be found to be both unexpectedly slow and woefully inadequate. The worldwide capabilities of ExxonMobil would mobilize huge quantities of equipment and personnel to respond to the spill, but not in the crucial first few hours and days when containment and cleanup efforts are at a premium. The US Coast Guard would demonstrate its prowess at ship salvage, protecting crews, and lightering operations, but would prove utterly incapable of oil spill containment and response. State and federal agencies would show differing levels of preparedness and command capability. The consequence—the waters of Prince William Sound, and eventually more than 1000 miles of beach in Southcentral Alaska, would be fouled by 10.8 million gallons of crude oil.

1.2 Key Findings

The NTSB concluded that the *Exxon Valdez* met all US and international segregated-ballast regulations, but that the standards at that time for segregated ballast and cargo tank size did not provide sufficient protection against oil spills by groundings or collisions.

Further it concluded that:

- Ice in Valdez Arm is a significant hazard to navigation and required closer monitoring and reporting. The monitoring of the amount and size of ice being calved from Columbia Glacier was inadequate for the safety of tankships transiting Prince William Sound.
- The master's judgment was impaired by alcohol during the critical period the vessel was transiting Valdez Arm. It found that the Exxon Shipping Company did not adequately monitor the master for alcohol abuse after his alcohol rehabilitation program. In addition, the company did not have a sufficient program to identify, and, if necessary remove from service or provide treatment for, employees who had substance abuse problems.
- The master's decision to depart from the TSS to avoid ice was probably reasonable, even though it required a heading toward shoal water.
- Navigating the *Exxon Valdez* between the ice field and Bligh Reef required a diligent, competent navigation watch capable of controlling the vessel, watching for ice, and fixing the vessel's position frequently to navigate the vessel safely; hence, two officers were required on the bridge, one with the navigation and ship handling experience to control the vessel and the other to frequently fix the vessel's position.
- The master's decision to leave the third mate in charge of the navigation watch was contrary to Federal regulations and company policy and was improper given the course of the vessel, the uncertain extent of the ice conditions, the proximity of a dangerous reef, and the fact that the third mate did not have the required pilotage endorsement.
- The performance of the third mate was deficient, probably because of fatigue, when he assumed supervision of the navigation watch from the master and that the third mate's failure to turn the vessel at the proper time and with sufficient rudder probably was the result of his excessive

workload and fatigued condition, which caused him to lose awareness of the location of Bligh Reef.

- There were no rested deck officers on the *Exxon Valdez* available to stand the navigation watch when the vessel departed from the Alyeska Terminal. Many conditions conducive to producing crew fatigue on the *Exxon Valdez* also existed on other Exxon Shipping Company vessels because many were three mate vessels and because the company had pursued reduced crewing procedures. Exxon Shipping had incentives and work requirements that could be conducive to fatigue and their crewing policies did not adequately consider the increase in workload caused by reductions in the number of crew. The Coast Guard was unduly narrow in its perspective when it evaluated reduced crewing requests for the *Exxon Valdez*; it based reductions primarily on the assumption that shipboard hardware and equipment might reduce the workload at sea, but did not consider the heavier workload associated with cargo operations in port and the frequency of such operations.
- Although moving the pilot station to Rocky Point was apparently based on a consideration for pilot safety, the move resulted in a reduction in pilotage services past Bligh Reef, where local knowledge was needed. Moving the pilot station to a position south of Bligh Reef would enhance navigation safety by ensuring the presence of an officer with local knowledge of the area on the bridge of each vessel transiting Valdez Arm past Bligh Reef.
- The Coast Guard had not maintained an effective vessel traffic service in Prince William Sound. The limited supervision of the Vessel Traffic Center probably contributed to the commanding officer's and operation officer's lack of awareness that tankers were departing from the TSS to avoid ice and were passing close to Bligh Reef. The VTS radar was operating satisfactorily, and the detection range of the radar was not significantly reduced by weather or sea conditions while the *Exxon Valdez* was transiting Valdez Arm. However, the VTC lost radar contact with the *Exxon Valdez* about 7.7 miles from the radar site, which is about 5.5 miles from the northern part of Bligh Reef, because the Center's watch stander did not use a higher range scale and not because of any limitation or malfunction of the radar. Had he used a higher range scale, the vessel probably could have been tracked as far as the site of the grounding, but no firm policy required him to do so. Monitoring the *Exxon Valdez* by radar as it transited Valdez Arm would have revealed to the VTC watch stander that the vessel had changed course to 180

degree, had departed the vessel TSS, and was headed for shoal water east of Bligh Reef. A query or warning from the VTC might have alerted the third mate to the impending danger from Bligh Reef. The monitoring of vessels in Valdez Arm was left to the discretion of the VTC watch stander because the senior watch stander decided to allow the Center's watch standers to monitor instead of plot the positions of vessels transiting Valdez Arm. A firm policy requiring the VTC to plot tankers transiting the full length of Valdez Arm could have alerted the commanding officer of the Marine Safety Office that tankships were departing from the TSS in the vicinity of Bligh Reef to avoid ice.

1.3 Lessons Learned

In the aftermath of the *Exxon Valdez* accident, ExxonMobil made major changes. It committed to safeguarding the environment, employees, and operating communities worldwide. As an example, to improve oil spill prevention, it has:

- Modified tanker routes.
- Instituted drug and alcohol testing programs for safety sensitive positions.
- Restricted safety sensitive positions to employees with no history of substance abuse.
- Implemented more extensive periodic assessments of ExxonMobil vessels and facilities.
- Strengthened training programs for vessel captains and pilots.
- And, applied new technology to improve vessel navigation and ensure the integrity of oil containment systems in the event a spill occurs and have improved their response capability.

Following the *Exxon Valdez* oil spill and against the background of a number of other disasters arising from the hazardous activities of companies other than Exxon and its affiliates, the company developed a framework for the safe and environmentally sound operation of its various undertakings. The framework was called the Operations Integrity Management Framework (OIMF). Within this framework, a series of expectations and guidelines were developed which included the Exxon Company International (ECI) Upstream OIMS Guidelines. The ECI guidelines contained 11 primary elements with associated expectations, and a series of guidelines for the achievement of these expectations. The company decreed that its affiliates should develop a management system in which all the expectations outlined in the OIMF and contained in the ECI Guidelines, were met. The elements referred to in the Guidelines are:

1. Management leadership, commitment, and accountability.
2. Risk assessment and management.
3. Facilities design and construction.
4. Information/documentation.
5. Personnel and training.
6. Operations and maintenance.
7. Management of change.
8. Third party services.
9. Incident investigation and analysis.
10. Community awareness and emergency preparedness.
11. Operations integrity assessment and improvement.

While there is no question that the *Exxon Valdez* spill was an unfortunate, but avoidable incident, it is also clear that it provided a necessary impetus to reexamine the state of oil spill prevention, response, and cleanup. In addition to the *Exxon Valdez* spill, the summer of 1989 experienced three oil spills that drained any resources left from the ongoing spill response in Alaska. Between 23 and 24 June, the T/V *World Prodigy* spilled 290,000 gallons of oil in Newport, Rhode Island; the T/V *Presidente Rivera* emptied 307,000 gallons of oil into the Delaware River; and the T/V *Rachel* hit Tank Barge *2514*, releasing 239,000 gallons of oil into Houston Ship Channel (Shigenaka, 2014). But these were not the only oil spills plaguing US waters during that time, and it resulted in action from the politicians.

In August of 1990, the US Congress voted unanimously to pass the Oil Pollution Act, which significantly improved measures to prevent, prepare for, and respond to oil spills in US waters. The shipping industry underwent a significant makeover in oil spill prevention, preparedness, and response. Examples of the result of the spills and legislation include the phasing out of tankers with single hulls, new regulations requiring the use of knowledgeable pilots, maneuverable tug escorts, and an appropriate number of people on the ship's bridge during transit.

But perhaps one of the most important elements of this law required those responsible for oil spills to foot the bill for both cleaning up the oil, and for economic and natural resource damages (NRD) resulting from it. This provision requires oil companies to pay into the Oil Spill Liability Trust Fund, a fund theoretically created by Congress in 1986 but not given the necessary authorization until the Oil Pollution Act of 1990. The fund helps the US Coast Guard pay for the upfront costs of responding to marine and coastal accidents that threaten to release hazardous materials such as oil into the environment. It also covers the assessment of potential environmental and cultural impacts, and implementing any restoration to make up for such impacts.

2. ESSO LONGFORD (1998)

Since 1969, the Gippsland Basin, located in the Bass Strait off the state of Victoria, Australia, supplied most of the state's gas requirements. It also supplies gas to New South Wales, Tasmania, and other locations. In the 50 years since the February 1965 discovery well, and the offshore facilities that followed, some 4 billion barrels of oil and 8 trillion cubic feet of gas were produced. In the half century since the first discovery, Esso Australia Resources Ltd (Esso), operator of the venture, and its partner BHP Billiton Petroleum (Bass Strait) Pty. Ltd. invested around $20 billion funding 17 platforms, associated subsea production systems, and other offshore installations feeding a network of about 370 miles of pipelines.

Onshore at Longford in south-eastern Victoria, Esso operates three gas plants to process gas flowing from wells in Bass Strait. It also operates a Crude Oil Stabilization Plant (CSP) at Longford to process oil flowing from other wells in Bass Strait. These four plants are interconnected, as the processing of gas produces some liquids which are further processed in the CSP. Similarly, the processing of crude oil in the stabilization plant produces some gas which is then fed to the gas plants for final processing before sale.

On the afternoon of Friday, September 25, 1998, a vessel in Gas Plant 1 (GP1) fractured, releasing hydrocarbon vapors and liquid. Explosions and a fire followed. Two Esso employees were killed and eight others were injured. Supplies of natural gas to domestic and industrial users in the state of Victoria were halted for 2 weeks.

The gas coming ashore from the Bass Strait platforms contains significant amounts of hydrocarbon liquids (condensate) and water. To meet the specified quality for sales gas, it is necessary to process the gas to remove all the water and most of the liquefiable components, and also to remove hydrogen sulfide, a noxious gas present in very small quantities.

The condensate arriving at Longford in the gas stream is removed in a system of large pipes called slug catchers and all traces of water and hydrogen sulfide are then removed by molecular sieves which preferentially extract these compounds from the gas stream. The liquefied petroleum gas (LPG) components then have to be removed.

In 1969 when the facility at Longford was established, there was only one gas plant (GPl) and a crude stabilization plant. The commissioning of Gas Plant 2 (GP2) in 1976 and Gas Plant 3 (GP3) in 1983 enhanced the site's capacity. GP2 and GP3 used newer technology then GP1 to process the gas, namely, a cryogenic process. This process uses a series of expansions and liquid separations followed by recompressions to remove the ethane and heavier components. Some sections of this cryogenic process are designed to operate at very low temperatures, well below the temperatures used in GPl.

GP1 uses a refrigerated lean oil absorption process to remove the LPG, so-called because lean oil (a light oil similar to aviation kerosene) is circulated at low temperature over trays in a tower (called an absorber) to extract the LPG components from the gas stream which is passing up the tower. The lean oil is enriched by LPG which it extracts and is then called rich oil. The processed gas from the top of the tower is piped away for sale and the cold, rich oil leaves the absorber, and is heated by passing through several heat exchangers before a two-stage distillation process to recover the LPG as a marketable product. Having the LPG components stripped from it, the rich oil becomes lean oil and is circulated back through the system of heat exchangers to return to the absorber as cold, lean oil.

2.1 Synopsis of the Event

The night before the incident there was a larger than usual flow of liquids into the plant from offshore. The result was a build-up of the level of condensate in the absorber. The volume of condensate could be controlled to some extent by raising the temperature. However, an automatic valve which controlled the temperature was not working properly and operators were using a manual by-pass valve. For various reasons, they did not keep the temperature high enough and the build-up of condensate continued. The outflow through the condensate outlet was too great for the downstream reprocessing so the overflow rate was automatically reduced. The level of condensate in the absorber tower then rose so high that it went off scale, that is, beyond the point where the operators could monitor it. In fact, it rose to the point where it overflowed into the rich oil stream.

The presence of condensate in the rich oil stream caused the rich oil to become much colder than normal. This caused an upset in the processing equipment downstream which in turn led to an automatic shutdown of the pumps which maintained the lean oil flow.

Operators were unable to restart these pumps and they remained shutdown for hours.

Because the circulation of warm lean oil had stopped, two of the heat exchangers became abnormally cold and a thick layer of frost formed on their exterior pipework. The temperature dropped below the design limit and the metal in one exchanger contracted to the point that it began to leak oil onto the ground. Unsuccessful attempts were made to fix this leak by tightening the bolts. At this point, operators decided to stop the flow into GP1 to try to deal with the situation. This stopped any further flow of cold condensate within the plant. But operators did not depressurize the plant. Rather, they decided to try again to restart the pumps to rewarm the heat exchanger. This was a critical error. The metal in the vessel by this time was so cold that it was brittle and it needed time to thaw out before being warmed. Operators succeeded in restarting the pumps and the reintroduction of warm liquid caused fracturing and catastrophic failure of one of the heat exchangers. A large quantity of volatile liquid and gas escaped and was ignited by a nearby ignition source.

The explosions and fire shutdown the three gas plants at Longford. The fire was not fully extinguished until September 27, 1998. The resumption of gas supply commenced on October 04, 1998 and was completed by October 14, 1998. The 10-day delay to restore gas supply was due to the need to extinguish the fires in GPl and to ensure the complete isolation of GPl, and the CSP, from GP2 and GP3.

On October 12, 1998, the Victoria Government announced its intention to establish a Royal Commission of Inquiry into the explosion and fire at Longford. The Commission submitted its report on June 28, 1999 (Dawson et al., 1999).

The Commission's key findings were divided among a number of categories.

2.1.1 Safety Management System

The Commission heard evidence that OIMS was a world class system and complied with world's best practice. It suggested that may be true of the expectations and guidelines upon which the system was based, but the same could not be said of the operation of the system in practice. Even the best management system is defective if it is not effectively implemented. Esso's OIMS, together with all the supporting manuals, comprised a complex management system. It was repetitive, circular, and contained unnecessary cross referencing. These characteristics made the system difficult to comprehend both by management and by operations personnel.

The Commission Further Concluded:

- The fundamental shortcoming was in the implementation of the OIMS, as seen in the inadequate state of knowledge of Esso personnel of the hazards associated with loss of lean oil circulation in GP1 and of the actions which could be taken to mitigate such hazards. As a result of this lack of knowledge, the Commission concluded that practices adopted by operations personnel fell far short of good operating practice and were counter to the safe operation of the plant on that day.
- The reliance placed by Esso on its OIMS for the safe operation of the plant was misplaced. The accident on September 25, 1998 demonstrated that important components of Esso's system of management were either defective or not implemented. If the implementation of OIMS by Esso was to be measured by the adequacy of its operating procedures, they were deficient and failed to conform with the ECI Upstream Guidelines or with the OIMS Systems Manual. If it was to be measured by reference to the actions and decisions of those persons who were attempting to resolve the process upsets on September 25, 1998, they were also deficient.

2.1.2 Lack of Knowledge

The Commission concluded there were deficiencies in the manner that the company dealt with the acquisition and retention of knowledge through its training system, its operating procedures, its documentation and data system, and its communication system. The Commission stated that the evidence from the operators and supervisors on the day of the incident indicated an apparent lack of their knowledge. Even if some aspects of that evidence could be criticized, the actual events which occurred on September 25, 1998 were a sure indication of a deficiency in the knowledge required to operate GPl safely.

2.1.3 Lack of Adequate Operation Procedures

The Commission noted an example of Esso's failure to implement OIMS was apparent from the state of the Longford Plant Operating Procedures Manual which contained the operating procedures for GPI and was located in the GPI control room. It was a controlled document and was identified by the OIMS Systems Manual as part of OIMS. The manual did not comply with the guidelines in critical respects. The operating procedures dealing with a lean oil absorption plant did not include any reference to the importance of maintaining lean oil flow in the operation of the plant. It did not contain any reference to the loss of lean oil flow and contained no procedures to deal with such an event. Nor did it contain any reference to GPl shutdown or startup procedures or the safe operating temperatures for the two heat exchangers.

The Commission concluded that the events leading up to the accident disclosed a number of instances where operators failed to adhere to basic operating practices. Some of these practices were written, for example, those relating to shift handover and operator log entries. Others would seem to be matters of common sense and include monitoring plant conditions, responding appropriately to alarms, reporting process upsets to supervisors, and undertaking appropriate checks before making adjustments to process variables.

2.1.4 Risk Assessment

Hazard and Operability (HAZOP) studies as part of the design process for new plant were a requirement of OIMS. OIMS also contained provision for retrospective HAZOP studies on existing plants. GP1 was constructed well before the introduction of OIMS and before the use of HAZOP studies became common practice. Following the introduction of OIMS, Esso recognized the need to undertake retrospective HAZOP studies of all its major facilities. Retrospective HAZOP studies were conducted for GP2 in September 1994, for GP3 in November 1994 and for the CSP in December 1995.

The Commission found that no formal hazard identification or structured risk assessment of any kind took place in GP1 after 1994. The Commission stated that the failure to carry out a HAZOP study for GP1 meant there was the risk that hazards would remain unidentified and uncontrolled. The events of September 25, 1998 demonstrated the existence of such hazards. Had a HAZOP study of GPl been conducted, as Esso initially believed it should, it would have acquired knowledge of those critical hazards. Knowledge would then have been disseminated by way of training, the development, and use of procedures, and the adoption of protective control systems. The Commission concluded that the failure to conduct a HAZOP study of GPI contributed to the disaster.

2.1.5 Control Room Log and Shift Handovers

The Commission noted that log book entries were not subjected to any examination either by Longford plant management or by management in Melbourne. They did not appear to have been used by management as a means of monitoring process conditions at the plant nor were they passed on to any person or group in Melbourne for plant surveillance purposes.

The shift supervisors' log was available to management personnel both at Longford and in Melbourne. However, process upsets were not generally included in that document due to the particular responsibilities of plant supervisors. It meant that the shift supervisors' log was not a substitute for a properly maintained control room log.

The Commission concluded that shift handovers and log book entries were used ineffectively in the lead up to the accident.

2.1.6 Operation in Alarm Mode

The Commission found evidence that in the GPl control room it was common for a large number of alarms to be active at any one time. Many of these alarms were nuisance alarms activated because the process variable monitored by the alarm was operated at the upper or lower end of its range and was constantly moving in and out of alarm range. This caused frequent and repetitive alarms. In the evidence, an operator said that nuisance alarms had the capacity to distract operators and frequently did. They could be very repetitive and could result in more important alarms not being picked up or noticed because their warning signals were lost among numerous other alarms.

The Commission found no evidence of any system to give priority to important alarms. Good operating practice would have dictated that critical alarms be identified and given priority. It would also have dictated that operators be informed of the correct way to respond to process upsets identified by the occurrence of critical alarms.

The lack of any system of priority for critical alarms may explain why the control room operator failed to respond promptly or adequately to the activation of the LFSD8 alarm at 8:20 a.m. on the morning of the accident. This alarm, which warned of a low flow shutdown of the lean oil pumps, was critical because it warned the operator of loss of the protective lean oil circulation system. Yet it was apparently ignored by the control room operator.

2.1.7 Monitoring of Operating Conditions

The Commission reported that a number of control room operators used charts, and to a lesser extent Process Information Data Acquisition System (PIDAS) records, to assist them to understand plant conditions during the course of their shift. They did not however appear to use such records as a means of monitoring the performance of the plant over an extended time. Panel operators did not, as a general practice, resort to charts or PIDAS records as a means of evaluating longer term process trends or longer term performance of equipment. They did, of course, evaluate the process from time to time by reference to the indicators on the controls. As with plant operators, plant supervisors had access to charts in the GPI control room. They also had access to PIDAS information through a computer terminal located on the plant supervisor's desk. When in the control room, supervisors used charts and computer records to understand and assess the workings of the plant during their shift and, to a lesser extent, in undertaking plant surveillance through the course of the shift. There was, however, no evidence to suggest that supervisors analyzed charts or used PIDAS recordings to monitor patterns in process variables or to conduct other forms of trend analysis. If supervisors did undertake such work, they did so only rarely, rather than as a matter of course. From 1997, plant supervisors were not expected to carry out this type of surveillance, nor was it their responsibility to monitor process operations in detail.

The Commission concluded that monitoring of PIDAS records for GPl in the weeks and months prior to the accident would have identified consistent deviations from normal operation in the absorbers in the form of high condensate levels and condensate flash drum temperature controller interference with the absorber level control. It would also have identified the practice of operating the absorbers in alarm. Had there been surveillance by qualified engineers, there would have been an opportunity to detect and correct the operating practices which led to the accident on September 25, 1998.

In the Commission's view, the failure to undertake ongoing analysis and evaluation of process trends within GPl, diminished the likelihood that upsets such as those which contributed to the accident on September 25, 1998 (operating conditions in the absorbers or condensate carryover) would be detected and avoided by appropriate responsive action. Had regular surveillance of operating conditions in GPI been undertaken by qualified engineers warning signals relevant to the accident; low absorber operating temperatures, high condensate levels, frequent condensate flash drum temperature control interference with level control of the absorber, the occurrence of condensate carryover, operation "in alarm"; would, in all likelihood, have been identified. This could have led to changes in operating practice for the absorbers. It could also have led to more rigorous monitoring of conditions in GPl.

2.1.8 Incident Reporting

The Commission heard evidence concerning an incident on August 28, 1998, a month before the accident that was not reported, but that the plant supervisor, and the panel operator conceded had a number of unusual features which warranted it being reported. These features included the fact that a critical spare lean oil pump was unavailable due to maintenance, with the consequence that the seal failure on the remaining pump required the shutdown of GP1 to effect repairs; the fact that such shutdown and subsequent restart had to be undertaken without the assistance of appropriate operating procedures; the fact that the incident involved a leak at one of the heat exchangers; and most importantly, the fact that, during the course of the incident, clear evidence emerged in the form of ice on piping and vessels that unusually cold temperatures were being experienced in vessels which usually operated hot, raising concerns about brittle fracture.

The Commission concluded that had the earlier incident been reported, as it should have been, the danger of equipment becoming subject to dangerously low temperatures upon the loss of lean oil flow for any length of time would, in all probability, have become known as would the steps available to avert the danger. The failure to report this incident thus stands as another example of a failure of the company to implement its management systems. Such failure deprived operations personnel of process information vital to the prevention of the incident on September 25, 1998.

3. EXXON ET AL. BLACKBEARD (2006)

In February 2005, ExxonMobil set out to drill the world's deepest offshore oil well. The project was ambitious, a well some 32,000 feet below the Gulf of Mexico seabed. It was hazardous because of the high pressure and high temperature at such depths, which could cause a blowout that would destroy the rig, put lives at risk, and create an ecological nightmare. ExxonMobil and its partners proceeded because Blackbeard's geology resembled that of Gulf of Mexico fields producing prolifically just 70 miles away. Data suggested the field could contain more than 1 billion barrels. If it did, Blackbeard would validate an entirely new oil frontier for ExxonMobil and the opportunity-strapped industry as a whole. Some trade journals called it the world's most watched oil play. After more than 500 days of drilling, the well had reached a depth of 30,067 feet, a record at the time, and was within striking distance of its target.

Seismic data suggested that 2000 feet further down was a giant prize: an "elephant" field of around a billion barrels of oil and gas. But 6 miles below the sea floor, the conditions were hellish, with high temperatures and pressures reaching 29,000 pounds per square inch.

The drilling team was getting nervous. The well had experienced a "kick." Attempts to relieve the pressure by pumping down heavy drilling mud were unsuccessful. Engineers worried that high-pressure gas might exceed the capacity of the blowout preventer (BOP).

"There was a pretty extensive discussion between the geoscientists, who wanted to keep going—here they were near their objectives—and the drillers, who were saying, 'We are just really not comfortable,'" recalled Rex W. Tillerson, ExxonMobil's chairman and chief executive officer, in an interview with the New York Times in July 2010 (Mouawad, 2010).

Tillerson eventually sided with the drillers. The well was capped and abandoned, and ExxonMobil wrote off Blackbeard as a $187 million dry hole.

"We were right at the ragged edge and they felt the risk was too great," Tillerson said.

At the time, the company was criticized for lacking courage. Some analysts were disparaging ExxonMobil. In this high-risk, high-reward industry, giant reservoirs go to those willing to gamble (Levine, 2009).

"ExxonMobil could have finished the well. They would have done fine," oil analyst George Froley was quoted as saying. "They just didn't have the guts."

Just over a year later, drilling resumed with a new operator. James R. Moffett, co-chair of the small Louisiana-based company McMoRan Exploration (MMR), said ExxonMobil had misread the pressure equation. He said that because of a quirk of such deep geology, the pressure underneath Blackbeard would drop within a few more feet of drilling, and thus become less perilous. Moffett went to work on Blackbeard in March 2008. Sure enough, the pressure eased. Moffett drilled another 2900 feet, and on 20 October, 7 months after drilling restarted, declared that he may have found "between a half-billion and several billion barrels of oil." Although 2 years later, they had yet to produce any oil from it.

After the BP Macondo blowout, the ExxonMobil decision takes on a different light. Paul Sankey, a Deutsche Bank analyst, explained in the New York Times.

"ExxonMobil's 'lack of guts' looks a lot more like justified conservatism and prudence, and a prescient awareness that safety, caution and catastrophic risk avoidance would be key themes as oil companies were forced to push the envelope in the search for new oil," he wrote in a report. "The fact is that Valdez pushed Exxon to the highest safety standards in the industry."

Before the DWH accident, the embodiment of a disastrous oil spill was the 1989 grounding of the *Exxon Valdez*. The incident, for which ExxonMobil was found responsible, led to a profound rethinking of safety management within the company. ExxonMobil followed up by developing a rigid system of rules for all its operations, from gas stations to offshore platforms, and it empowered everyone, even contractors, to speak up about safety problems.

4. BP TEXAS CITY 2005

4.1 CSB BP Texas City

In 1998, BP had one refinery in North America. In early 1999, BP merged with Amoco and then acquired ARCO in 2000, ending up with five refineries in North America. Prior to 1999, Amoco owned the Texas City refinery, the third-largest oil refinery in the United States with 475,000 barrels per day (bpd) refining capacity and had been in operation since 1934. Cost-cutting and failure to invest in the 1990s by Amoco and then BP, left the Texas City refinery's infrastructure and process equipment in disrepair. Operator training and staffing were also downsized. BP replaced the centralized HSE management systems of Amoco and Arco with a decentralized HSE management system. The effect of decentralizing HSE in the new organization resulted in a loss of focus on process safety management.

On March 23, 2005, at 1:20 p.m., the BP Texas City Refinery experienced explosions and fires that killed 15 people and injured another 180, alarmed the surrounding community, and resulted in financial losses exceeding $1.5 billion. The incident occurred during the startup of an isomerization (ISOM) unit when a raffinate splitter tower was overfilled; pressure relief devices opened, resulting in a release of flammable liquid from a blowdown stack that was not equipped with a flare. The release of flammables led to the explosion and fire. All the fatalities occurred in or near office trailers located close to the blowdown drum. A shelter-in-place order was issued that required 43,000 people to remain indoors. Houses were damaged as far away as three-quarters of a mile from the refinery.

4.2 Synopsis of the Event

For 2 years prior to the incident, BP had used a rigorous prestartup procedure that required all startups after turnarounds to go through a Pre-Startup Safety Review which included completing maintenance work, performing required safety reviews, checking equipment, and ensuring that utilities, control valves, and other equipment were functioning and correctly aligned. The process safety coordinator responsible for an area of the refinery that included the ISOM was unfamiliar with the process, and no Pre-Startup Safety Review procedure was conducted.

BP supervision decided to initiate the startup of the ISOM unit raffinate section during the night shift on March 22, 2005. However, after the startup was begun, it was stopped and the raffinate section was shutdown to be restarted during the next shift. Starting and then stopping the unit was unusual, and not covered in the startup procedures, which only addressed one continuous startup.

The Night Lead Operator controlled filling the raffinate section from the satellite control room because it was close to the process equipment. The Night Board Operator controlled the other two process units from the central control room. The Night Lead Operator did not use the startup procedure or record completed steps for the process of filling the raffinate section equipment, which left no record for the operators on the next shift.

When the Day Board Operator changed shifts in the central control room with the Night Board Operator shortly after 6:00 a.m., he received very little information on the state of the unit other than what was written in the logbook.

On the morning of 23 March, the raffinate tower startup began with a series of miscommunications. The early morning shift directors' meeting discussed the raffinate startup, and Day Supervisor B, who lacked ISOM experience, was told that the startup could not proceed because the storage tanks that received raffinate from the splitter tower were believed to be full. The Shift Director stated in postincident interviews that the meeting ended with the understanding that the raffinate section would not be started. However, that was not communicated to the ISOM operations personnel.

Day Supervisor A told the operations crew that the raffinate section would be started but did not distribute or review the applicable startup procedure with the crew, despite being required to do so in the procedure. Because the startup procedure that should have provided information on the progress of the startup by the night shift was not filled out and did not provide instructions for a noncontinual startup, the Day Board Operator had no precise information of what steps the night crew had completed and what the day shift was to do.

The Day Board Operator, acting on what he believed were the unit's verbal startup instructions and his understanding of the need to maintain a higher level in the tower to protect downstream equipment, closed the level control valve. The level sight glass, used to visually verify the tower level, had been reported by operators as unreadable because of a buildup of dark residue; the sight glass had been nonfunctional for several years. Knowing the condition of the sight glass, the Day Board Operator did not ask the outside crew to visually confirm the level. Even though the tower level control valve was not at 50% in "automatic" mode, as required by the startup procedure, the Day Board Operator said he believed the condition was safe as long as he kept the level within the reading range of the transmitter. The Day Board Operator observed a 97% level when he started circulation and thought that this level was normal. He said he did not recall observing a startup where the level was as low as 50%. At 10:10 a.m., 20,000 bpd of raffinate feed was being pumped into the tower and 4100 bpd was erroneously indicated as leaving the tower through the level control valve. The Day Board Operator said he was aware that the level control valve was shut.

As the unit was being heated, the Day Supervisor A, an experienced ISOM operator, left the plant at 10:47 a.m. due to a family emergency. The second Day Supervisor was devoting most of his attention to the final stages of the Aromatics Recovery Unit (ARU) startup; he had very little ISOM experience and, therefore, did not get involved in the ISOM startup. No experienced supervisor or ISOM technical expert was assigned to the raffinate section startup after the Day Supervisor A left, although BP's safety procedures required such oversight.

The Day Board Operator continued the liquid flow to the splitter tower, but was unaware that the actual tower level continued to rise. At 9:55 a.m., two burners were lit in the raffinate furnace, which preheated the feed flowing into the splitter tower and served as a reboiler, heating the liquid in the tower bottom. At 11:16 a.m., operators lit two additional burners in the furnace. While the transmitter indicated that the tower level was at 93% (8.65 feet) in the bottom 9 feet of the tower, the U.S. Chemical Safety and Hazard Investigation Board (CSB) determined from postincident analysis that the actual level in the tower was 67 feet. The fuel to the furnace was increased at 11:50 a.m., at which time the actual tower level was 98 feet, although the transmitter indicated that the level was 88% (8.4 feet) and decreasing.

At 12:41 p.m., the tower's pressure rose to 33 pounds (psig) (228 kPa), due to the significant increase in the liquid level compressing the remaining nitrogen in the raffinate system. The operations crew, however, believed the high pressure to be a result of the tower bottoms overheating, which was not unusual in previous startups. In response to the high pressure, the outside operations crew opened the 8-in. chain-operated valve that vented directly to the blowdown drum, which reduced the pressure in the tower.

The Day Board Operator and the Day Lead Operator agreed that the heat to the furnace should be reduced, and at 12:42 p.m. fuel gas flow was reduced to the furnace. At this time the raffinate splitter level transmitter displayed 80% (8 feet), but the actual tower level was 140 feet.

From 10 a.m. to 1 p.m. the transmitter showed the tower level declining from 97% to 79%. The Day Board Operator thought the level indication was accurate, and believed it was normal to see the level drop as the tower heated up. At the time of the pressure upset, the Day Board Operator became concerned about the lack of heavy raffinate flow out of the tower, and discussed with the Day Lead Operator the need to remove heavy raffinate from the raffinate splitter tower. None of the ISOM operators knew the tower was overfilling. At 12:42 p.m., the Day Board Operator opened the splitter level control to 15% output, and over the next 15 min opened the valve five times until, at 1:02 p.m., it was 70% open. However, heavy raffinate flow had not actually begun until 12:59 p.m.

The heavy raffinate flow out of the tower matched the feed into the tower was 20,500 bpd at 1:02 p.m. and by 1:04 p.m. had increased to 27,500 bpd. Unknown to the operators, the level of liquid in the 170 foot tower at this time was 158 feet, but the level transmitter reading had continued to decrease and now read 78% (7.9 feet). Although the total quantity of material in the tower had begun to decrease, heating the column contents caused the liquid level at the top of the column to continue increasing until it completely filled the column and spilled over into the overhead vapor line leading to the column relief valves and condenser.

At 1:14 p.m., hydrocarbon liquid flowed out of the top of the raffinate splitter tower and into the vertical overhead vapor line, due to overfilling and rapid heating of the column.

As the liquid filled the overhead line, the resulting hydrostatic head in the line increased. The tower pressure (which remained relatively constant) combined with the hydrostatic head exceeded the set pressures of the safety relief valves. The valves opened and discharged liquid raffinate into the raffinate splitter disposal header collection system. Both the Day Board Operator in the central control room and the outside operators in the satellite control room saw the splitter tower pressure rising rapidly to 63 psig (434 kPa); however, interviews revealed that the outside operators did not hear the three splitter tower relief valves open.

Once the blowdown system filled, flammable liquid discharged to the atmosphere from its stack and fell to the ground. Shortly after the ISOM operators began troubleshooting the pressure spike, they received, via radio, the first notification that the blowdown drum was overflowing. In response to the radio message, the Board Operator and Lead Operator used the computerized control system to shut the flow of fuel to the heater, while the other operators left the satellite control room and ran toward an adjacent road, to redirect traffic away from the blowdown drum, as required by BP's "Emergency Response Procedure A-7."

The ISOM operators stated they had insufficient time to sound the emergency alarm before the explosion. Approximately 15 s after hearing the radio message, both the Board Operator and the Lead Operator said they started the process of shutting off the fuel to the furnace using the computerized control system. Their testimony is substantiated by the computerized control system data, which showed that the fuel gas flow control valve was shut 5 s before the explosion. Hundreds of alarms registered in the computerized control system at 1:20:04 p.m., including the high level alarm on the blowdown drum; the flood of alarms indicates when the explosion occurred. Consequently, ISOM operations personnel did not have sufficient time to assess the situation and sound the emergency alarm prior to the explosion.

The released volatile liquid evaporated as it fell to the ground and formed a flammable vapor cloud. The most likely source of ignition was an idling diesel pickup truck located about 25 feet from the blowdown drum. The 15 employees killed in the explosion were contractors working in and around temporary trailers that had been previously sited by BP as close as 121 feet from the blowdown drum.

4.3 Key Findings

The Texas City disaster was caused by organizational and safety deficiencies at all levels of the BP Corporation. Warning signs of a possible disaster were present for several years, but company officials did not intervene effectively to prevent it. The extent of the serious safety culture deficiencies was further revealed when the refinery experienced two additional serious incidents just a few months after the March 2005 disaster. In one, a pipe failure caused a reported $30 million in damage; the other resulted in a $2 million property loss. In each incident, community shelter-in-place orders were issued.

The following findings are taken from BP's Fatal Accident Investigation (Mogford, 2005):

- Over the years, the working environment had eroded to one characterized by resistance to change, and lacking trust, motivation, and a sense of purpose. Coupled with unclear expectations around supervisory and management behaviors this meant that rules were not consistently followed, rigor was lacking and individuals felt disempowered from suggesting or initiating improvements.
- Process safety, operations performance, and systematic risk reduction priorities had not been set and consistently reinforced by management.
- Many changes in a complex organization led to the lack of clear accountabilities and poor communication, which together resulted in confusion in the workforce over roles and responsibilities.
- A poor level of hazard awareness and understanding of process safety on the site resulted in people accepting levels of risk that were considerably higher than comparable installations. One consequence was that temporary office trailers were placed within 150 feet of a blowdown stack which vented heavier than air hydrocarbons to the atmosphere without questioning the established industry practice.
- Given the poor vertical communication and performance management process, there was neither adequate early warning system of problems, nor any independent means of understanding the deteriorating standards in the plant.

The following findings are taken from the Investigation Report of the CSB (U.S. Chemical Safety and Hazard Investigation Board, 2007).

The ISOM startup procedure required that the level control valve on the raffinate splitter tower be used to send liquid from the tower to storage. However, this valve was closed by an operator and the tower was filled for over 3 h without any liquid being removed. This led to flooding of the tower and high pressure, which activated relief valves that discharged flammable liquid to the blowdown system. Underlying factors involved in overfilling the tower included:

- The tower level indicator showed that the tower level was declining when it was actually overfilling. The redundant high level alarm did not activate, and the tower was not equipped with any other level indications or automatic safety devices.

- The control board display did not provide adequate information on the imbalance of flows in and out of the tower to alert the operators to the dangerously high level.
- A lack of supervisory oversight and technically trained personnel during the startup, an especially hazardous period, was an omission contrary to BP safety guidelines. An extra board operator was not assigned to assist, despite a staffing assessment that recommended an additional board operator for all ISOM startups.
- Supervisors and operators poorly communicated critical information regarding the startup during the shift turnover; BP did not have a shift turnover communication requirement for its operations staff.
- ISOM operators were likely fatigued from working 12-h shifts for 29 or more consecutive days.
- The operator training program was inadequate. The central training department staff had been reduced from 28 to 8, and simulators were unavailable for operators to practice handling abnormal situations, including infrequent and high hazard operations such as startups and unit upsets.
- Outdated and ineffective procedures did not address recurring operational problems during startup, leading operators to believe that procedures could be altered or did not have to be followed during the startup process.
- BP Texas City managers did not effectively implement their prestartup safety review policy to ensure that nonessential personnel were removed from areas in and around process units during startups, an especially hazardous time in operations. The process unit was started despite previously reported malfunctions of the tower level indicator, level sight glass, and a pressure control valve.
- Occupied trailers were sited too close to a process unit handling highly hazardous materials. All fatalities occurred in or around the trailers.
- The size of the blowdown drum was insufficient to contain the liquid sent to it by the pressure relief valves. The blowdown drum overfilled and the stack vented flammable liquid to the atmosphere, which fell to the ground and formed a vapor cloud that ignited. A relief valve system safety study had not been completed.
- Neither Amoco nor BP replaced blowdown drums and atmospheric stacks, even though a series of incidents warned that this equipment

was unsafe. In 1992, OSHA cited a similar blowdown drum and stack as unsafe, but the citation was withdrawn as part of a settlement agreement and therefore the drum was not connected to a flare as recommended. Amoco, and later BP, had safety standards requiring that blowdown stacks be replaced with equipment such as a flare when major modifications were made. In 1997, a major modification replaced the ISOM blowdown drum and stack with similar equipment, but Amoco did not connect it to a flare. In 2002, BP engineers proposed connecting the ISOM blowdown system to a flare, but a less expensive option was chosen.

- The BP Board of Directors did not provide effective oversight of BP's safety culture and major accident prevention programs. The Board did not have a member responsible for assessing and verifying the performance of BP's major accident hazard prevention programs. Cost-cutting, failure to invest and production pressures from BP Group executive managers impaired process safety performance at Texas City.
- Reliance on the low personal injury rate at Texas City as a safety indicator failed to provide a true picture of process safety performance and the health of the safety culture.
- Deficiencies in BP's mechanical integrity program resulted in the "run to failure" of process equipment at Texas City.
- A "check the box" mentality was prevalent at Texas City, where personnel completed paperwork and checked off on safety policy and procedural requirements even when those requirements had not been met.
- BP Texas City lacked a reporting and learning culture. Personnel were not encouraged to report safety problems and some feared retaliation for doing so. Therefore, the lessons from incidents and near misses, were generally not captured or acted upon. Important relevant safety lessons from a British government investigation of incidents at BP's refinery in Grangemouth, Scotland, were also not incorporated at Texas City.
- The BP Texas City site had a number of reporting programs, yet serious near misses and other critical events were often unreported. In the eight previous ISOM blowdown system incidents, three were not reported in any BP database, five were reported as environmental releases, and only two were investigated as safety incidents. In the 5 years prior to the 2005 disaster, over three-quarters of the raffinate splitter tower startups' level ran above the range of the level transmitter and in nearly half, the level was out of range for more than 1 h. These operating deviations were not reported by operations personnel or reviewed in the computerized history by Texas City managers. During the March 2005 ISOM startup, the operating deviations were more serious in degree but similar in kind to

past startups. Yet the operating envelope program designed to capture and report excursions from safe operating limits was not fully functional and did not capture high distillation tower level events in the ISOM to alert managers to the deviations.
- Safety campaigns, goals, and rewards focused on improving personal safety metrics and worker behavior rather than on process safety and management safety systems. While compliance with many safety policies and procedures was deficient at all levels of the refinery, Texas City managers did not lead by example regarding safety.
- BP Texas City did not effectively assess changes involving people, policies, or the organization that could impact process safety.
- Beginning in 2002, BP Group and Texas City managers received numerous warning signals about a possible major catastrophe at Texas City. In particular, managers received warnings about serious deficiencies regarding the mechanical integrity of aging equipment, process safety, and the negative safety impacts of budget cuts and production pressures.

On August 17, 2005, following two further safety incidents at Texas City, the CSB issued an urgent safety recommendation to the BP Group Executive Board of Directors that it convene an independent panel of experts to examine BP's corporate safety management systems, safety culture, and oversight of the North American refineries. BP accepted the recommendation and commissioned the BP US Refineries Independent Safety Review Panel, chaired by former Secretary of State James Baker, III (Baker Panel). The scope of the Baker Panel's work did not include determining the root causes of the Texas City ISOM incident.

The following is taken from the Baker Panel Report (Baker, 2007) that was issued on January 16, 2007. The Panel found that "significant process safety issues exist at all five US refineries, not just Texas City," and that BP had not instilled "a common unifying process safety culture among its US refineries." The report found "instances of a lack of operating discipline, toleration of serious deviations from safe operating practices, and [that an] apparent complacency toward serious process safety risk existed at each refinery." The Panel concluded that "material deficiencies in process safety performance exist at BP's five US refineries."

The Baker Panel Report stated that BP's corporate safety management system "does not effectively measure and monitor process safety performance" for its US refineries. The report also found that BP's overreliance on personal injury rates impaired its perception of process safety risks, and that BP's Board of Directors had "not ensured, as a best practice, that BP's management has implemented an integrated, comprehensive, and effective process safety management system for BP's five US refineries." The report's findings covered three broad themes: corporate safety culture, process safety management systems, and performance evaluation, corrective action, and corporate oversight.

4.4 Corporate Safety Culture

4.4.1 Process Safety Leadership

Based on its review, the Baker Panel believed that BP had not provided effective process safety leadership and had not adequately established process safety as a core value across its five US refineries. While BP had an aspirational goal of "no accidents, no harm to people," BP had not provided effective leadership in making certain its management and US refining workforce understood what was expected of them regarding process safety performance. BP had emphasized personal safety and had achieved significant improvement in personal safety performance, but BP did not emphasize process safety. BP mistakenly interpreted improving personal injury rates as an indication of acceptable process safety performance at its US refineries. BP's reliance on this data, combined with an inadequate process safety understanding, created a false sense of confidence that BP was properly addressing process safety risks. The Baker Panel further found that process safety leadership appeared to have suffered as a result of high turnover of refinery plant managers.

4.4.2 Employee Empowerment

A good process safety culture requires a positive, trusting, and open environment with effective lines of communication between management and the workforce, including employee representatives. At Texas City, BP had not established a positive, trusting, and open environment with effective lines of communication between management and the workforce, although the safety culture appeared to be improving.

4.4.3 Resources and Positioning of Process Safety Capabilities

BP had not always ensured that it identified and provided the resources required for strong process safety performance at its US refineries. Despite having numerous staff at different levels of the organization that support process safety, BP did not have a designated, high-ranking leader for process safety dedicated to its refining business. In addition, BP's corporate management mandated numerous initiatives that applied to the US refineries and that, while well intentioned, had overloaded personnel at BP's US refineries.

This "initiative overload" may have undermined process safety performance at the US refineries. Also, operations and maintenance personnel in BP's five US refineries sometimes worked high rates of overtime, and that could impact their ability to perform their jobs safely and increased process safety risk.

4.4.4 Incorporation of Process Safety Into Management Decision Making

The Baker Panel also found that BP did not effectively incorporate process safety into management decision making. BP tended to have a short-term focus, and its decentralized management system and entrepreneurial culture delegated substantial discretion to US refinery plant managers without clearly defining process safety expectations, responsibilities, or accountabilities. In addition, while accountability was a core concept in BP's Management Framework for driving desired conduct, the company had not demonstrated that it effectively held executive management and refining line managers and supervisors, both at the corporate level and at the refinery level, accountable for process safety performance at its five US refineries.

4.4.5 Process Safety Cultures at BP's US Refineries

BP had not instilled a common, unifying process safety culture among its US refineries. Each refinery had its own separate and distinct process safety culture. While some refineries were far more effective than others in promoting process safety, significant process safety culture issues existed at all five US refineries, not just Texas City. Although the five refineries did not share a unified process safety culture, each exhibited similar weaknesses. The Baker Panel found instances of a lack of operating discipline, toleration of serious deviations from safe operating practices, and apparent complacency toward serious process safety risks at each refinery.

4.5 Process Safety Management Systems

The Baker Panel's findings to the effectiveness of process safety management systems that BP utilized for its five US refineries relate to its process risk assessment and analysis, compliance with internal process safety standards, implementation of external good engineering practices, process safety knowledge and competence, and general effectiveness of BP's corporate process safety management system.

4.5.1 *Process Risk Assessment and Analysis*

While all of BP's US refineries had active programs to analyze process hazards, the system as a whole did not ensure adequate identification, and rigorous analysis of those hazards. The Baker Panel's examination also indicated that the extent and recurring nature of this deficiency was not isolated, but systemic.

4.5.2 *Compliance With Internal Process Safety Standards*

The Baker Panel's technical consultants and the Baker Panel observed that BP did have internal standards and programs for managing process risks. However, the Baker Panel found that BP's corporate safety management system did not ensure timely compliance with internal process safety standards and programs at its five US refineries. This finding relates to several areas that were addressed by BP internal standards: rupture disks under relief valves, equipment inspections, critical alarms and emergency shutdown devices, area electrical classification, and near miss investigations.

4.5.3 *Implementation of External Good Engineering Practices*

The Baker Panel also found that BP's corporate safety management system did not ensure timely implementation of external good engineering practices that support and could improve process safety performance at BP's five US refineries. Such practices play an important role in the management of process safety in refineries operating in the United States.

4.5.4 *Process Safety Knowledge and Competence*

Although many members of BP's technical and process safety staff had the capabilities and expertise needed to support a sophisticated process safety effort, the Baker Panel believed that BP's system for ensuring an appropriate level of process safety awareness, knowledge, and competence in the organization relating to its five US refineries had not been effective in a number of respects. First, BP had not effectively defined the level of process safety knowledge or competency required of executive management, line management above the refinery level, and refinery managers. Second, BP had not adequately ensured that its US refinery personnel and contractors had sufficient process safety knowledge and competence. The information that the Baker Panel reviewed indicated that process safety education and training needed to be more rigorous, comprehensive, and integrated. Third, the Baker Panel found that at most of BP's US refineries, the implementation of and over-reliance on BP's computer-based training contributed to inadequate process safety training of refinery employees.

4.5.5 Effectiveness of BP's Corporate Process Safety Management System

BP had an aspirational goal and expectation of "no accidents, no harm to people, and no damage to the environment," and was developing programs and practices aimed at addressing process risks. These programs and practices included the development of new standards, engineering technical practices, and other internal guidance, as well as the dedication of substantial resources. Despite these positive changes, the Baker Panel's examination indicated that BP's corporate process safety management system did not effectively translate corporate expectations into measurable criteria for management of process risk or define the appropriate role of qualitative and quantitative risk management criteria.

4.5.6 Panel Observations Relating to Process Safety Management Practices

The Baker Panel observed several positive notable practices or, in the case of BP's process safety minimum expectation program, an excellent process safety management practice. The notable practices relate to creation of an engineering authority at each refinery and several other refinery-specific programs.

4.6 Performance Evaluation, Corrective Action, and Corporate Oversight

Maintaining and improving a process safety management system requires the periodic evaluation of performance, the identification of deficiencies, and the measures to be taken to correct the deficiencies. Significant deficiencies existed in BP's site and corporate systems for measuring process safety performance, investigating incidents and near misses, auditing system performance, addressing previously identified process safety-related action items, and ensuring sufficient management and Board oversight. Many of the process safety deficiencies were not new, but were identifiable to BP based upon lessons from previous process safety incidents, including process incidents that occurred at BP's facility in Grangemouth, Scotland in 2000.

4.6.1 Measuring Process Safety Performance

BP primarily used injury rates to measure process safety performance at its US refineries before the Texas City accident. Although BP was not alone in this practice, BP's reliance on injury rates significantly hindered its perception of process risk. It also tracked some other metrics relevant to process safety at its US refineries. However, it became apparent that BP did not understand or accept what this data indicated about the risk of a major accident or the overall performance of its process safety management systems. As a result, BP's corporate safety management system for its US refineries did not effectively measure and monitor process safety performance.

4.6.2 Incident and Near Miss Investigations

BP acknowledged the importance of incident and near miss investigations, and it employed multiple methods at different levels of the organization to distribute information regarding incidents and lessons learned. Although BP was improving aspects of its incident and near miss investigation process at the time of the accident, BP had not instituted effective root cause analysis procedures to identify systemic causal factors that may contribute to future accidents. When true root or system causes are not identified, corrective actions may address immediate or superficial causes, but not likely the true root causes. The Baker Panel also believed that BP had an incomplete picture of process safety performance at its US refineries because its process safety management system resulted in under reporting of incidents and near misses.

4.6.3 Process Safety Audits

The Baker Panel found that BP had not implemented an effective process safety audit system for its US refineries based on the Baker Panel's concerns about auditor qualifications, audit scope, reliance on internal auditors, and the limited review of audit findings. The Baker Panel was also concerned that the principal focus of the audits was on compliance and verifying that required management systems were in place to satisfy legal requirements. It did not appear that BP used the audits to ensure that the management systems were delivering the desired safety performance or to assess a site's performance against industry best practices.

4.6.4 Timely Correction of Identified Process Safety Deficiencies

BP promptly expended significant efforts to identify deficiencies. However, the Baker Panel found that sometimes the company failed to address such issues promptly, nor did it track to completion process safety deficiencies identified during hazard assessments, audits, inspections, and incident investigations. The Baker Panel's review, for example, found repeated audit findings at the US refineries, suggesting that true root causes were not being identified and corrected. This problem was especially apparent with overdue mechanical integrity inspection and testing. Although BP regularly conducted various assessments, reviews, and audits within the company, the follow through after these reviews repeatedly fell short. This failure to follow through compromised the effectiveness of even the best audit program or incident investigation. In addition, BP did not take full advantage of opportunities to improve process operations at its US refineries and its process safety management systems. BP did not effectively use the results of its operating experiences, process hazard analyses, audits, near misses, or accident investigations to improve process operations and process safety management systems.

4.6.5 Corporate Oversight

BP acknowledged to the Baker Panel, the importance of ensuring that the company-wide safety management system functioned as intended. However, the company's system for assuring process safety performance used a bottom-up reporting system that originates with each business unit, such as a refinery. As information was reported up, data were aggregated. By the time information was formally reported at the Refining and Marketing segment level, for example, refinery-specific performance data were no longer presented separately.

The Baker Panel's examination indicates that BP's executive management either did not receive refinery-specific information that suggested process safety deficiencies at some of the US refineries or did not effectively respond to the information that it did receive. According to annual reports on health, safety, security, and environmental assurance that BP management provided to the Environment and Ethics Assurance Committee of BP's Board of Directors for 1999 through 2005, management was monitoring process safety matters, including plant and operational integrity issues. The reports identified safety and integrity management risks that various levels of the organization confronted and described management actions proposed to address and mitigate those risks. However, the reports and other documents that the Baker Panel examined indicate that issues persisted relating to assurance of effective implementation of BP's policies and expectations relating to safety and integrity management.

For these reasons, the Baker Panel believed that BP's process safety management system was not effective in evaluating whether the steps that BP took were actually improving the company's process safety performance. The Baker Panel found that neither BP's executive management nor its refining line management had ensured the implementation of an integrated, comprehensive, and effective process safety management system.

BP's Board of Directors were monitoring process safety performance of operations based on information that corporate management presented to it. A substantial gulf appeared to exist between the actual performance of BP's process safety management systems and the company's perception of that performance. Although the executive and refining line management was responsible for ensuring the implementation of an integrated, comprehensive, and effective process safety management system, the Baker Panel found the Board had not ensured, as a best practice, that management did so. In reviewing the conduct of the Board, the Baker Panel was guided by its chartered purpose to examine and recommend any needed improvements. In the Baker Panel's judgment, this purpose did not call for an examination of legal compliance, but called for excellence. It is in this context and in the context of best practices that the Baker Panel believed that BP's Board could and should have done more to improve its oversight of process safety at the five US refineries.

4.7 Lessons Learned

Simply targeting the mistakes of BP's operators and supervisors misses the underlying and significant cultural, human, and organizational causes of the incident that have a greater preventative impact. One underlying cause was that BP used inadequate methods to measure safety conditions at Texas City. For instance, a very low personal injury rate at Texas City gave BP a misleading indicator of process safety performance. In addition, while most attention was focused on the injury rate, the overall safety culture, and process safety management program had serious deficiencies. Despite numerous previous fatalities at the Texas City refinery (23 deaths in the 30 years prior to the 2005 disaster) and many hazardous material releases, BP did not take effective steps to stem the progression to a catastrophic event.

Further evidence of a systemic problem with BP's management systems occurred in July 2005 and March 2006. In July 2005, *Thunder Horse*, BP's giant new production platform in the Gulf of Mexico, nearly sank during a hurricane. In their rush to finish the $1 billion platform, workers had installed a valve backwards, allowing the ballast tanks to flood. Inspections revealed other shoddy work. Repairs costing hundreds of millions would keep *Thunder Horse* out of commission for 3 years.

Then, in March 2006, corrosion in BP's Prudhoe Bay pipelines caused a 267,000-gallon oil leak, the worst spill ever on Alaska's North Slope. This accident was a result of the company's failure to properly test and clean miles of aging pipe.

As part of a strategic plan to restructure BP's US refining portfolio, the company completed the sale of the Texas City Refinery on February 01, 2013 after reporting it spent over $1 billion in modernizing and improving the plant. It said Texas City lacked strong integration into any of its marketing assets.

5. BP ET AL. MACONDO (2010)

On April 20, 2010, a multiple-fatality incident occurred at the Macondo oil well approximately 50 miles off the coast of Louisiana in the Gulf of Mexico during temporary well-abandonment activities on the DWH drilling rig. Control of the well was lost, resulting in a blowout, the uncontrolled release of oil and gas (hydrocarbons) from the well. The hydrocarbons found an ignition source on the rig. The resulting explosions and fire killed 11 people, seriously injured 17 others, forced the evacuation of 115 from the rig, resulted in the sinking of the DWH, and caused massive marine and coastal damage from a reported 4 million barrels of released hydrocarbons.

BP was the main operator/lease holder responsible for the well design, and Transocean was the drilling contractor that owned and operated the DWH. On the day of the incident, the crew was completing temporary abandonment of the well so that it could be left in a safe condition until production could begin later using another offshore facility.

Abandonment activities are meant to safely and securely plug the well using cement barriers. In the case of the Macondo well, a critical cement barrier intended to keep the hydrocarbons in the reservoir had not been effectively installed at the bottom of the well. BP and Transocean personnel misinterpreted a test to assess the cement barrier integrity, leading them to erroneously believe that the hydrocarbon bearing zone at the bottom of the well was sealed. When the crew removed drilling mud from the well in preparation to install an additional cement barrier, the open Blowout Preventer (BOP) was the only physical barrier that could have potentially prevented hydrocarbons from reaching the rig and surrounding environment. The ability of the BOP to act as this barrier was contingent primarily upon human detection of the kick and timely activation and closure of the BOP.

In the case of the Macondo, removing drilling mud after the test allowed hydrocarbons to flow past the failed cement barrier toward the rig. The hydrocarbons continued to flow from the reservoir for almost an hour without human detection or the activation of the automated controls to close the BOP. Eventually, oil and gas passed above the BOP and forcefully released to the rig. In response, the well operations crew manually closed the BOP. Oil and gas that had already flowed past the BOP continued to gush onto the rig, igniting, and exploding. The explosion likely activated an automatic emergency response system designed to shear the drill pipe passing through the BOP and seal the well, but it was unsuccessful.

The Macondo blowout was the subject of multiple official investigations and perspectives, including those by the National Commission (January 2011), National Academy of Engineering (2012), Department of Interior—Joint Investigation Team (US Coast Guard and Bureau of the Ocean Energy Management, Regulation and Enforcement) (September 2011), Deepwater Horizon Study Group, BP (September 2010), and Transocean (June 2010). But the potential legal implications from the severity of the Macondo blowout limited the flow of information from BP and Transocean, both to the public and the entities investigating the incident. This became apparent as new documents and depositions controlled by the US District Court for the Eastern District of Louisiana were released under the multidistrict litigation (MDL) docket and when Transocean complied with the CSB's subpoena requests years after they were originally submitted to the company.

For example, the major investigation reports, except Transocean's, either were published before BOP testing was completed or did not have access to the full set of postincident BOP data. Details that emerged in the final phase of BOP testing were imperative, as they revealed latent failures in the DWH BOP before it was deployed to the wellhead. Also, the 2013 MDL and Transocean records shed light on the operator/drilling contractor relationship between BP and Transocean. This relationship ultimately led to vaguely established safety roles and responsibilities that affected human performance and major accident risk management at Macondo. Finally, a 2016 trial provided testimony from rig personnel who previously evoked their Fifth Amendment right, revealing additional insights into the decisions and actions of the well operations crew leading up to the blowout.

5.1 Key Findings

BP formed an investigation team that was charged with gathering the facts surrounding the accident, analyzing available information to identify possible causes, and making recommendations to enable prevention of similar accidents in the future. BP's investigating team identified eight key findings related to the causes of the accident (BP, 2011). These findings are briefly described below.

- *The annulus cement barrier did not isolate the hydrocarbons.* The day before the accident, cement had been pumped down the production casing and up into the wellbore annulus to prevent hydrocarbons from entering the wellbore from the reservoir. The annulus cement that was placed across the main hydrocarbon zone was a light, nitrified foam cement slurry. This cement probably experienced nitrogen breakout and migration, allowing hydrocarbons to enter the wellbore annulus. *The investigation team concluded that there were weaknesses in cement design and testing, quality assurance, and risk assessment.*
- *The shoe track barriers did not isolate the hydrocarbons.* Having entered the wellbore annulus, hydrocarbons passed down the wellbore and entered the 9 7/8 in. × 7 in. production casing through the shoe track, installed in the bottom of the casing. Flow entered into the casing rather than the casing annulus. For this to happen, both barriers in the shoe track must have failed. The first barrier was the cement in the shoe track, and the second was the float collar, a device at the top of the shoe track designed to prevent fluid ingress into the casing. *The investigation team concluded that hydrocarbon ingress was through the shoe track, rather than through a failure in the production casing itself or up the wellbore annulus and through the casing hanger seal assembly. The investigation team identified potential failure modes that could explain how the shoe track cement and the float collar allowed hydrocarbon ingress into the production casing.*
- *The negative pressure test was accepted although well integrity had not been established.* Prior to temporarily abandoning the well, a negative pressure test was conducted to verify the integrity of the mechanical barriers (the shoe track, production casing, and casing hanger seal assembly). The test involved replacing heavy drilling mud with lighter seawater to place the well in a controlled underbalanced condition. In retrospect, pressure readings and volume bleed at the time of the negative pressure test were indications of flow-path communication with the reservoir, signifying that the integrity of these barriers had not been achieved. *The Transocean*

rig crew and BP well site leaders reached the incorrect view that the test was successful and that well integrity had been established.

- *Influx was not recognized until hydrocarbons were in the riser.* With the negative pressure test accepted, the well was returned to an overbalanced condition, preventing further influx into the wellbore. Later, as part of normal operations to temporarily abandon the well, heavy drilling mud was again replaced with seawater, underbalancing the well. Over time, this allowed hydrocarbons to flow up through the production casing and pass through the BOP. Indications of influx with an increase in drill pipe pressure are discernable in real-time data from approximately 40 min before the rig crew took action to control the well. The rig crew's first apparent well control actions occurred after hydrocarbons were rapidly flowing to the surface. *The rig crew did not recognize the influx and did not act to control the well until hydrocarbons had passed through the BOP and into the riser.*
- *Well control response actions failed to regain control of the well.* The first well control actions were to close the BOP and diverter, routing the fluids exiting the riser to the rig's mud gas separator (MGS) system rather than to the overboard diverter line. *If fluids had been diverted overboard, rather than to the MGS, there may have been more time to respond, and the consequences of the accident may have been reduced.*
- *Diversion to the MGS resulted in gas venting onto the rig.* Once diverted to the MGS, hydrocarbons were vented directly onto the rig through the 12 in. goosenecked vent exiting the MGS, and other flow lines also directed gas onto the rig. This increased the potential for the gas to reach an ignition source. *The design of the MGS system allowed diversion of the riser contents to the MGS vessel although the well was in a high flow condition. This overwhelmed the MGS system.*
- *The fire and gas system did not prevent hydrocarbon ignition.* Hydrocarbons migrated beyond areas on the DWH that were electrically classified to areas where the potential for ignition was higher. *The heating, ventilation, and air conditioning system probably transferred a gas-rich mixture into the engine rooms, causing at least one engine to overspeed, creating a potential source of ignition.*
- *The BOP emergency mode did not seal the well.* Three methods for operating the BOP in the emergency mode were unsuccessful in sealing the well.
  - The explosions and fire very likely disabled the emergency disconnect sequence, the primary emergency method available to the rig

personnel, which was designed to seal the wellbore and disconnect the marine riser from the well.

- The condition of critical components in the yellow and blue control pods on the BOP very likely prevented activation of another emergency method of well control, the automatic mode function (AMF), which was designed to seal the well without rig personnel intervention upon loss of hydraulic pressure, electric power, and communications from the rig to the BOP control pods. An examination of the BOP control pods following the accident revealed that there was a fault in a critical solenoid valve in the yellow control pod and that the blue control pod AMF batteries had insufficient charge; these faults likely existed at the time of the accident.
- Remotely operated vehicle intervention to initiate the autoshear function, another emergency method of operating the BOP, likely resulted in closing the BOP's blind shear ram (BSR) 33 h after the explosions, but the BSR failed to seal the well.

Through a review of rig audit findings and maintenance records, the investigation team found indications of potential weaknesses in the testing regime and maintenance management system for the BOP.

The team did not identify any single action or inaction that caused this accident. Rather, a complex and interlinked series of mechanical failures, human judgments, engineering design, operational implementation, and team interfaces came together to allow the initiation and escalation of the accident.

The Joint Investigation Team of the Bureau of Ocean Energy Management, Regulation and Enforcement (BOEMRE) (formerly the Minerals Management Service or "MMS") and the United States Coast Guard (the Panel) identified a number of causes of the Macondo blowout (The Bureau of Ocean Energy Management Regulation and Enforcement, 2011).

The Panel found that a central cause of the blowout was failure of a cement barrier in the production casing string (a high-strength steel pipe set in a well to ensure well integrity and to allow future production). The failure of the cement barrier allowed hydrocarbons to flow up the wellbore, through the riser and onto the rig, resulting in the blowout. The precise reasons for the failure of the production casing cement job were not known, but the Panel concluded that the failure was likely due to: (1) swapping of cement and drilling mud (referred to as "fluid inversion") in the shoe track (the section of casing near the bottom of the well); (2) contamination of the shoe track cement; or (3) pumping the cement past the target location in the well, leaving the shoe track with little or no cement (referred to as "over-displacement").

The loss of life at the Macondo site on April 20, 2010 and the subsequent pollution of the Gulf of Mexico through the Summer of 2010 were the result of poor risk management, last-minute changes to plans, failure to observe and respond to critical indicators, inadequate well control response, and insufficient emergency bridge response training by companies and individuals responsible for drilling at the Macondo well and for the operation of the DWH.

BP, as the designated operator under BOEMRE regulations, was ultimately responsible for conducting operations at Macondo in a way that ensured the safety and protection of personnel, equipment, natural resources, and the environment. Transocean, the owner of the DWH, was responsible for conducting safe operations and for protecting personnel onboard. Halliburton, as a contractor to BP, was responsible for conducting the cement job, and, through its subsidiary (Sperry Sun), had certain responsibilities for monitoring the well. Cameron was responsible for the design of the DWH BOP stack.

At the time of the blowout, the rig crew was engaged in "temporary abandonment" activities to secure the well after drilling was completed and before the rig left the site. In the days leading up to 20 April, BP made a series of decisions that complicated cementing operations, added incremental risk, and may have contributed to the ultimate failure of the cement job. These decisions included:

- *The use of only one cement barrier*. BP did not set any additional cement or mechanical barriers in the well, even though various well conditions created difficulties for the production casing cement job.
- *The location of the production casing*. BP decided to set production casing in a location in the well that created additional risk of hydrocarbon influx.
- *The decision to install a lock-down sleeve*. BP's decision to include the setting of a lock-down sleeve (a piece of equipment that connects and holds the production casing to the wellhead during production) as part of the temporary abandonment procedure at Macondo increased the risks associated with subsequent operations, including the displacement of mud, the negative test sequence and the setting of the surface plug.
- *The production casing cement job*. BP failed to perform the production casing cement job in accordance with industry-accepted recommendations.

The Panel concluded that BP failed to communicate these decisions and the increasing operational risks to Transocean. As a result, BP and Transocean personnel onboard the DWH on the evening of April 20, 2010, did not fully identify and evaluate the risks inherent in the operations that were being conducted.

On 20 April, BP and Transocean personnel onboard the DWH missed the opportunity to remedy the cement problems when they misinterpreted anomalies encountered during a critical test of cement barriers called a negative test, which seeks to simulate what will occur at the well after it is temporarily abandoned and to show whether cement barrier(s) will hold against hydrocarbon flow.

The rig crew conducted an initial negative test on the production casing cement job that showed a pressure differential between the drill pipe and the kill line (a high-pressure pipe leading from the BOP stack to the rig pumps). This was a serious anomaly that should have alerted the rig crew to potential problems with the cement barrier or with the negative test. After some discussion among members of the crew and a second negative test on the kill line, the rig crew explained the pressure differential away as a "bladder effect," a theory that later proved to be unfounded. Around 7:45 p.m., after observing for 30 min that there was no flow from the kill line, the rig crew concluded that the negative test was successful. At this point, the rig crew most likely concluded that the production casing cement barrier was sound.

The cement in the shoe track barrier, however, had in fact failed, and hydrocarbons began to flow from the Macondo reservoir into the well. Despite a number of additional anomalies that should have signaled the existence of a kick or well flow, the crew failed to detect that the well was flowing until 9:42 p.m. By then it was too late. The well was blowing drilling mud up into the derrick and onto the rig floor. If members of the rig crew had detected the hydrocarbon influx earlier, they might have been able to take appropriate actions to control the well. There were several possible reasons why the crew did not detect the kick:

- *The rig crew had experienced problems in promptly detecting kicks.* The DWH crew had experienced a kick on March 08, 2010 that went undetected for approximately 30 min. BP did not conduct an investigation into the reasons for the delayed detection of the kick. Transocean personnel admitted to BP that individuals associated with the 08 March kick had "screwed up by not catching" it. Ten of the 11 people on duty on 08 March, who had well control responsibilities, were also on duty on 20 April.
- *Simultaneous rig operations hampered the rig crew's well-monitoring abilities.* The rig crew's decision to conduct simultaneous operations during

the critical negative tests—including displacement of fluids to two active mud pits and cleaning the pits in preparation to move the rig—complicated well-monitoring efforts.

- *The rig crew bypassed a critical flow meter.* At approximately 9:10 p.m., the rig crew directed fluid displaced from the well overboard, which bypassed the Sperry Sun flow meter, which is a critical kick detection tool that measures outflow from the well. The DWH was equipped with other flow meters, but the Panel found no evidence that these meters were being monitored prior to the blowout.

Once the crew discovered the hydrocarbon flow, it sent the flow to a MGS, a piece of equipment not designed to handle high flow rates. The MGS could not handle the volume of hydrocarbons, and it discharged a gas plume above the rig floor that ignited.

The Panel found evidence that the configuration of the DWH general alarm system and the actions of rig crew members on the bridge of the rig contributed to a delay in notifying the entire crew of the presence of very high gas levels on the rig. Transocean had configured the DWH general alarm system in "inhibited" mode, which meant that the general alarm would not automatically sound when multiple gas alarms were triggered in different areas on the rig. As a result, personnel on the bridge were responsible for sounding the general alarm. Personnel on the bridge waited approximately 12 min after the sounding of the initial gas alarms to sound the general alarm, even though they had been informed that a "well control problem" was occurring. During this period, there were approximately 20 alarms indicating the highest level of gas concentration in different areas on the rig.

The BOP stack, a massive 360-ton device installed at the top of the well, was designed to allow the rig crew to handle numerous types of well control events. However, on 20 April, the BOP stack failed to seal the well to contain the flow of hydrocarbons. The explosions likely damaged the DWH's multiplex cables and hydraulic lines, rendering the crew unable to activate the BOP stack.

The BOP stack was equipped with an "automatic mode function," which upon activation would trigger the BSR, two metal blocks with blades on the inside edges that are designed to cut through the drill pipe and seal the well during a well control event. The Panel concluded that there were two possible ways in which the BSR might have been activated: (1) on 20 April, by the AMF, immediately following loss of communication with the rig or (2) on 22 April, when a remotely operated vehicle triggered the "autoshear" function, which is designed to close the BSR if the lower marine riser package disconnects from the rest of the BOP stack. Regardless of how the BSR was activated, it did not seal the well.

A forensic examination of the BOP stack revealed that elastic buckling of the drill pipe had forced the drill pipe up against the side of the wellbore and outside the cutting surface of the BSR blades. As a result, the BSR did not completely shear the drill pipe and did not seal the well. The buckling of the drill pipe, which likely occurred at or near the time when control of the well was lost, was caused by the force of the hydrocarbons blowing out of the well; by the weight of the 5000 feet of drill pipe located in the riser above the BOP forcing the drill pipe down into the BOP stack; or by a combination of both. As a result of the failure of the BSR to completely cut the drill pipe and seal the well, hydrocarbons continued to flow after the blowout.

Prior to the events of 20 April, BP and Transocean experienced a number of problems while conducting drilling and temporary abandonment operations at Macondo. These problems included:

- *Recurring well control events and delayed kick detection.* At least three different well control events and multiple kicks occurred during operations at Macondo. On 08 March, it took the rig crew at least 30 min to detect a kick in the well. The delay raised concerns among BP personnel about the DWH crew's ability to promptly detect kicks and take appropriate well control actions. Despite these prior problems, BP did not take steps to ensure that the rig crew was better equipped to detect kicks and to handle well control events. As of 20 April, Transocean had not completed its investigation into the 08 March incident.
- *Scheduling conflicts and cost overruns.* At the time of the blowout, operations at Macondo were significantly behind schedule. BP had initially planned for the DWH to move to BP's Nile well by March 08, 2010. In large part as a result of this delay, as of 20 April, BP's Macondo operations were more than $58 million over budget.
- *Personnel changes and conflicts.* BP experienced a number of problems involving personnel with responsibility for operations at Macondo. A reorganization that took place in March and April 2010 changed the roles and responsibilities of at least nine individuals with some responsibility for Macondo operations. In addition, the Panel found evidence of a conflict between the BP drilling and completions operations manager and the BP wells team leader and evidence of a failure to adequately delineate roles and responsibilities for key decisions.

At the time of the blowout, both BP and Transocean had extensive procedures in place regarding safe drilling operations. BP required that its drilling and completions personnel follow a "documented and auditable risk management process." The Panel found no evidence that the BP Macondo team fully evaluated ongoing operational risks, nor did it find evidence that BP communicated with the Transocean rig crew about such risks.

Transocean had a number of documented safety programs in place at the time of the blowout. Nonetheless, the Panel found evidence that Transocean personnel questioned whether the DWH crew was adequately prepared to independently identify hazards associated with drilling and other operations.

Everyone on board the DWH was obligated to follow the Transocean "stop work" policy that was in place on 20 April. It provided that "[e]ach employee has the obligation to interrupt an operation to prevent an incident from occurring." Despite the fact that the Panel identified a number of reasons that the rig crew could have invoked stop work authority, no one on the DWH did so that day.

The Panel's recommendations sought to improve the safety of offshore drilling operations in a variety of different ways:

- *Well design.* Improved well design techniques for wells with high flow potential, including increasing the use of mechanical and cement barriers, to decrease the chances of a blowout.
- *Well integrity testing.* Better well integrity test practices (e.g., negative test practices) to allow rig crews to identify possible well control problems in a timely manner.
- *Kick detection and response.* The use of more accurate kick detection devices and other technological improvements to help ensure that rig crews can detect kicks early and maintain well control. Also, better training to allow rig crews to identify situations where hydrocarbons should be diverted overboard.
- *Rig engine configuration (air intake locations).* Assessment and testing of safety devices, particularly on rigs where air intake locations create possible ignition sources, to decrease the likelihood of explosions and fatalities in the event of a blowout.
- *Blowout preventers.* Improvements in BOP stack configuration, operation, and testing to allow rig crews to be better able to handle well control events.

- *Remotely-operated vehicles (ROVs)*. Standardization of ROV intervention panels and intervention capabilities to allow for improved response during a blowout.

The CSB builds on previously published investigation reports by analyzing evidence that, in some respects, became available only following their publication, offering technical, human, organizational, and regulatory perspectives beyond those of previous reports (U.S. Chemical Safety and Hazard Investigation Board, 2016).

The CSB Macondo investigation identified several safety gaps and worthy lessons:

- Testing limitations masked latent failures of the DWH BOP, affecting its operation on the day of the incident, and these latent failures will continue to exist for similarly designed BOPs unless modifications are made to current standard industry testing protocols.
- Pressure conditions in a well can cause the drill pipe to buckle (or bend) in a BOP even after a crew has initially sealed a well, potentially incapacitating emergency functions of the BOP intended to cut the drill pipe and seal the well.
- Industry is challenged to effectively assess the human performance expectations and human factor implications of the barriers and safety systems meant to control or mitigate the hazards of safety-critical well operations.
- Cognitive and social skills training, in conjunction with technical competencies, can be valuable for combating cognitive biases and other mental traps that may influence decision making within complex systems.
- Gaps between work-as-imagined by well designers, managers, or regulatory authorities and work-as-done by the well operations crew must be continually identified, managed, and minimized by building a resilient process that can sustain desirable operations during both expected and unexpected conditions.
- Obstacles continue to exist that not only limit sharing of lessons from incident investigations within individual companies, but also across the operator/drilling contractor boundary and across international geographical regions.
- An equal focus and effort to collect, measure, and improve process safety performance indicators to that currently dedicated to *personal* safety statistics is necessary to reduce the potential for a major accident event.
- Corporate Boards of Directors' oversight, shareholder activism, and US Securities and Exchange Commission (SEC) reporting requirements

have the potential to influence an organization's focus on major accident risk.

- Incongruities among proclaimed values, actual practices, and unstated basic assumptions within an organization's culture impacts its focus on safety, necessitating efforts to effectively assess, monitor and modify all three cultural components for safety change to occur.
- Complexities of multi-party risk management between an operator and drilling contractor in the US offshore industry necessitate more explicit and established safety roles and responsibilities, as well as oversight.

5.2 Lesson Learned

In 2011, the Secretary of the Department of the Interior (DOI) redefined the responsibilities previously performed by the MMS and reassigned the functions of the offshore energy program among three separate organizations: the Bureau of Ocean Energy Management (BOEM), the Bureau of Safety and Environmental Enforcement (BSEE), and the Office of Natural Resources Revenue (ONRR) (Ramseur, 2015).

As one of the responsible parties, BP is reported to have spent over $14 billion in cleanup operations (U.S. District Court for Eastern District of Louisiana, 2016). In addition, BP has paid over $15 billion to the federal government, state and local governments, and private parties for economic claims and other expenses, including reimbursements for response costs related to the oil spill. BP and other responsible parties have agreed to civil and/or criminal settlements with the Department of Justice (DOJ). Settlements from various parties, to date, total almost $6 billion. BP is to pay the United States a civil penalty of $5.5 billion under the Clean Water Act (CWA), payable over 15 years. BP will pay $7.1 billion to the United States and the five Gulf states over 15 years for NRD. This is in addition to the $1 billion already committed for early restoration. BP will also set aside an additional amount of $232 million to be added to the NRD interest payment at the end of the payment period to cover any further NRD that are unknown at the time of the agreement. A total of $4.9 billion will be paid over 18 years to settle economic and other claims made by the five Gulf Coast states. Up to $1 billion will be paid to resolve claims made by more than 400 local government entities. The principal payments arising from the agreements will be made over a period of 18 years.

In July 2010, IOGP established the Global Industry Response Group (GIRG) to identify, learn from and apply the lessons of Macondo and similar well accidents. The GIRG brought together more than 100 technical experts and managers from some 20 companies around the world. Most of them worked full time on the project for the better part of a year. They pooled their knowledge and experience to create three dedicated teams focused on oil spill prevention, intervention, and response.

Their recommendations led to:

- An industry-wide well control incident database.
- A task force on BOP reliability.
- Improved human factors training and competencies.
- The development and implementation of key international standards for well design and operations management.
- The creation of the Subsea Well Response Project—known as SWRP—to improve intervention capabilities.
- The creation of the Oil Spill Response Joint Industry Project—known as the OSR-JIP—to improve oil spill response capabilities.
- Mutual aid agreements and framework to enable operators to access additional resources in the event of an major oil spill.

As a result of GIRG, industry can show that it has learned from events such as Macondo and has worked to reduce the likelihood and consequences of future incidents.

REFERENCES

Baker, J. A., III (2007). *The report of the BP U.S. Refineries independent safety review panel.*

BP. (2011). The report of the internal BP incident investigation team. *Deepwater Horizon: Accident Investigation Report.*

Dawson, D. M., Brooks, B. J., & Longford Royal Commission (1999). *The Esso Longford gas plant accident—Report of the Longford Royal Commission.*

Khan, F., Rathnayaka, S., & Ahmed, S. (2015). Methods and models in process safety and risk management: Past, present and future. *Process Safety and Environmental Protection, 98,* 116–147.

Levine, S. (2009). EXXON—Juggernaut or Dinosaur? *Bloomberg/BusinessWeek* (February 5)

Mogford, J. (2005). *BP, fatal accident investigation report, isomerization unit explosion.* Texas City, Texas (December 9, 2005).

Mouawad, J. (2010). New culture of caution at Exxon after Valdez. *The New York Times* (July 12)

National Transportation Safety Board. (1990). Marine accident report. *Grounding of the U.S. Tankship Exxon Valdez on Bligh Reef, Prince William Sound near Valdez, Alaska: Report No. NTSB/MAR-90/04, (July 31, 1990).*

Parker, W. B. (February 1990). *Alaska oil spill commission, spill, the wreck of the Exxon Valdez, implications for safe transportation of oil.*

Ramseur, J. L. (April 2015). *Deepwater Horizon oil spill: Recent activities and ongoing developments, congressional research service.*

Shigenaka, G. (2014). *Twenty-five years after the Exxon Valdez oil spill: NOAA's scientific support, monitoring, and research.* Seattle: NOAA Office of Response and Restoration 78.

The Bureau of Ocean Energy Management. (September 2011). *Regulation and enforcement, U.S. Department of interior*. Report regarding the cause of the April 20, 2010 Macondo Well Blowout.

U.S. Chemical Safety and Hazard Investigation Board. (2007). *Investigation report: refinery. Explosion and fire, Texas*.

U.S. Chemical Safety and Hazard Investigation Board. (2016). *Investigation report executive summary: Drilling Rig explosion and fire at the Macondo well*. Mississippi Canyon, Gulf of Mexico: Deepwater Horizon Rig.

U.S. District Court for Eastern District of Louisiana (2016). Multi-District Litigation No 2179, Section J, Judge Barbier and Magistrate Judge Shushan, case 2:10-md-02179-CJB-SS, Document 16022-1, filed 03/22/16.

CHAPTER THREE

Elements of Process Safety Management

Paul R. Amyotte[1], Cathleen S. Lupien
Dalhousie University, Halifax, NS, Canada
[1]Corresponding author: e-mail address: Paul.Amyotte@Dal.Ca

Contents

1. INTRODUCTION

Process—any activity involving a highly hazardous chemical including using, storing, manufacturing, handling, or moving such chemicals at the site, or combination of these activities.

OSHA (2000)

Methods in Chemical Process Safety, Volume 1
ISSN 2468-6514
http://dx.doi.org/10.1016/bs.mcps.2017.01.004

Process Safety—the prevention and mitigation of process-related injuries and damage arising from process incidents involving fire, explosion and toxic release.

Amyotte (2013a)

Process Safety Management (PSM)—the application of management principles and systems to the identification, understanding and control of process hazards to prevent process-related injuries and incidents.

CSChE (2012a, 2012b)

The scope of this chapter is embodied in the above suite of definitions—i.e., management actions directed at the reduction of risk arising from flammable, reactive, and toxic materials found in the process industries. The specific objective is to examine key process safety management (PSM) systems and their elements, as well as process safety concepts requiring integration within a PSM system for them to be truly effective. Fig. 1 indicates the structure of the chapter which progresses from a brief look at management systems and safety management systems (SMSs) in general, through to a more detailed analysis of PSM systems including regulatory, performance measurement, and improvement aspects. The penultimate section preceding the conclusion gives an overview of the relationship between a PSM system and several important process safety concepts: (i) inherently safer design (ISD), (ii) recognition of warning signs, (iii) process safety culture, and (iv) dynamic risk assessment.

In terms of research methodology, a search was conducted of the process safety literature dealing mostly with management system features over the past decade. The search was, again for the most part, restricted to industrial

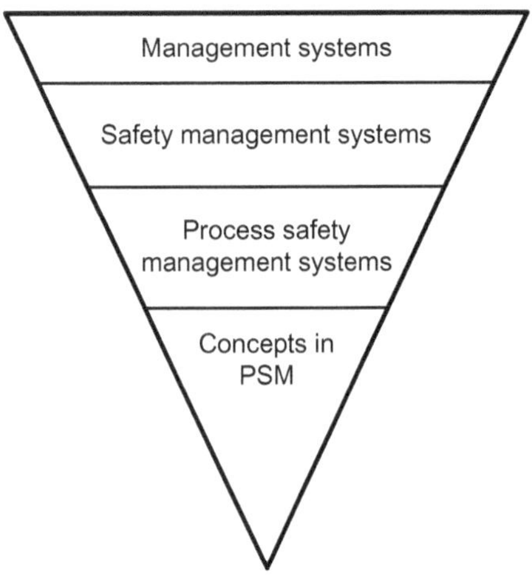

Fig. 1 Structure of the current chapter on elements of process safety management.

and process safety books, internet sources, and journals (e.g., *Journal of Loss Prevention in the Process Industries*, *Process Safety and Environmental Protection*, *Process Safety Progress*, and *Safety Science*), with some coverage of conference proceedings. We make no claim that the search design and results are exhaustive, but we are confident that the sampling is well representative of the body of knowledge in this area. There are undoubtedly national and perhaps international PSM systems we have not discussed; the systems and elements described here are, however, indicative of the global consensus for comprehensive PSM coverage.

Motivation for the current work—i.e., why attach special significance to the use of management systems in assuring process safety?—comes from a number and variety of sources. In their review of the occurrence of major process incidents, Amyotte et al. (2016) describe seven core concepts for the prevention of such events: (i) the creation of paradigm-enhancing organizations (e.g., the Center for Chemical Process Safety in the United States), (ii) ISD, (iii) awareness of the total cost of major accidents, (iv) consideration of the broader societal and cultural aspects of major accidents, (v) process safety culture, (vi) process safety competency, and (vii) dynamic risk assessment. As illustrated throughout this chapter, each of these concepts is related in some fashion to the success or lack thereof in employing a PSM system. They are also key contributors, either implicitly or explicitly, to the evolution of process safety practice that is illustrated in Fig. 2.

The following general works also reference the relevance of PSM systems, elements, and concepts (as discussed in this chapter):

- MKOPSC (2012)—An international panel drawn largely from academia identified 19 focus areas for future research, including PSM knowledge transfer, standardization of process safety methods, safety culture, ISD, and risk management.
- Knegtering and Pasman (2009)—The authors' proposed improvement foci for modern PSM include an adequate process safety measurement system and a continuous learning system.
- Vaughen and Kletz (2012)—Fig. 3 shows how many of our current process safety elements (e.g., management of change) and concepts (e.g., ISD) have been developed in response to major process incidents such as Flixborough, Seveso, and Bhopal. The authors describe our present situation of managing risk, the unexpected, and complex systems, as well as a future in which managing information and addressing public perception are paramount.

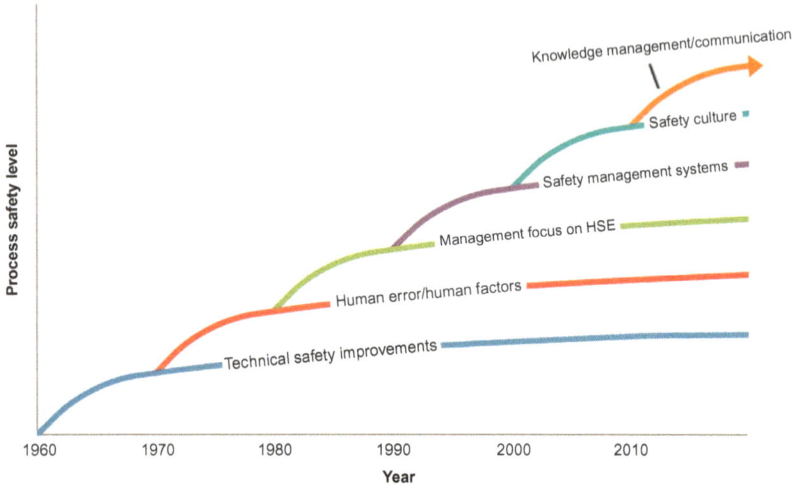

Fig. 2 Significant contributions to the evolution of process safety. *Amyotte, P. R., Berger, S., Edwards, D. W., Gupta, J. P., Hendershot, D. C., Khan, F. I., et al. (2016). Why major accidents are still occurring.* Current Opinion in Chemical Engineering, 14, *1–8; updated from MKOPSC (Mary Kay O'Connor Process Safety Center), Texas A&M University System, A Frontiers of Research Workshop. (2012). Process safety research agenda for the 21st century. A policy document developed by a representation of the global process safety academia. October 21–22, 2011, College Station, TX. College Station, TX: Mary Kay O'Connor Process Safety Center, Texas A&M University System; De Rademaeker, E., Suter, G., Pasman, H. J., & Fabiano, B. (2014). A review of the past, present and future of the European loss prevention and safety promotion in the process industries.* Process Safety and Environmental Protection, 92, *280–291; Amyotte, P., Margeson, A., Chiasson, A., Khan, F., & Ahmed, S. (2014). There is no such thing as a black swan process incident.* Proceedings of Hazards 24, IChemE, *Edinburgh, UK (May 7–9, 2014).*

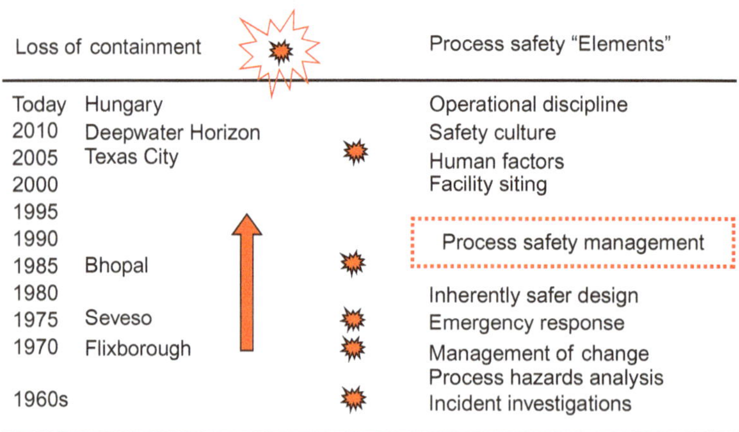

Fig. 3 Evolution of process safety elements by application of lessons learned (Vaughen & Kletz, 2012).

- Pasman, Knegtering, and Rogers (2013)—Central to the authors' holistic approach to process risk reduction are the concepts of dynamic risk assessment, monitoring safety performance indicators for PSM systems, and recognizing weak signals to improve situational awareness—all of which are critical in enhancing facility resilience.
- De Rademaeker, Suter, Pasman, and Fabiano (2014)—This paper (along with MKOPSC, 2012) provides the foundation for Fig. 2 in the current work. The authors also describe the major themes for the *14th International Symposium on Loss Prevention and Safety Promotion in the Process Industries* held in 2013 and organized by the European Federation of Chemical Engineering (EFCE); two of these themes were human factors and management systems, and learning from accidents and knowledge transfer.
- Khan, Rathnayaka, and Ahmed (2015)—Safety management and SMSs figure prominently in this comprehensive review of the state-of-the-art with respect to process safety and process risk management.

In closing this section, we present a quote from one of the last papers written by the late Professor Trevor Kletz—a mentor and teacher to all in the field of process safety:

> *As a consultant I knew the importance of talking to employees at all levels and checking details from time to time. After promotion many people take a helicopter approach, leaving the detail to others. From a helicopter all we see are forests. If we want to see if the forest is healthy, we have to land the helicopter and look at the leaves and twigs.*
>
> ***Kletz (2012)***

To continue with Trevor's analogy, we will now board the helicopter for an overview of management and SMSs. We will then disembark and walk among the details of PSM systems and concepts.

2. MANAGEMENT SYSTEMS

> *A management system describes the set of procedures an organization needs to follow in order to meet its objectives.*
>
> ***ISO (2016)***
>
> *Management is doing things right; leadership is doing the right things.*
>
> ***Peter F. Drucker, as quoted in Kletz and Amyotte (2010)***

According to Schein (2010), *management* is a largely North American construct for which there is no comparable word in several other languages.

He cites the example of the German language, which—although conveying the meaning of concepts such as *leading* and *directing*—does not have a counterpart for the act of *managing* (Schein, 2010). Perhaps then, as remarked by Kletz and Amyotte (2010), it would be more appropriate to speak of process safety *leadership* rather than process safety *management*. We will, however, abide with convention and use the familiar term PSM; it should come as no surprise though that later sections of this chapter examine the key role of corporate leaders in managing process safety efforts. (On a personal note from one of the authors (P.R.A.), a highly educational and entertaining treatise on leadership from a completely different field can be found in the book by Ferguson and Moritz, 2015.)

Meyer and Reniers (2016) identify two main categories of management systems: business management systems and risk management systems. Regardless of the classification, the *ISMEC* progression of steps proposed by Bird and Germain (1996) for management control is appropriate: (1) ***I****dentification* of the work to be performed, (ii) ***S****tandards* for the work at all levels, (iii) ***M****easurement* of performance to standards, (iv) ***E****valuation* of performance level compared to standards, and (v) ***C****ommendation* for compliance to standards and constructive correction of substandard work. With only minor tweaking of the *ISMEC* framework, one easily arrives at the currently accepted view of the essential elements of any management system, often expressed as *plan, do, check, act* (Arntz-Gray, 2016; Dougherty, 1999—with emphasis on the *check* function by means of independent auditing).

The thoughts expressed above are in accordance with Asfahl and Rieske (2010) who remark that a manager assumes an enlarged scope of responsibility (and therefore accountability) that involves hazard analysis, standards compliance, and capital investment planning. Brauer (2006) states that management involves planning, organizing, and directing the elements integral to an organization achieving its goals. These elements include activities, people, equipment, materials, facilities, regulations, time, cost, and the physical, social, and management environments (Brauer, 2006).

Analogous to Meyer and Reniers (2016), Hammer and Price (2001) comment that several different, overlapping management systems will typically be functioning within a given organization. These could be core management functions unique to the enterprise, as well as aspects related to financial, personnel, quality, environmental, and health and safety aspects (Hammer & Price, 2001). Meyer and Reniers (2016) give several examples of international management systems for dealing with quality and business performance, and also environmental, integrity, and occupational health and safety matters. The website of the International Organization for Standardization (ISO, 2016) provides information on standards for energy, environmental, quality, food safety, information security, sustainable events, occupational health and safety, and antibribery management systems.

Goetsch (2015) refers to total quality management (TQM) in the context of this system being a forerunner to total safety management (TSM), in which a holistic approach is taken to enhance the safety of employees, products, and processes when establishing safe operating procedures and practices. One recognizes in this approach the modern integrated view of preventing and mitigating loss in the categories of people, property, process, and environment (Wilson & McCutcheon, 2003). In a similar vein, Coulter (1995) deals with the application of TQM to the health, safety, and environmental spheres; Kontogiannis, Leva, and Balfe (2016) give a comprehensive and contemporary review of TSM principles and methodologies. For coverage of systems dealing with general and safety management, quality and safety management, and health, safety, and environmental management, readers are directed to the papers by Swuste et al. (2016), Celik (2009), and Gholami, Nassiri, Yarahmadi, Hamidi, and Mirkazemi (2015), respectively.

3. SAFETY MANAGEMENT SYSTEMS

The objective of a SMS [Safety Management System] is to reduce injuries and to preserve the environment and the productive lives of assets. An effective SMS is recognized by many businesses as an essential requirement for remaining in business.

Lutchman, Maharaj, and Ghanem (2012)

Having established the industrial relevance of management systems in general, we now turn our attention to those systems that focus on management of safety concerns. Lutchman et al. (2012) argue the business case for SMSs by virtue of the profitability and sustainability benefits accruing to an organization, as shown in Table 1.

The archival journal literature reviewed in the current work is replete with papers dealing with various aspects and applications of SMSs. Here we have separated them into three categories of papers dealing primarily with: (i) fundamental features, (ii) sectoral examples, and (iii) transportation examples. There is, of course, some overlap in the categories in that a given paper

Table 1 Benefits of an Effective Safety Management System (Lutchman et al., 2012)

Quantifiable Benefits	Less Quantifiable Benefits
• Fewer incidents and injuries • Less severe injuries • Fewer work-related fatalities and diseases • Reduced absenteeism • Reduced loss time away from work • Reduced health care treatment cost • Fewer employee turnovers • Reduction in insurance cost • Lowered operating and production cost	• A more motivated workforce • Ability to attract and retain the best and brightest • Greater stakeholder commitment • More community support and engagement • Higher trust and credibility among all stakeholders

could involve, for example, performance measurement (fundamental feature) of a management system for occupational safety in the construction industry (sectoral example).

3.1 Fundamental Features

Grote (2012) provides an interesting analysis of SMSs in which the issue of system similarity for various high-risk activities is addressed. She establishes three attributes for customizing a purpose-fit SMS: (i) the types of safety being managed, (ii) how uncertainty is managed, and (iii) the regulatory regime applicable to the safety management efforts (Grote, 2012). Grote (2012) speaks of the desirability of knowledge transfer across high-risk sectoral boundaries involving the nuclear industry, chemical process industries, and aviation domains.

In a somewhat similar vein, Pillay (2015) gives an analysis of published articles to demonstrate the relationship between accident causation factors in different industries and the safety management strategies typically employed in those industries. Examples given include sociotechnical systems in the aviation industry, technological and cultural systems in the oil and gas industry, and behavioral systems in road transportation (Pillay, 2015).

Wahlstrom and Rollenhagen (2014) take a creative look at SMSs by using a metaphor based on requirements from control theory:

- a *system model* to enable prediction of outcomes for specific actions (e.g., SMS elements),

- *observability* of the system so its state can be determined (e.g., system performance indicators),
- *controllability* of the system so specific actions can be taken to effect changes in the system state (e.g., risk reduction measures), and
- a *preference relation* to distinguish between desired and undesired system states (e.g., cost/benefit analysis).

A different sort of comparison is drawn by Moorkamp, Kramer, van Gulijk, and Ale (2014) in their examination of the relationship between SMS theory and resilience engineering theory. They consider the former theory to be one in which uncertainty is *minimized*, while the latter involves *coping* with uncertainty (Moorkamp et al., 2014). One can see clear overtones of dynamic operational risk management—as discussed later in this chapter—in their comment that resilience engineering attempts to manage safety by *accounting for the constantly changing nature of dynamic operational conditions* (Moorkamp et al., 2014). There is thus a role for probabilistic as well as deterministic measures in managing an organization's safety efforts; this observation is in accordance with the earlier discussion of the work of Grote (2012).

Fernandez-Muniz, Montes-Peon, and Vazquez-Ordas (2007) performed a study involving 455 Spanish companies in an attempt to develop a scale for measurement of the extent to which SMSs had been implemented. The developed tool consists of 43 items spread over eight groupings: safety policy, worker incentives, safety training, communication related to prevention, preventive planning, emergency planning, internal control, and benchmarking techniques (Fernandez-Muniz et al., 2007).

Bianchini, Donini, Pellegrini, and Saccani (2017) describe their use of an *Efficacy Index* for the same purpose—i.e., quantitatively evaluating the effectiveness of implementation of a SMS (in this case, for occupational health and safety). The index relates the costs arising from a loss-producing event or near-miss to the estimated costs to prevent and protect with respect to the unintended event.

Finally, there is clear evidence in the general safety literature of research into concepts that are also key to the success of PSM (as discussed later in this chapter). These include knowledge management and communication (Vinodkumar & Bhasi, 2010), safety leadership (Pilbeam, Doherty, Davidson, & Denyer, 2016; Sheehan, Donohue, Shea, Cooper, & De Cieri, 2016), and leading and lagging indicators for safety performance measurement (Sheehan et al., 2016).

3.2 Sectoral Examples

As demonstrated in Table 2, the breadth of system applications for managing different types of safety in different industrial/activity sectors throughout the world is quite comprehensive. While the depth of subject matter coverage in these papers is substantial, comments here are limited to three brief points. First, it is interesting to note the adoption of a risk-based approach for occupational health and safety management in the work of Sousa et al. (2014,

Table 2 Examples of Safety Management Systems in Different Industrial/Activity Sectors

Reference	Sector	Management Concern
Antonsen, Skarholt, and Ringstad (2012)	Oil and gas industry offshore Norway	Safety
Boustras, Hadjimanolis, Economides, Yiannaki, and Nicolaides (2015)	Microfirms (<10 employees) in Cyprus	Health and safety
Newslow (2014)	Food industry	Food safety
Sousa, Almeida, and Dias (2014, 2015)	Construction industry	Occupational health and safety
Zhou, Goh, and Li (2015)	Construction industry	Safety
Lingard, Hallowell, Salas, and Pirzadeh (2017)	Infrastructure construction project in Australia	Safety
Battaglia, Passetti, and Frey (2015)	Municipal waste companies in Italy	Occupational health and safety
Jeon, Lee, Shin, and Park (2009)	Dams in Korea	Dam safety
Zhang, Wu, Chen, Skibniewski, and Hsu (2014)	Buildings adjacent to tunneling excavation in China	Building safety
Qian and Lin (2016)	Underground (tunnel) engineering in China	Safety/risk
Wu, Xu, Zhou, Peng, and Yu (2014)	Coal mines in China	Mine safety
Donaldson, Borys, and Finch (2013)	Community sporting organizations in Australia	Sports injury safety

2015); this is similar to developments over the past decade in the field of PSM (as subsequently explained in this chapter). Second, the role of a SMS in relation to input parameters, other core functions, and management objectives is clearly illustrated in Fig. 4 from the work of Qian and Lin (2016). Third, the coal mine SMS from the work of Wu et al. (2014), as presented in Fig. 5, can be seen to explicitly embody the cycle of plan/do/check/act that was previously described as being essential to the effective functioning of any management system.

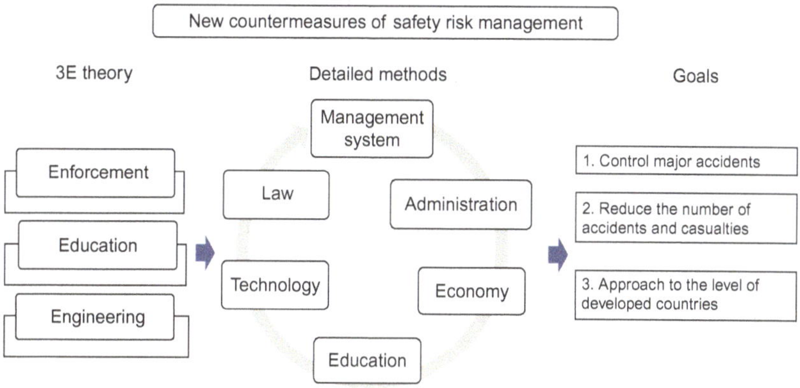

Fig. 4 Framework for safety risk management of underground engineering in China (Qian & Lin, 2016).

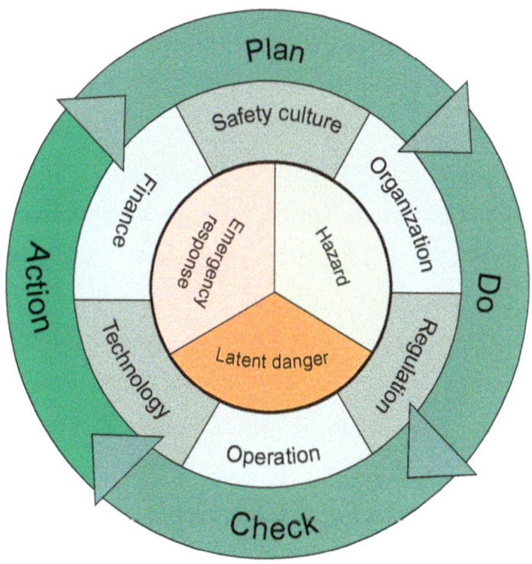

Fig. 5 Framework for a coal mine safety management system (Wu et al., 2014).

3.3 Transportation Examples

Literature examples of SMS applications to the transportation industry exist for each of the land, sea, and air sectors. According to Warmerdam, Newnam, Sheppard, Griffin, and Stevenson (2017), work-related vehicles account for 30% of traffic volume in Australia; additionally, drivers engaged in work-related activities are more likely to suffer injury than drivers on the road for purposes unrelated to work. With this risk-based justification for their research, Warmerdam et al. (2017) present their findings on the need for improvements in organization accountability, communication practices, vehicle-related risk reduction, driver competency, and incident investigation. These improvement areas are equally applicable to the management of risks posed by flammable, explosible, and toxic materials.

Similarly, Mooren, Grzebieta, Williamson, Olivier, and Friswell (2014) cite statistical evidence for their risk-based examination of safety management of heavy vehicle transport. Heavy trucks comprise 3% of registered vehicles in the United States, account for 7% of total vehicular mileage, and are involved in 11% of all driver fatalities (Mooren et al., 2014). Again, their findings on the importance of factors such as management commitment, safety training, and worker participation (Mooren et al., 2014) are equally applicable to the field of PSM.

In his study of safety and risk at sea (specifically with respect to tankers), Havold (2010) gives an extensive discussion of safety culture—a concept that underpins all SMSs. His description of safety culture being composed of values, attitudes, perceptions, and competencies (Havold, 2010) gives a good preview of our discussion later in this chapter on process safety culture. The theme of management system measurement requirements is also evident in the call for safety culture metrics that can be employed as leading performance indicators (Havold, 2010).

The Civil Aviation Safety Authority in Australia has provided a SMS guide (CASA, 2014) for aviation operators and organizations. The guide distinguishes between a SMS and a business/quality management system, and discusses principles such as safety culture that are broadly applicable (CASA, 2014). The document also considers human factors within a SMS and presents the familiar Swiss cheese model (Reason, 1990) in the context of flight operations of a Boeing 737-300 aircraft (CASA, 2014).

Cacciabue, Cassani, Licata, Oddone, and Ottomaniello (2015) also present the management system approach to safety from an aviation perspective, but again with broad applicability. They use terminology instantly recognizable to process safety practitioners—such as *bow-tie analysis* and *risk matrix*—and when describing the purpose of a SMS as being to prevent, control, and contain the consequences of hazardous events (Cacciabue et al., 2015). Cacciabue et al. (2015) further comment on the significant cultural changes that must accompany the SMS approach. This recurring notion of the critical role of safety culture in managing aviation safety is also apparent in the work of Remawi, Bates, and Dix (2011) and Liao (2015), with the latter study invoking the concepts of just, reporting, and learning cultures (as discussed later in this chapter with respect to PSM).

Before moving on to PSM, we conclude the review in this section with mention of two additional research studies on aviation safety management. Stroeve, Som, van Doorn, and Bakker (2016) present a holistic, risk-based approach for examination of a specific aviation hazard (runway incursions). The work of Nascimento, Majumdar, Ochieng, Schuster, and Studic (2016) describes the development of a SMS for a specific mode of air transport (helicopters).

4. PROCESS SAFETY MANAGEMENT

Process safety is about keeping it in the pipes (as expressed by numerous process safety practitioners).

Process safety is about understanding the chemistry and physics of the manufacturing process.

Hendershot (2016)

Given the above quotes (and notwithstanding the more formal definitions found in Section 1), process safety can be said to be about *containment*—i.e., keeping hazardous materials (the *chemistry*), and energy (the *physics*), within their designed process boundaries. PSM is then about planning, doing, and checking the work needed to ensure containment, and also acting on the results of these efforts. And a PSM system is about assisting company personnel (especially managers) in achieving the organization's containment objectives.

In this section, we examine PSM with a focus on various PSM systems and their elements. The starting point is a brief look at the regulation of PSM; the section concludes with an analysis of performance metrics associated with PSM systems, as well as suggestions from the literature for the continuous improvement of these systems. Case study examples from both the archival literature and the investigations of the US Chemical Safety Board (CSB) are also given.

4.1 Regulatory Aspects

As noted earlier in reference to the work of Grote (2012), the regulatory regime is one of the determinants in designing a SMS. Thus, our objective here is to touch on the topic of PSM regulation; a detailed review of the subject is, however, outside the analytical scope of this chapter. For this latter purpose, readers are directed to the recent paper by Sreenevasan (2015) and the comprehensive review undertaken by the Canadian Association of Petroleum Producers (CAPP, 2014).

We first comment that it would seem well advised to view the regulation of PSM—whether by prescription or based on performance (as explained in the next paragraph)—in the manner expressed by the following authors and corresponding quotes:

> *... it must be noted that regulations themselves cannot improve process safety performance; instead, regulations should be considered as the minimum standards that can provide the motivation for improvement.*
>
> ***Mannan (2015)***
>
> *Thoughtful companies recognize that meeting the minimum legal requirements may not be sufficient, and also audit for conformance with their internal company standards and procedures.*
>
> ***Gill (2015)***
>
> *... regulatory compliance should be an outcome of a good process safety management program, not an objective.*
>
> ***Attributed to M. Broadribb in Hendershot (2016).***

Mannan (2015) also remarks that regulations should be based on science and an understanding of risk, and that compliance must be enforced by the appropriate regulatory body.

The second point made here is that when PSM regulations have been introduced in various parts of the world, it has usually been in response to major process incidents (see, for example, Shin, 2013; Sreenevasan, 2015; recall also Fig. 3). Maher, Long, Cromartie, Sutton, and Steinhilber (2016) describe the US regulatory response to the 2010 Deepwater Horizon (Macondo) incident, and the resulting modified SMS requirements for offshore operations. They also distinguish between prescriptive regulations and goal- or performance-based regulations in the following manner (Maher et al., 2016):

- Prescriptive regulations state-specific actions and requirements that must be met to achieve compliance.
- Goal- or performance-based regulations state the required outcome but leave the manner in which the outcome is achieved to the implementer.

Engineering educators will recognize in these definitions an analogy with undergraduate engineering accreditation systems based on input measurements (e.g., lecture and tutorial hours), and those based on outcomes (i.e., graduate attributes such as proficiency in design and use of engineering tools).

Luo (2010) examines the relationship between PSM inspection citations in the United States (as regulated by OSHA, the Occupational Safety and Health Administration), and management system deficiencies identified in investigations conducted by the US CSB for 19 major chemical accidents. The correlation between the two datasets was found to be significant and enabled suggestions for improvements in enforcement and implementation of the OSHA-regulated PSM system. Kwon, Lee, Seo, and Moon (2016) provide another case study related to experiences with PSM regulation in a particular country (in this case, Korea).

Our third comment on regulatory aspects is that in most industrialized countries, PSM is indeed heavily regulated. A notable exception is Canada, which relies mostly on voluntary compliance and initiatives led largely by industry and industry/technical associations (e.g., the PSM Standard (Piette, 2012) developed by the Canadian Society for Chemical Engineering (CSChE, 2012b)). The CAPP (2014) report terms this voluntary compliance more perception than reality, given the bits and pieces of PSM-type requirements in existing legislation; one such example is the emergency response planning requirements for chemical accidents as specified in the Canadian Environmental Protection Act (Di Menna, 2012).

Macza (2008) provides a historical perspective for the Canadian PSM regulatory scene in terms of constitutional authority over most process industry plants resting with the provinces rather than the federal government. A complicating factor for those in favor of PSM regulation in Canada is that the vast majority of safety regulation at the provincial level deals explicitly only with occupational safety. Although some authors (e.g., Sreenevasan, 2015) have called for national PSM regulation in Canada, this sentiment is not shared by all practitioners (Di Menna, 2012).

In their study of the effect of complex occupational health and safety rule sets on the behavior of managers and corporations, Hale, Borys, and Adams (2015) state that their findings are applicable to any externally imposed regulations (e.g., those promulgated by government). They further comment on the likely applicability of their analysis to internal company standards, and the desirability of extending their research to rules that are not legally binding such as in the case of industry standards (Hale et al., 2015).

The current situation in Canada with respect to PSM regulation would seem to afford an excellent opportunity in this regard.

In summary, PSM regulations are best viewed as minimum standards for which compliance should be an outcome not an objective. Major process incidents are a key driving force for governments to introduce or alter a PSM regulatory framework. PSM regulation, although prevalent globally, is not universal as evidenced by the largely voluntary compliance regime in Canada.

4.2 Systems

We now present a number of PSM systems utilized in countries and regions throughout the world. The elements for each system are given in a series of tables and figures drawn from pertinent references. The primary objective in this section is to present the various frameworks of PSM elements, with the elements themselves being the subject of the next section.

Table 3 gives both the elements, and the components for each element, in the PSM system recommended by the Canadian Society for Chemical Engineering (CSChE). Details can be found in the form of a 4th edition Guide (CSChE, 2012a) and a 1st edition Standard (CSChE, 2012b). As noted in CSChE (2012a), the purpose of the Guide is to provide an overview of PSM and an introduction to the Standard; the objective of the Standard is to identify performance requirements that are auditable for continuous improvement purposes (CSChE, 2012b).

The PSM approach in Canada (Table 3) is based on an earlier system shown in Table 4 that was originally developed by the Center for Chemical Process Safety (CCPS) in the United States. Table 5 gives the PSM Standard regulated by the US Occupational Safety and Health Administration (OSHA) under 29 CFR 1910.119; descriptions of this system are given in groupings of the first 3 elements by Mason (2001a) and the remaining 11 elements by Mason (2001b). PSM elements regulated by the US Environmental Protection Agency (EPA) under 40 CFR 68 in the Risk Management Plan (RMP) Rule are shown in Table 6 (CFR = Code of Federal Regulations).

Table 7 illustrates the newer risk-based PSM (RBPS) system now recommended by CCPS; the arrangement is by accident prevention pillar (4) and PSM element (20). The evolutionary thinking behind the development of this system is shown in Fig. 6 from CCPS (2014), which gives an updated overview of the risk-based approach. The fit of RBPS within

Table 3 PSM Elements and Components: Canadian Society for Chemical Engineering (CSChE, 2012a, 2012b)

Element	Component
1. Accountability: Objectives and goals	• Continuity of operations • Continuity of systems • Continuity of organization • Quality process • Control of exceptions • Alternative methods • Communications • Company expectations
2. Process knowledge and documentation	• Chemical and occupational health hazards • Process definition/design criteria • Process and equipment design • Protective systems • Normal and upset conditions • Process risk management decisions • Company memory
3. Capital project review and design procedures	• Appropriation request procedures • Hazard reviews • Siting • Plot plan • Process design and review procedures • Project management procedures and controls
4. Process risk management	• Hazard identification • Risk analysis of operations • Reduction of risk • Residual risk management • Process management during emergencies • Encouraging client and supplier companies to adopt similar risk management practices • Selection of businesses with acceptable risk
5. Management of change	• Change of process technology • Change of facility • Organizational changes • Variance procedures • Permanent changes • Temporary changes
6. Process and equipment integrity	• Reliability engineering • Materials of construction • Fabrication and inspection procedures • Installation procedures • Preventative maintenance

Continued

Table 3 PSM Elements and Components: Canadian Society for Chemical Engineering (CSChE, 2012a, 2012b)—cont'd

Element	Component
	• Process, hardware, and systems inspection and testing • Maintenance procedures • Alarm and instrument management • Decommissioning and demolition procedures
7. Human factors	• Operator—process/equipment interface • Administrative control versus engineering control • Human error assessment
8. Training and performance	• Definition of skills and knowledge • Design of operating and maintenance procedures • Initial qualifications assessment • Selection and development of training programs • Measuring performance and objectives • Instructor program • Records management • Ongoing performance and refresher training
9. Incident investigation	• Major incidents • Third party participation • Follow-up and resolution • Communication • Incident recording, reporting and analysis • Near-miss reporting
10. Company standards, codes, and regulations	• External codes/regulations • Internal standards
11. Audits and corrective actions	• PSM systems audit • Process safety audits • Compliance reviews • Internal/external auditors • Corrective actions
12. Enhancement of process safety knowledge	• Quality control programs and process safety • Professional and trade association programs • Technical association programs • Research, development, documentation, and implementation • Improved predictive system • Process safety resource center and reference library

Table 4 PSM Elements: Center for Chemical Process Safety, American Institute of Chemical Engineers (CCPS, 1992)

1. Accountability: Objectives and goals
2. Process knowledge and documentation
3. Capital project review and design procedures
4. Process risk management
5. Management of change
6. Process and equipment integrity
7. Human factors
8. Training and performance
9. Incident investigation
10. Standards, codes, and laws
11. Audits and corrective actions
12. Enhancement of process safety knowledge

CCPS's Vision 20/20 is explained in the following passage quoted from CCPS (ccpsonline.org) in Amyotte, Berger, et al. (2016): [Vision 20/20…] *looks into the not-too-distant future to demonstrate what perfect process safety will look like when it is championed by industry; driven by five tenets of culture, standards, competency, management systems and lessons learned; and enhanced by community passion and four global societal themes*. The American Institute of Chemical Engineers publication *Chemical Engineering Progress* has recently introduced a new column, *Process Safety Values*, to communicate the Vision 20/20 tenets and themes—e.g., a committed process safety culture (PSV, 2016). (See also McCavit, Berger, & Nara, 2014; McCavit, Berger, Grounds, & Nara, 2015 for discussion of CCPS's Vision 20/20.)

Rosen (2015) provides an interesting and personal look at risk-based PSM in the form of *ten commandments* (several of which—in particular I, IV, and VI—will be quite familiar by now). Thou shalt (Rosen, 2015):

I. Always honor thy container.
II. Always maintain a sense of vulnerability.
III. Eliminate normalization of deviation (i.e., not accept operation outside set limits).
IV. Know thy chemistry.

Table 5 PSM Elements and Relevant 29 CFR 1910.119 Articles: US Occupational Safety and Health Administration (OSHA, 2000, 2013)

Element	Article
Employee participation	29 CFR 1910.119(c)
Process safety information	29 CFR 1910.119(d)
Process hazard analysis	29 CFR 1910.119(e)
Operating procedures	29 CFR 1910.119(f)
Training	29 CFR 1910.119(g)
Contractors	29 CFR 1910.119(h)
Prestartup safety review	29 CFR 1910.119(i)
Mechanical integrity	29 CFR 1910.119(j)
Hot work permit	29 CFR 1910.119(k)
Management of change	29 CFR 1910.119(l)
Incident investigation	29 CFR 1910.119(m)
Emergency planning and response	29 CFR 1910.119(n)
Compliance audits	29 CFR 1910.119(o)
Trade secrets	29 CFR 1910.119(p)

V. Educate, train, and drill employees.
VI. Create and nurture a strong risk-based process safety culture.
VII. Recognize those who exemplify process and occupational safety.
VIII. Not tolerate omissions in documentation.
IX. Not manage from behind thy desk.
X. Not violate rules.

There are clearly many similarities among the PSM systems displayed in Table 3 (CSChE), Table 4 (original CCPS), Table 5 (OSHA), Table 6 (EPA), and Table 7 (RBPS CCPS). There are also differences in the naming, number, and arrangement of elements in each of the systems. Detailed comparisons are given in: (i) CAPP (2014) for the CSChE PSM system (Table 3) and the OSHA PSM (Table 5)/EPA RMP (Table 6) systems, (ii) CCPS (2007) for the original CCPS PSM system (Table 4), the OSHA PSM (Table 5)/EPA RMP (Table 6) systems, and the RBPS CCPS system (Table 7), and (iii) CCPS (2016a) for the OSHA PSM (Table 5)/EPA RMP (Table 6) systems and the RBPS CCPS system (Table 7).

Table 6 PSM Elements and Relevant 40 CFR 68 Sections: US Environmental Protection Agency (EPA, 2004)

Element	Section
Process safety information	40 CFR 68.65
Process hazard analysis	40 CFR 68.67
Operating procedures	40 CFR 68.69
Training	40 CFR 68.71
Mechanical integrity	40 CFR 68.73
Management of change	40 CFR 68.75
Prestartup review	40 CFR 68.77
Compliance audits	40 CFR 68.79
Incident investigation	40 CFR 68.81
Employee participation	40 CFR 68.83
Hot work permit	40 CFR 68.85
Contractors	40 CFR 68.87
Emergency response program	40 CFR 68.95

The comparison given in CCPS (2016a) demonstrates that at least in terms of element identification, the following concepts are made explicit in the risk-based approach to PSM management promoted in CCPS (2007, 2014):

- process safety culture,
- process safety competency,
- conduct of operations,
- measurement and metrics, and
- management review and continuous improvement.

This move by professional industry organizations to identifying more system elements and grouping them in thematic areas is also reflected in the PSM system recently developed by the Energy Institute in the United Kingdom. One again sees in Table 8 the explicit naming of key process safety concepts—e.g., *competency*. Defining and ensuring process safety competency at all levels in a company is critical to the success of PSM efforts (CCPS, 2015).

By regulation in Australia, major hazard facilities (MHFs) are required to establish and implement a comprehensive system to safely manage the

Table 7 Accident Prevention Pillars and Risk-Based PSM Elements: Center for Chemical Process Safety, American Institute of Chemical Engineers (CCPS, 2007)

Pillar	Element
Commit to process safety	• Process safety culture • Compliance with standards • Process safety competency • Workforce involvement • Stakeholder outreach
Understand hazards and risk	• Process knowledge management • Hazard identification and risk analysis
Manage risk	• Operating procedures • Safe work practices • Asset integrity and reliability • Contractor management • Training and performance assurance • Management of change • Operational readiness • Conduct of operations • Emergency management
Learn from experience	• Incident investigation • Measurement and metrics • Auditing • Management review and continuous improvement

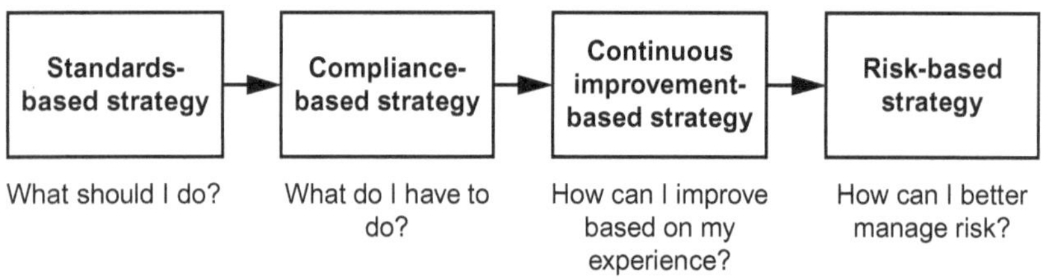

Fig. 6 Evolution of accident prevention and process safety strategies (CCPS, 2014).

hazards and risks encountered during operations (SWA, 2012). Table 9 shows guidance on such systems provided by Safe Work Australia (SWA, 2012). Concepts such as *leadership*, *competency*, and *improvement* are again present, along with the familiar technical requirements for hazard identification, risk management, permit-to-work systems, confined space entry procedures, etc.

As a direct result of the 1976 dioxin release in Seveso, Italy, the European Union (EU) operates in a highly regulated environment for the control of

Table 8 PSM Focus Areas and Elements: Energy Institute (Energy Institute, 2010)

Focus Area	Element
Process safety leadership	• Leadership commitment and responsibility • Identification and compliance with legislation and industry standards • Employee selection, placement and competency, and health assurance • Workforce involvement • Communication with stakeholders
Risk identification and assessment	• Hazard identification and risk assessment • Documentation, records, and knowledge management
Risk management	• Operating manuals and procedures • Process and operational status monitoring, and handover • Management of operational interfaces • Standards and practices • Management of change and project management • Operational readiness and process startup • Emergency preparedness • Inspection and maintenance • Management of safety critical devices • Work control, permit to work, and task risk management • Contractor and supplier, selection, and management
Review and improvement	• Incident reporting and investigation • Audit, assurance, management review, and intervention

major-accident hazards involving dangerous substances (Meyer & Reniers, 2016). Table 10 gives a listing of the primary requirements in this regard for upper tier (Seveso high or tier 2) facilities under the EU Seveso III Directive; Table 11 identifies the broad issues that must be addressed by the SMS referenced in Table 10. Implementation of the Seveso III Directive in Great Britain is accomplished by the Control of Major Accident Hazards (COMAH) Regulations (HSE, 2015).

Of particular note in Table 10 is the obligation to consider intersite domino effects. A case in point here is the 1986 Sandoz/Schweizerhalle chemical warehouse fire in Basel, Switzerland, which resulted in far-reaching pollution of the Rhine River (Meyer & Reniers, 2016). Process safety incidents do not respect facility or even national boundaries.

Table 9 Common Safety Management System Elements Found at Most Major Hazard Facilities (MHFs): Safe Work Australia (SWA, 2012)

Leadership, management, accountability, and commitment
Hazard and risk management
• Hazard identification, risk assessment, control measures
Information and documentation
• Includes regulatory compliance
Design and construction
• Throughout facility life cycle
Incident management
• Reporting, investigation, analysis, follow-up
Management of change
Contractor management
Emergency preparedness and response
• Includes community communication/consultation
Purchasing
• Materials, equipment, standards
Systems of work/operations and maintenance
• Includes permit to work and confined space
Personnel
• Includes training and competency
Monitoring, auditing, review, and improvement
Possible inclusion of other management system elements:
• Health and fitness for work
• Environment and waste management
• Quality management

Three examples from Asia help to illustrate the efficacy and widespread adoption of the core principles embodied in the PSM systems found in Tables 3–6. Fig. 7 shows the PSM elements mandated by the State Administration of Work Safety (SAWS) in China. Tables 12 and 13 present similar information for PSM in Korea and Singapore, respectively. As part of a special issue of the journal *Process Safety and Environmental Protection* to commemorate the 30th anniversary of the Bhopal disaster, Goh, Tan, and Lai (2015) reviewed what happened at Bhopal in relation to the practice of PSM in Singapore. This is an excellent example of leadership in driving the continuous improvement loop critical to the success of any PSM system.

In a document intended primarily for senior leaders in high-hazard industries, the Organization for Economic Cooperation and Development (OECD) provides helpful advice to ensure an appropriate process safety

Table 10 Principal Obligations for Upper Tier Establishments and Relevant Directive 2012/18/EU Articles Concerning the Control of Major Accident Hazards Involving Dangerous Substances: Seveso III Directive of the European Union

Obligation	Article
Notification to competent authority	2012/18/EU Article 7
Development of major-accident prevention policy (MAPP) and safety management system (SMS) for implementation	2012/18/EU Article 8
Consideration of intersite domino effects	2012/18/EU Article 9
Provision of safety report	2012/18/EU Article 10
Development of internal emergency plan	2012/18/EU Article 12
Provision of information to competent authority to enable development of external emergency plan	2012/18/EU Article 12
Provision of information to land-use planning authorities	2012/18/EU Article 13
Provision of information to the public	2012/18/EU Article 14

Adapted From Leonard, T. (2013). Seveso III directive. Implications for the Irish industry. Presentation to Engineers Ireland, Dublin, Ireland (November 28, 2013). Available at: https://www.engineersireland.ie/EngineersIreland/media/SiteMedia/groups/Divisions/fire-safety/Seveso-III-Directive-implications-for-Irish-Industry.pdf?ext=.pdf (last accessed November 16, 2016); Seveso III. (2012). Directive 2012/18/EU of the European Parliament and of the Council of 4 July 2012 on the control of major-accident hazards involving dangerous substances, amending and subsequently repealing Council Directive 96/82/EC. *Official Journal of the European Union, L 197*, 1–37.

Table 11 Issues to be Addressed by the Safety Management System Implemented for the Control of Major Accident Hazards Involving Dangerous Substances: Seveso III Directive of the European Union (Annex III of Seveso III, 2012)

Organization and personnel
Identification and evaluation of major hazards
Operational control
Management of change
Planning for emergencies
Monitoring performance
Audit and review

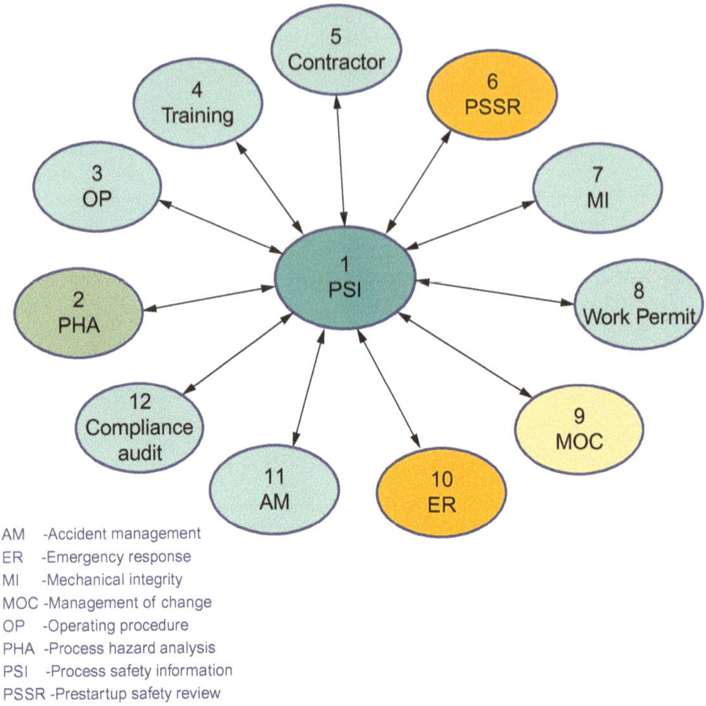

Fig. 7 Elements of the PSM system regulated in China by the State Administration of Work Safety (SAWS) (Zhao, Suikkanen, & Wood, 2014).

Table 12 Contents of the Process Safety Report Mandated for Submission by Hazardous Installations to the Regulator: Korea (Kwon et al., 2016)

Process safety information
Hazard analysis and risk assessment report
Procedure and planning for safe operations for installations • Procedure and manual for safe operation • Procedure and planning for mechanical integrity • Procedure for hot work permit • Safety control procedure for contractor work • Education and training plan • Procedure for management of change • Procedure for prestartup • Audit procedure • Procedure for incident investigation • Others related to safety management
Emergency planning and response

Table 13 PSM Elements: Singapore Standard SS 506 Part 3 (Huat, 2012)

Process safety information
Hazard identification, risk assessment and risk control
Training, awareness and competence
Operating procedures and safe work practices
Management of change
Prestartup safety
Contractors
Mechanical integrity and reliability
Control of hazardous substances
Emergency preparedness and response
Incidents, accidents, nonconformity, corrective action, and preventive action

Fig. 8 OECD essential elements of corporate governance for process safety. *OECD. (2012). Corporate governance for process safety: Guidance for senior leaders in high hazard industries.*

culture lives in the corporate boardroom as well as the facility workspaces (OECD, 2012). The notion from the previous paragraph that leadership and culture sit at the center of a continuous improvement cycle from risk-awareness through to action is also illustrated in Fig. 8.

We close our discussion of PSM systems with a brief mention of the *safety case* approach which is prevalent in regulation of the nuclear, aerospace, and offshore oil and gas industries. AcuTech (2012) describes the Safety and Environmental Management Program (SEMP), developed in 1993 by the offshore oil and gas industry, as a process safety-like program that was published in API RP 75 (American Petroleum Institute, Recommended Practice 75). Current US SEMS (Safety and Environmental Management System) offshore regulations have incorporated SEMP/API RP 75 by reference (AcuTech, 2012).

Sutton (2014) identifies the use of a safety case as another manner in which offshore safety can be managed. He defines a safety case as *the case that the designers and operators of a facility make to all interested parties that the facility is safe* (Sutton, 2014), and gives three principles upon which a safety case is built:

1. Risk control is the responsibility of the people who create the risk.
2. Setting and achieving goals, not following prescriptive regulations, is how safe operations are achieved.
3. Risk must be reduced below an acceptable threshold.

Sutton (2014) argues that the difference between the SMS and safety case approaches is not as great as one might think, with both sharing the first two principles in the above list. He comments that the heart of a safety case is in fact a SMS, and that the safety report required under the Seveso III Directive (Table 10) is essentially a safety case (Sutton, 2014).

The same statement concerning the Seveso safety reports is made by the US CSB in the safety case analysis conducted as part of its regulatory review following the 2012 Chevron refinery fire in Richmond, CA (CSB, 2014a). Key among the CSB's recommendations (CSB, 2014a) is the recommended requirement within PSM for continuous risk reduction to ALARP (as low as reasonably practicable). This, coupled with the need for more goal-based and fewer activity-based or compliance-driven regulations, addresses the second and third safety case principles in Sutton's list (Sutton, 2014).

Readers are referred to Sutton (2014) and CSB (2014a) for further ideas on the relationship between safety cases and PSM (regulated or otherwise). Both documents offer clearly written, authoritative information.

4.3 Elements

There is, of course, no substitute for careful examination of the relevant technical documentation if one wishes to learn the theoretical underpinning for a given PSM element (e.g., CSChE (2012a, 2012b) with respect to the elements in Table 3). Then follows the incubation period of practical experience with respect to the element and how it interacts in an integrated manner with its counterparts in the entire system. This experience can be personal and can also be learned from others in the form of case studies.

Table 14 Matrix Showing the Interrelationship Among Critical Elements in the OSHA PSM Standard; the Symbol • Indicates a Significant Correlation Between Elements (Aziz et al., 2016)

Element	PSI	PHA	OP	TNG	PSSR	MI	MOC
Employee participation	•	•	•	•	•	•	•
Process safety information (PSI)	N/A	•	•	•	•	•	•
Process hazard analysis (PHA)	•	N/A	•		•	•	•
Operating procedures (OP)	•	•	N/A	•	•	•	•
Training (TNG)	•	•	•	N/A	•	•	•
Contractors	•	•	•	•		•	•
Prestartup safety review (PSSR)				•	N/A	•	•
Mechanical integrity (MI)		•			•	N/A	•
Hot work permit	•		•				
Management of change (MOC)	•	•	•		•	•	N/A
Incident investigation		•	•	•		•	
Emergency planning and response	•	•				•	
Compliance audits	•	•	•	•	•	•	•
Trade secrets	•	•	•				

In this section, therefore, we review the recent process safety literature for guidance and practical examples related to individual PSM elements. We first draw attention to an innovative study by Aziz, Shariff, and Rusli (2016) in which the interrelationship among the OSHA PSM Standard elements (Table 5) is examined. Table 14 gives the interrelations among the 14 PSM elements and a subset of seven elements deemed critical on the basis of OSHA citations. While the authors recognized the importance of all elements working together as an integrated whole, process hazard analysis (PHA) and mechanical integrity (MI) were the most highly correlated with other elements in their analysis (Aziz et al., 2016).

Table 15 provides references for papers related to elements of the RBPS system (CCPS, 2007) shown in Table 7. The analysis was limited to papers having an explicit management focus; thus, the numerous works available on topics such as specific *hazard identification and risk analysis* techniques are not referenced here.

Table 15 References From the Recent Process Safety Literature With Respect to Specific Elements in the CCPS Risk-Based PSM System (Table 7)

References	Representative PSM Element
Frank (2007)	Process safety culture
Hendershot (2012)	Process safety culture
King (2013)	Process safety culture
Olewski, Ahammad, Quraishy, Gan, and Vechot (2016)	Process safety culture
Baybutt (2016a)	Process safety competency
Puyosa (2012)	Process knowledge management
Aziz, Shariff, and Rusli (2014)	Process knowledge management
Rowe and Francois (2016)	Process knowledge management
Scholtz and Maher (2014)	Operating procedures
Hayes (2015)	Asset integrity and reliability
Majid, Shariff, and Rusli (2015)	Contractor management
Philley (2002)	Management of change
Kelly (2013)	Management of change
Wincek, Sousa, Myers, and Ozog (2015)	Management of change
Gerbec (2016)	Management of change
Haesle, Devlin, and McCavit (2009)	Conduct of operations
Forest (2012)	Conduct of operations
Forest (2014)	Conduct of operations
Majid, Shariff, and Loqman (2016a)	Emergency management
Baybutt (2015)	Auditing
Allford (2016)	Auditing

As an example of how a given reference citation was categorized in Table 15, consider the title of Dennis Hendershot's paper (Hendershot, 2012): *Process Safety Management—You Can't Get It Right Without a Good Safety Culture*. It seems clear that this reference relates primarily to the element *process safety culture*, and this is indeed the case. Note, however, that the second column in Table 15 is headed *Representative PSM Element.* When reading Hendershot (2012), one sees additional references to the influence of safety culture on activities such as PHA and incident investigation, as well as the following section heading: *A Good Culture—Critical to all PSM Activities* (Hendershot, 2012). This is in accordance with the previously described work of Aziz et al. (2016).

Our presentation now touches on a sampling of Table 15 references to illustrate some features of specific PSM elements. Olewski et al. (2016) describe the importance of safety culture from a university perspective—in their case, managing a major research project on the consequences of ground spills of LNG (liquefied natural gas). As recent events have demonstrated, serious incidents involving hazardous materials are not restricted to industry (CSB, 2010a).

Scholtz and Maher (2014) provide helpful advice on the development of effective operating procedures. They suggest consideration of the following points (Scholtz & Maher, 2014):

- approval of the procedure template by all stakeholders,
- use of a user-friendly procedure format that is consistently applied across the facility,
- breaking down processes into separate units to facilitate referencing of procedures, and
- establishing an appropriate depth of information to be included in procedures, with subsequent adjustment of the training program.

Hayes (2015) gives an illustrative case study of the importance of effective asset integrity management by examining a gas transmission pipeline rupture. This 2010 incident in San Bruno, CA caused eight fatalities (members of the public), and resulted from the failure of a longitudinal seam weld on a line that had not been inspected or tested since installation in 1956 (Hayes, 2015). Among the many excellent points made by the author is the following key observation (Hayes, 2015): *The primary strategy for ensuring public safety was the management of system integrity by means of compliance with regulations. The key question in people's minds was "does it comply?," rather than "is it safe?"* One is reminded of the earlier quote from Hendershot (2016) that PSM regulatory compliance should be an outcome not an objective (quote attributed to M. Broadribb in Hendershot, 2016).

MOC is a powerful system that enables continuous improvement through change. If it is not properly applied, however, it can cripple an organization and inhibit such progress (Kelly, 2013). These statements by Brian Kelly aptly summarize the multifaceted nature of MOC—management of change. Change is inevitable for progress, yet it must be effectively managed so as not to impede the very thing it enables; Kelly (2013) refers to MOC as one of the more fundamental elements of PSM. It must be well understood that risk can be heightened not only by technical changes involving valves, pumps, and compressors, but also by organizational changes that occur in the normal course of business and in major events such as mergers and acquisitions (Philley, 2002; Wincek et al., 2015).

As identified in CCPS (2007, 2016a), conduct of operations is one of the new PSM elements appearing in the risk-based PSM system shown in Table 7. CCPS (2014) describes this element as *the execution of operational and management tasks in a deliberate and structured manner*. Further details can be found in CCPS (2011a, 2012) for both conduct of operations and its subcomponent operational discipline (defined in CCPS (2012) as *displaying behaviors within a system of checks and balances that help ensure that things are done correctly and consistently*). Conduct of operations has strong ties to the cultural dimension of an organization (CCPS, 2014), as well as other PSM elements such as operating procedures, training and performance assurance, and management of change (Haesle et al., 2009). Forest (2012) offers a timely reminder in Fig. 9 that effective conduct of operations requires discipline not only by process operators (operational discipline), but also by

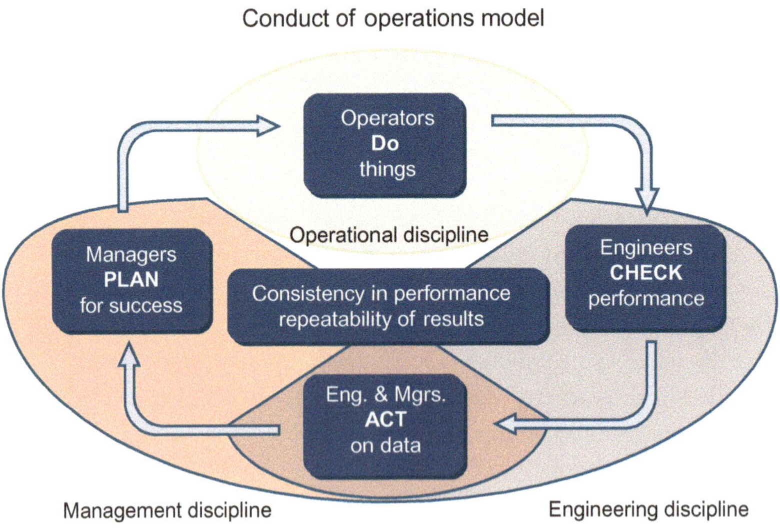

Fig. 9 Conduct of operations model involving operational, engineering, and management discipline (Forest, 2012).

engineering and management personnel (engineering discipline and management discipline, respectively).

There are other examples in the literature that do not correspond directly by element name to the RBPS system given in Table 7—such as the work of Abu-Khader (2004) on human factors and Majid, Shariff, Rusli, and Azman (2016b) on trade secrets. These papers do, however, correlate by specific element name to the CSChE system (Table 3) and the OSHA Standard (Table 5), respectively; they would also be relevant to risk-based PSM (Table 7) by virtue of the various subcomponents of the 20 primary elements. For example, Forest (2012) comments on how conduct of operations helps to reduce the likelihood of human error.

Before moving on to the measurement of PSM performance, we conclude this section by making note of the usefulness of CSB investigation reports and case studies for learning about the functioning of a PSM system and, in particular, the individual elements. This is a theme that reappears later in this chapter when discussing important process safety concepts and their relation to PSM. In a similar vein, Amyotte (2013b) remarks on how CSB reports and videos have significant value in teaching process safety to undergraduate engineering students.

The cover page of each CSB report (which are freely available at csb.gov) typically contains a listing of key issues related to some aspect of incident root causation (management system deficiency, process safety concept, regulatory concern, etc.). The reports therefore afford excellent learning opportunities and should constitute a critical component of a company's efforts to enhance process safety knowledge. The final component (process safety resource center and reference library) under this PSM element (no. 12) in Table 3 includes case histories concerning incidents illustrating PSM principles (CSChE, 2012a). There is of course no reason for library resources to be restricted to books on shelves. Table 16 provides a snapshot of what is possible using resources available in the public domain.

4.4 Metrics

You don't improve what you don't measure.

CCPS (2011b)

Process Safety Performance Indicators (PSPIs)—key performance metrics that indicate when a process safety accident is most likely to occur.

Azizi (2016)

Leading Indicators Indicators that proactively measure the effectiveness of risk controls.

WSHCouncil (2012)

Table 16 Sampling of US Chemical Safety Board (CSB) Reports and Key Issues Identified

Reference	Incident	Key Issue
CSB (2016a)	• Williams Geismar Olefins Plant • Geismar, LA • Reboiler rupture, explosion, fire • Two fatalities, 167 reported injured	• Overpressure protection • Process hazard analysis • Management of change • Prestartup safety review • Operating procedures • Hierarchy of controls • Process safety culture
CSB (2014b)	• Tesoro Anacortes Refinery • Anacortes, WA • Heat exchanger rupture, explosion, fire • Seven fatalities	• Inherently safer design • Tesoro process safety culture • Control of nonroutine work • Mechanical integrity industry standard deficiencies • Regulatory oversight of petroleum refineries
CSB (2010b)	• Xcel Energy Hydroelectric Plant • Cabin Creek, CO • Penstock fire • Five fatalities, three injured	• Safe limits for working in confined space flammable atmospheres • Prejob safety planning of hazardous maintenance work • Contractor selection and oversight • Emergency response and rescue

Lagging Indicators—Indicators that monitor reactively the effectiveness of risk controls.

WSHCouncil (2012)

Much has been written over the past 5–10 years on the subject of process safety performance indicators (PSPIs)—or more generally, safety performance indicators (SPIs), and what would have been referred to earlier as key performance indicators (KPIs). Table 17 provides a listing of research and case study references taken from the recent process safety literature on PSM performance measurement.

We also draw attention to the comprehensive review of the literature on process safety indicators given by Swuste, Theunissen, Schmitz, Reniers, and Blokland (2016), and the very accessible treatment of PSPIs written by Azizi (2016). These two papers collectively form an excellent entry point to the world of PSM metrics for novice readers, as well as a thoughtful refresher for those who are more experienced in this field. Hence, our further discussion here on PSM performance measurement is brief.

Table 17 References From the Recent Process Safety Literature With Respect to Performance Measurement of Process Safety Management

Reference	Overview
Rosenthal, Kleindorfer, and Elliott (2006)	Examination of survey factors likely to predict low-probability/high-consequence (LP/HC) accidents given the limited LP/HC accident database available at the time
Chosnek and Clifton (2008)	PSM system implementation at Gulf Coast Waste Disposal Authority and use of metrics related to: (i) incidents and near-misses, (ii) management of change, (iii) training, (iv) enforcement actions, and (v) LEL (lower explosive limit) control actions
Chang and Liang (2009)	Model development for performance evaluation of PSM systems used at paint manufacturing facilities in Taiwan
Cummings (2009)	Review of the development, current state, and future possibilities for PSM metrics based on the author's experience with DuPont
Payne, Bergman, Rodriguez, Beus, and Henning (2010)	Study of the leading/lagging effects of process safety climate on incidents by means of surveying a multinational corporation working with hazardous materials
Khan, Abunada, John, and Benmosbah (2010)	Development of risk-based process safety indicators using the UK HSE barrier approach with consideration of the relationship between leading and lagging indicators
Knijff, Allford, and Schmeizer (2013)	Guidance from the European Process Safety Centre (EPSC) concerning leading indicators for process safety performance
Wang, Mentzer, Gao, Richardson, and Mannan (2013)	Exploration of denominators (normalization factors) for lagging performance indicators which offer an alternative to the commonly used "work hours"
Mendeloff, Han, Fleishman-Mayer, and Vesely (2013)	Study conducted on behalf of the US Chemical Safety Board to critically examine the usefulness of ANSI/API RP 754 Tier 1 and Tier 2 process safety events as indicators of PSM performance

Continued

Table 17 References From the Recent Process Safety Literature With Respect to Performance Measurement of Process Safety Management—cont'd

Reference	Overview
Pasman and Rogers (2014)	Application of Bayesian networks to relate process safety performance indicators to facility risk level
Kadri et al. (2014)	Presentation of cases from Air Products and Chemicals, Inc. in which CCPS and ANSI/API RP 754 leading and lagging indicators were used to drive process safety performance improvement programs
Kenan and Kadri (2014)	CCPS review and update on the use and effectiveness of leading indicators of process safety performance
Vaughen, Downes, Fox, and Belonger (2015)	Overview of CCPS (2016b): *Guidelines for Integrating Management Systems and Metrics to Improve Process Safety Performance*, which is aimed at helping companies consolidate process safety metrics throughout their safety, health, environmental, quality, and security (SHEQ&S) efforts
Leveson (2015)	Identification of system-specific leading safety indicators based on the author's accident causation model STAMP (system-theoretic accident model and processes)
Kerin (2016)	Exposition of leading process safety metrics developed by the IChemE Safety Centre (ISC)

From a fundamental perspective, Fig. 10 outlines a general, four-step methodology for program operation with respect to process safety performance measurement. The deliverables from the second step are the sets of leading and lagging indicators for identified safety critical activities (WSHCouncil, 2012). [As an aside—it is well established that occupational safety indicators are typically not appropriate for process safety purposes. For example, near-miss indicators must be directly related to process events such as pressure or temperature excursions, not occupational safety-type activities such as working at height.]

Azizi (2016) describes two methods for PSPI selection which rely either on the use of *barriers* (Fig. 11) or *tiers* (Fig. 12). The barrier-based approach utilizes Reason's Swiss cheese model (Reason, 1990), whereas the tier-based

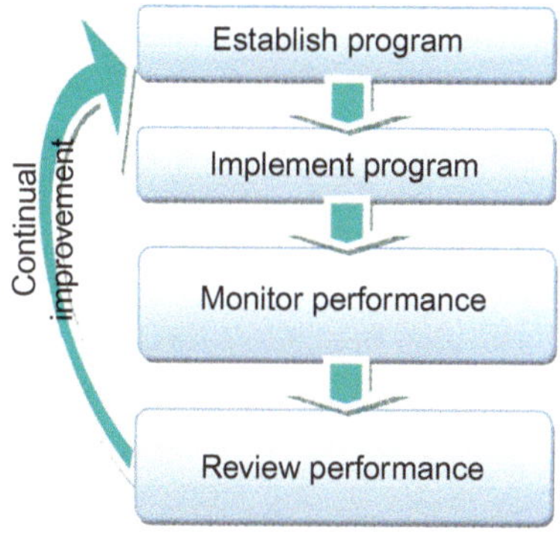

Fig. 10 Methodology for operation of a program to measure process safety performance (WSHCouncil, 2012).

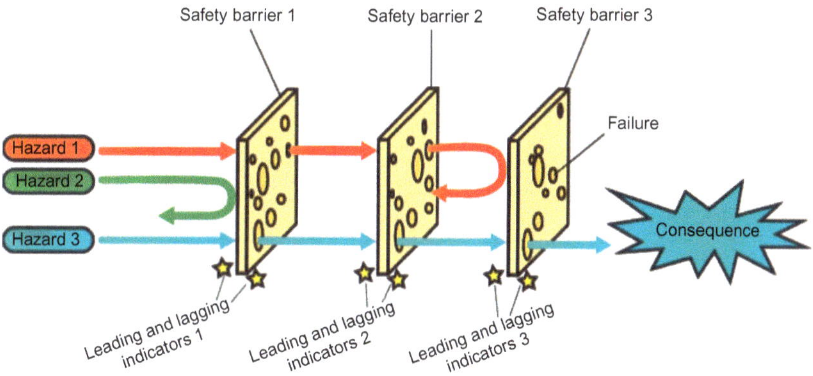

Fig. 11 Barrier-based approach in which leading and lagging indicators are defined for each barrier (Azizi, 2016).

approach comes from the accident pyramid concept of consequences escalating from base to tip (Azizi, 2016). Azizi (2016) identifies the barrier-based approach with the UK Health and Safety Executive (HSE, 2006), and the tier-based approach with the American Petroleum Institute (API, 2010), International Association of Oil and Gas Producers (OGP, 2011), and Center for Chemical Process Safety (CCPS, 2011b).

Examples of PSPIs drawn from ANSI/API Recommended Practice (RP) 754 (API, 2010) include the following:

- Tier 1 (Lagging)
  - hospital admission and/or fatality of a third party
  - officially declared community evacuation or shelter-in-place
  - fire or explosion resulting in direct company cost of $25,000 or more

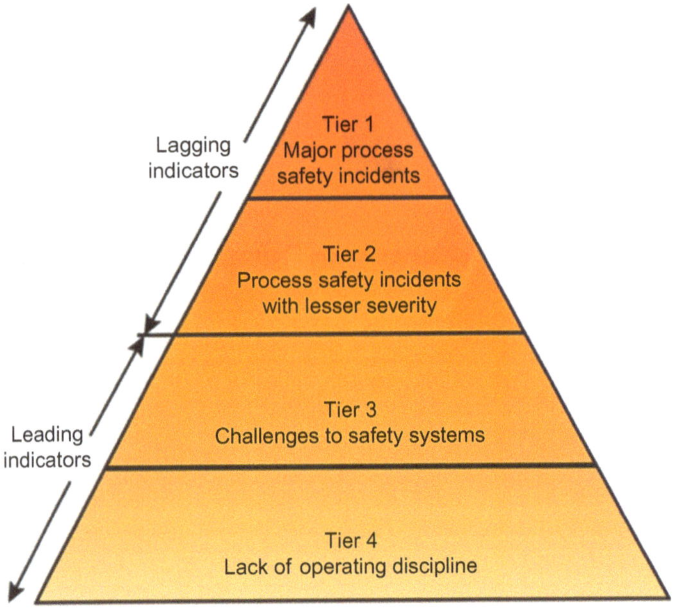

Fig. 12 Tier-based approach in which leading and lagging indicators are defined for each tier (Azizi, 2016).

- Tier 2 (Lagging)
  - employee, contractor, or subcontractor recordable injury
  - fire or explosion resulting in direct company cost of $2500 or more
  - discharge of a pressure relief device to atmosphere resulting in liquid carryover
- Tier 3 (Leading)
  - safe operating limit excursions
  - primary containment inspection results outside acceptable limits
  - demands on safety systems
- Tier 4 (Leading)
  - completion of process hazard evaluations
  - work permit compliance
  - management of change and prestartup safety review compliance

It should be noted that the above examples are for illustration only. ANSI/API RP 754 (API, 2010) and HSE (2006) should be consulted for all other purposes related to the development and use of tier-based and barrier-based PSPIs, respectively. For information on SPI development related to chemical accident prevention, preparedness, and response—in this case for public authorities and the general public—OECD (2008) provides guidance in relation to activities such as land-use planning and emergency coordination with facilities processing hazardous materials.

4.5 Improvements

The need for continual improvement has been mentioned several times to this point in our discussion of PSM systems and their constituent elements. Here we present practical advice on this point with reference to the recent process safety literature. First though, we make note of a recently published CCPS book: *Guidelines for Implementing Process Safety Management, 2nd edition* (CCPS, 2016c). Although not yet reviewed by the current authors, it is expected that this new addition to the CCPS guideline series will provide helpful information on many PSM-related topics including the subject of this section. Chapter 6 in CCPS (2016c) is titled *Improving an Existing PSM Element or System.*

While the need to improve PSM elements and systems is perhaps self-evident, additional motivation can be found in the review by Shariff, Aziz, and Majid (2016) who remark on the continued occurrence of major process accidents decades after the formal adoption of PSM by industry. Having said this, the study by Bottani, Monica, and Vignali (2009) reminds us that companies operating without a SMS at all are at a distinct disadvantage with respect to activities like risk analysis, corrective actions, and employee training.

Nevertheless, PSM systems do sometimes fail. Kelly (2011) gives 10 contributing reasons for these failures (in no particular order of priority): (i) failure by senior management to understand and support the goals of a process safety program, (ii) focus of PSM activities primarily on regulatory compliance, (iii) inappropriate assignment of resources to support process safety, (iv) discipline of workers involved in an incident for reasons other than policy/protocol violations, (v) mismatch between the type of operation and the process safety framework selected, (vi) competition among process safety and other management programs and initiatives, (vii) ineffective risk communication to management resulting in poor understanding of risk by corporate decision makers, (viii) lack of engagement/commitment to process safety by all staff, (ix) management failure to learn from previous incidents, and (x) failure to hold middle managers accountable for process safety deliverables.

On the final item in the above list, a recent paper by Rezvani and Hudson (2016) highlights the indispensable role of middle management in effectively managing safety in the process industries. In an overall sense, Arendt (2006) provides practical suggestions for strategies to improve PSM (e.g., adding new PSM activities to existing PSM elements and creating new PSM elements), as well as sources of information for improvement (e.g., sharing best practices within industry groups and benchmarking within a peer group). Several authors have written on the development and improvement of PSM systems in small companies (Bragatto, Ansaldi, & Agnello, 2015; Goddard, 2012; Herber, 2012; Louvar, 2008).

Klein and Dharmavaram (2012) write on achieving higher levels of PSM performance by focusing on improvements in, among other areas, maintaining a sense of vulnerability, risk management practices, and operational discipline. Young and Hodges (2012) suggest establishing a mentoring program to enhance competencies with regard to the CCPS risk-based PSM system (Table 7).

Paradies (2011) makes an interesting comparison between safety management in the process industries and in the US nuclear navy. He identifies a number of PSM gaps that can be viewed as potential areas for PSM improvement (Paradies, 2011): (i) assuming of ultimate responsibility for research, design, operations, and maintenance of process facilities by senior management, (ii) appropriate management resolution of conflicts between PSM priorities and production/budget issues, (iii) requirements for advanced technical training and competence of PSM-facility leadership, (iv) strict standards for operational and supervisory personnel training, both on hiring and on a continuing basis, (v) performance auditing and direct reporting requirements of audit results to senior leadership, (vi) emphasis on regular self-assessments, (vii) strict enforcement of compliance with procedure and management standards, and (viii) emphasis on design to prevent plant upsets and accidents.

Table 18 gives several case studies that afford excellent learning opportunities on PSM improvement needs and techniques. (Although presented in a previous section as avenues for understanding specific PSM elements, industry examples such as Hayes (2015) and the CSB reports in Table 16 have additional value as improvement motivators.) While only one representative PSM improvement topic was selected for most of the entries in Table 18, the source references contain a wealth of information in this regard. For example, Cazabon and Erickson (2010) comment on process safety information, PHA, and prestartup safety review in addition to MOC. The exception to this single-entry rule for the third column in Table 18 is the paper on Bhopal by Vaughen (2015). Given the extreme

Table 18 References From the Recent Process Safety Literature With Respect to Case Study Lessons for PSM Improvements

Reference	Incident or Area of Emphasis	Representative Topic for PSM Improvement
Bloch and Wurst (2010)	Spent caustic tank explosion in a refinery	Recognition of precursor warning signs
Cazabon and Erickson (2010)	Natural gas explosion in a facility making glass fiber mat	Management of change
Howell (2010)	Explosion in a reactor at a chemical company	Operating procedures
Brackey (2013)	Various over three decades, including a dust explosion, reactor explosion, and explosion and fire in a compressor train	Process hazard analysis
Rashid, Ramzan, Iqbal, Yasin, and Yousaf (2013)	Fertilizer plant; no single incident, although examples are given in relation to specific PSM elements (startup of an ammonium carbamate pump for the element given in the next column)	Incident investigation
Koivupalo, Sulasalmi, Rodrigo, and Vayrynen (2015)	Global steel company; emphasis is on managing health and safety as a complete entity in a frequently changing organization	Consistency of PSM practices and tools among various plant sites within an organization
Vaughen (2015)	Bhopal	Fig. 13
Wold and Laumann (2015)	Oil and gas producing company; emphasis is on the role of a safety management system as a communication system	Use of work procedures and safety standards as both tools and communication avenues within an organization
Lee, Kwon, Cho, Kim, and Moon (2016)	Hydrogen fluoride release at a chemical plant in Gu-mi City, Korea	Emergency management

Fig. 13 Weaknesses in the PSM program at Bhopal (Vaughen, 2015).

nature of Bhopal, and its global and enduring impact, we have included Fig. 13 from the work of Vaughen (2015). Readers interested in further consideration of the PSM legacy of Bhopal are referred to the paper by Amyotte, Berger, et al. (2016) in the journal *Current Opinion in Chemical Engineering*.

The final piece of advice in this section comes from John Bresland—a former member and chair of the US CSB, and also someone with a long and continuing career in the process industries. He comments in Bresland (2016) that he saw three types of companies during his time at the CSB: (i) those that do not understand the hazards in their operations, lack an appropriate process safety program, and thus experience major accidents, (ii) those that do understand the hazards they face and the pertinent regulations, have an excellent process safety program driven by qualified people, yet still experience process incidents having minor and also severe consequences, and (iii) those that understand the hazards and regulations, have an excellent safety program and well-qualified people, and do not have serious process incidents.

The third category above is clearly the desired state for a process industry company. To help achieve this goal, Bresland offers his top 10 rules for process safety success (Bresland, 2016):

1. Having leadership that is committed to process safety, including CEOs, managers, and supervisors—anyone in a leadership position.
2. Attracting the best possible people from senior managers to control room operators by means of strict hiring procedures.

3. Ensuring equipment reliability through an effective mechanical (asset) integrity program.
4. Being passionate about attention to detail.
5. Carefully monitoring operations with process safety metrics.
6. Taking a long-term view on risk; shutting down when necessary to resolve problems.
7. Preparing for injuries and off-site consequences of process incidents.
8. Investigating all accidents in a comprehensive manner.
9. Refusing to let complacency set in.
10. Developing and nurturing a strong process safety culture.

5. PROCESS SAFETY CONCEPTS IN PSM

For a long time, people were saying that most accidents were due to human error and this is true in a sense but it's not very helpful. It's a bit like saying that falls are due to gravity.

Professor Trevor Kletz, as quoted in Edwards (2014).

Earlier, we briefly observed that attention to human factors (and the accompanying reduction in human error) is either an explicit element or an underlying component of successful management of process safety. There are, in fact, a number of overarching process safety concepts that achieve maximum benefit when interwoven through the fabric of a PSM system. We therefore offer a brief coverage of: (i) ISD, (ii) recognition of warning signs, (iii) process safety culture, and (iv) dynamic risk assessment. These and other important concepts are covered in detail in other chapters of this volume; our intention here is to examine the elements within a PSM system where each finds particular relevance. Examples drawn from the work of the US CSB are again given to reinforce the theoretical treatment.

5.1 Inherently Safer Design

ISD, or simply inherent safety, is a maturing area within the broader field of process safety. Book treatments are available (e.g., CCPS, 2009; Kletz & Amyotte, 2010) and research is actively underway (e.g., Abidin, Rusli, Shariff, & Khan, 2016; Roy et al., 2016).

How then does ISD dovetail with PSM? Amyotte, Goraya, Hendershot, and Khan (2007) comment that *explicit incorporation of the principles of inherent safety in the basic definition and functional operation of the various PSM elements can help to improve the quality of the safety management effort.* They then give several illustrations of ISD incorporation in the elements of the

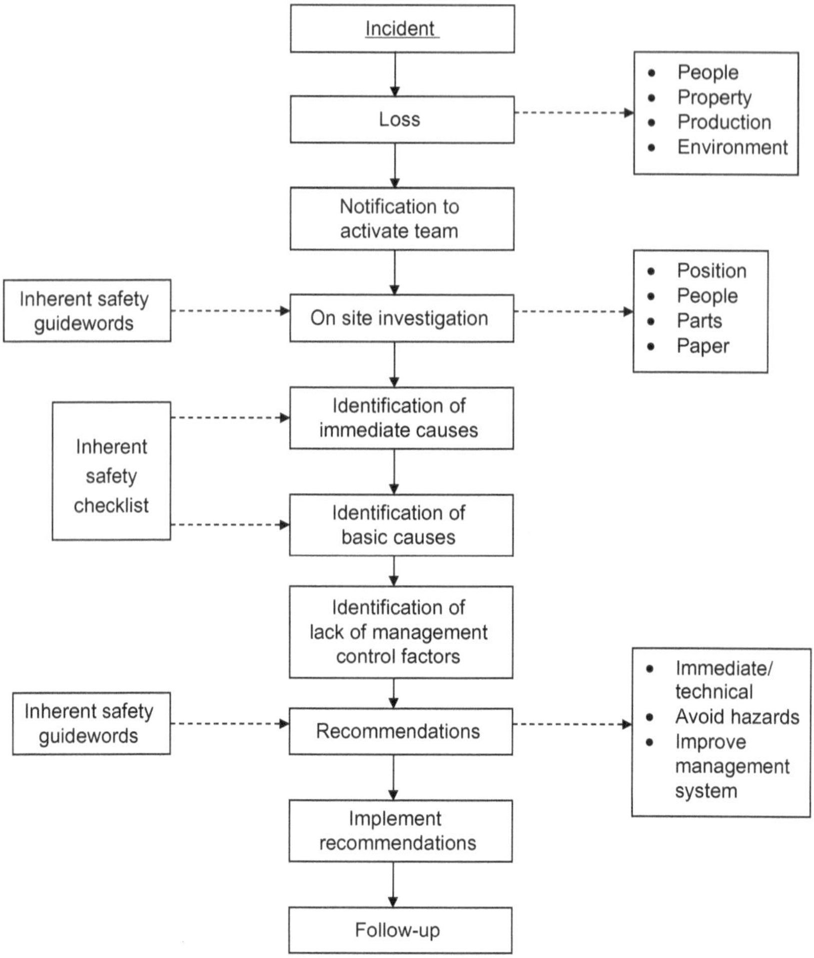

Fig. 14 An inherent safety-based incident investigation methodology. *Adapted from Goraya, A., Amyotte, P. R., & Khan, F. I. (2004). An inherent safety-based incident investigation methodology.* Process Safety Progress, 23, *197–205.*

CSChE-recommended PSM system shown in Table 3 (Amyotte et al., 2007). An example from an earlier paper by Goraya, Amyotte, and Khan (2004) is given in Fig. 14, which demonstrates how inherent safety guidewords and checklist items can enhance a traditional incident investigation protocol. The guidewords are simply the primary ISD principles of minimization, substitution, moderation, and simplification; these principles are identified in CSChE (2012a) under the *reduction of risk* component of the *process risk management* element.

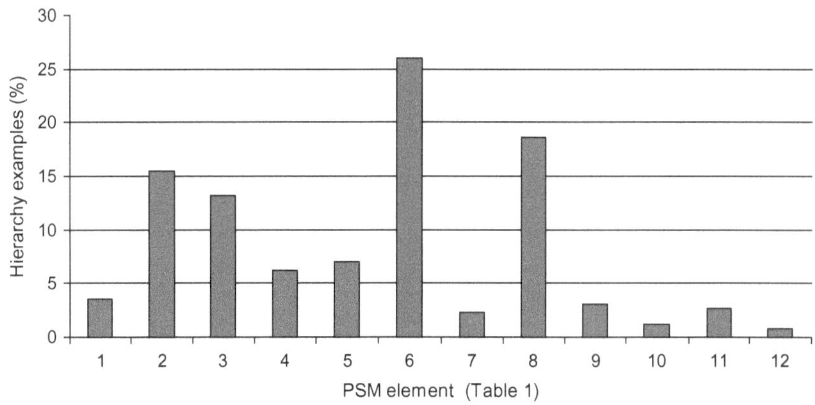

Fig. 15 Breakdown of overall hierarchy examples from review of CSB reports according to applicable PSM element. Table 1 (*x*-axis) in Fig. 15 = Table 3 in the current work; i.e., the element numbers correspond to the PSM system recommended by the Canadian Society for Chemical Engineering (Amyotte et al., 2011).

The usefulness of CSB investigation reports as a learning tool for PSM enhancement is demonstrated once again in the work of Amyotte, MacDonald, and Khan (2011). They analyzed a set of 63 CSB reports and identified over 200 examples from the hierarchy of controls for risk reduction: inherent safety, passive engineered safety, active engineered safety, and procedural safety (Amyotte et al., 2011). Fig. 15 shows that it is possible to assign each safety measure example to a specific PSM element, thus establishing a quantitative link between a process safety concept—ISD, or more broadly, the hierarchy of controls—and PSM (Amyotte et al., 2011). A similar study has recently been conducted for the period 2011–16 (Irvine, Amyotte, & Khan, 2016) and will be the subject of a future archival journal submission.

5.2 Recognition of Warning Signs

There are—or rather, there should be—no *black swan* (unforeseen and unpredictable) process incidents (Amyotte, Margeson, Chiasson, Khan, & Ahmed, 2014). In theory there are always warning signs of an impending undesirable event; in practice these precursor signals can be weak and conceptual or cultural, as opposed to strong and material or physical. A slow decline in a critical feature such as a willingness to report near-misses might be more difficult to identify than a mechanical integrity issue with a visibly corroded process vessel or pipeline. Nevertheless, both are important to the assurance of process safety.

As with ISD, numerous references are available on the subject of warning sign recognition. The publications of Andrew Hopkins (e.g., Hopkins, 2000, 2009a) are especially helpful in understanding the relationship between warning signs and the elements of an effective safety culture and SMS (Amyotte et al., 2014). The previous discussion on PSPIs and the PSPI references cited should also be viewed as relevant to the topic of warning signals for process incidents.

CCPS (2012) deals with the following areas as potentially beneficial in the identification of warning signs: (i) leadership and safety culture, (ii) training and competency, (iii) process safety information, (iv) procedures (operating and maintenance), (v) asset (mechanical) integrity, (vi) risk analysis and management of change, (vii) audits, (viii) learning from experience, and (ix) near-miss and incident reporting/investigation. The close correspondence of this list to the CCPS risk-based PSM system pillars and elements (Table 7) is not a coincidence. Like ISD, warning sign recognition is an integral PSM concept.

In his own review of CSB investigation reports, Baybutt (2016b) concludes that his analysis can be used by companies to improve their PSM performance by focusing on areas in which other companies have experienced difficulty. This epitomizes the concept of learning from experience as expressed in the above list from CCPS (2012). The final item in this list is the subject of a recent study by Gnoni and Saleh (2017) in which consideration of near-misses as accident precursors was shown to generate opportunities for system redesign, improved operational procedures, and worker training.

5.3 Process Safety Culture

In general, the problem is not that we don't know what to do, but rather that we do not always do what we already know how to do, and what we know we should do.

Hendershot (2015)

Several compilations of thoughts and advice from the literature have been given in this chapter: (i) Rosen (2015) and his 10 commandments for risk-based PSM, (ii) Kelly (2011) on why PSM systems fail, (iii) Paradies (2011) with a comparison between PSM and safety management in the US nuclear navy, and (iv) Bresland (2016) and his top 10 rules for process safety success. Common to all is the concept of process safety culture—either explicitly in the case of Rosen (2015) and Bresland (2016), or clearly evident as the motivating force to effect the changes advocated by Kelly (2011) and Paradies (2011).

Text references on safety culture abound—e.g., Goetsch (2010) on developing a safety-first corporate culture; the fundamental treatise on safety culture by Hopkins (2005); and the compendium of lessons from high reliability organizations also provided by Hopkins (2009b). Whether a separately defined and named element as in Table 7 (RBPS), or an implicit component throughout a given system, process safety culture considerations have a significant impact on PSM success.

Hopkins (2005) is an excellent resource for understanding the basic principles of safety culture. He considers three concepts—safety culture, collective mindfulness, and risk awareness—and explains the similarities and overlaps in each approach (Hopkins, 2005). The breakdown of safety culture into just, reporting, learning, and flexible (decision-making) cultures (Hopkins, 2005) helps relate the overall concept to specific PSM elements (e.g., incident investigation). Collective mindfulness involves several features such as preoccupation with failure and sensitivity to operations (Hopkins, 2005); these correlate well with avoiding complacency and maintaining a sense of vulnerability, respectively, as expressed earlier in this chapter. Risk awareness incorporates the avoidance of normalizing evidence (Hopkins, 2005), sometimes called normalization of deviance or deviation. This is an extremely undesirable state in which abnormal events (such as "small" process fires) are accepted as the norm, eventually leading to a process incident with much more severe consequences.

Table 19 is a summary of the analysis involving ISD, warning signs, and process safety culture conducted by Amyotte, Khan, and Lupien (2016), which in turn builds on the work of Amyotte and Khan (2016). Six CSB-investigated incidents were reviewed with respect to these three common causation factors as referenced in Table 19 and Figs. 16–21. Each hazard is defined in terms of loss or generation of containment, and both hazards and incident causes are related to a specific PSM element from Table 3. Inadequate consideration of ISD, warning signs, and process safety culture increases risk by permitting the existence of hazards and creating management system deficiencies.

5.4 Dynamic Risk Assessment

Just as ISD is most effective when considered at the earliest possible stage in the design phase of a process plant, there is a concurrent need for dynamic risk assessment during the operational phase (Amyotte, Berger, et al., 2016).

Table 19 Hazard and Causation Factor Correlation for CSB-Investigated Incidents With Elements of the CSChE-Recommended PSM System Shown in Table 3

Reference	Figure	Hazard	PSM Element (Table 3)
CSB (2010c)	16	Gas blow	Training and performance
CSB (2011)	17	Iron dust accumulations	Process knowledge and documentation
CSB (2013)	18	Sulfidation corrosion	Process and equipment integrity
CSB (2016b)	19	Inappropriate storage	Process risk management
CSB (2010b)	20	Confined space	Company standards, codes, and regulations
CSB (2014c)	21	Closure of the explosion pentagon	Process and equipment integrity

Causation Factor	PSM Element (Table 3)
Lack of consideration of inherently safer design	Process risk management
Poor recognition of warning signs	Incident investigation
Inadequate process safety culture	Accountability: Objectives and goals

Adapted From Amyotte, P., Khan, F., & Lupien, C. (2016b). Different hazards, similar causes, same results. *Loss Prevention Bulletin.*

Fig. 16 Cleaning of fuel gas piping by gas blow at Kleen Energy site 1 week before incident (CSB, 2010c).

Fig. 17 Iron dust deposits on elevated surface at Hoeganaes Corporation facility (CSB, 2011).

Fig. 18 Pipeline degradation due to sulfidation corrosion at Chevron Richmond refinery (CSB, 2013).

Fig. 19 Plywood bin walls in which FGAN (fertilizer-grade ammonium nitrate) and other fertilizer products were inappropriately stored at West Fertilizer Company (CSB, 2016b).

Fig. 20 Access door cut into confined space (penstock tunnel) at Xcel Energy hydroelectric plant (CSB, 2010b).

Fig. 21 Blender wall deformation and indications of blade scraping (thus effectively completing the explosion pentagon) at AL Solutions facility (CSB, 2014c).

Khan et al. (2016) define dynamic risk assessment as *a method that updates estimated risk of a deteriorating process according to the performance of the control system, safety barriers, inspection and maintenance activities, the human factor, and procedures.*

Fig. 22 schematically represents the potential evolution of safety standards enabled by the familiar PDCA (plan, do, check, adjust/act) cycle

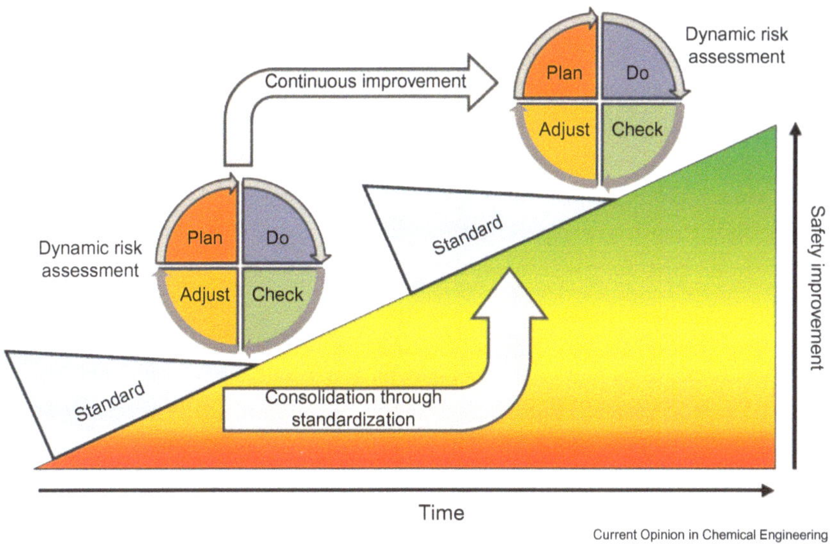

Fig. 22 Representation of continuous safety improvement through dynamic risk management (Khan et al., 2016).

embodied in both dynamic risk assessment and PSM. Based on name alone, dynamic risk assessment should be expected to interface with any PSM activities related to hazard identification and risk analysis, assessment, and management. Because one of the focal points of dynamic risk assessment is the potential degradation of safety barriers, its use is also relevant to the analysis of PSPIs (whether developed from a barrier- or tier perspective). Readers are referred to Khan et al. (2016) for a review of the current state-of-the-art with respect to dynamic risk assessment.

6. CONCLUDING REMARKS

SMSs are widely employed in numerous industrial, transportation, and recreational sectors throughout the world. We have attempted to show in this chapter how the professional management of safety—whatever the specific safety objectives might be—is critical to the advancement of both industry and society.

PSM systems typically consist of about 10–20 separate, yet intertwined, elements. Representative PSM systems from Asia, Australia, Europe, and North America have been presented to illustrate both their common features (e.g., management of change) and their differences (e.g., number and organization of elements). There has been a trend in recent years to systems with more elements and thematic groupings of elements, although earlier systems having fewer elements continue to be used globally (often in a regulatory context). Regardless of the system, there is a renewed emphasis in PSM on fundamental concepts such as competency and conduct of operations—essentially, doing what one is supposed to do and doing it well. It will be interesting to see whether the safety case approach, which is a mainstay of offshore oil and gas processing, makes further inroads into onshore PSM efforts.

For the most part, we have stayed out of the PSM regulatory debate. We have observed, however, that regulations are usually driven by the occurrence of major process incidents and that regulation, while prevalent, is not universal. There is a strong feeling among process safety practitioners that where they exist, PSM regulations need to be viewed as minimum standards requiring compliance as an outcome not an objective.

As with any management endeavor, measurement of process safety performance is paramount. While leading indicators are intuitively attractive as PSPIs, lagging indicators also have a role to play in avoiding future incidents. Given the current emphasis on barrier- and tier-based PSPIs, research on user-friendly tools for quickly translating process excursion/upset data into predictive indicators would seem valuable.

Continuous improvement is at the core of a PSM system by virtue of the plan/do/check/act (PDCA) cycle. Learning from the experience of others is key to continuous improvement. It is therefore essential that case studies continue to be presented and published to ensure wide dissemination of PSM lessons learned to the global process safety community. The investigations conducted by the US CSB and the ensuing reports are an invaluable resource in this regard.

Core process safety concepts such as ISD, recognition of warning signs, process safety culture, and dynamic risk assessment should not be viewed as stand-alone concepts. They are most effective in preventing and mitigating process incidents when considered as integral components of PSM. Ongoing research on these topics should include the PSM applications of the techniques and methodologies being developed.

The following statement by David Guss from his presentation *Implementing PSM—Where is the Finish Line?* seems a fitting end to this chapter: *So where is the finish line? Maybe when we are able to successfully operate our facilities without having a serious process safety event every day, not just for a week, a month, a year, or even a decade, but for the life of the facility* (Guss, 2015).

ACKNOWLEDGMENTS

The authors gratefully acknowledge the funding assistance of the Natural Sciences and Engineering Research Council of Canada (Discovery Grant) and the Province of Nova Scotia (Cooperative Education Incentive).

REFERENCES

Abidin, M. Z., Rusli, R., Shariff, A. M., & Khan, F. I. (2016). Three-stage ISD matrix (TIM) tool to review the impact of inherently safer design implementation. *Process Safety and Environmental Protection, 99*, 30–42.

Abu-Khader, M. M. (2004). Impact of human behaviour on process safety management in developing countries. *Process Safety and Environmental Protection, 82*, 431–437.

AcuTech. (2012). *Safety and environmental management systems (SEMS) background summary.* McLean, VA: AcuTech Consulting Group. Available at: http://www.ourenergypolicy.org/wp-content/uploads/2012/05/SEMSWhitePaper4-17-12byAcutech.pdf. last accessed November 18, 2016.

Allford, L. (2016). The auditing of process safety. *Journal of Loss Prevention in the Process Industries, 43*, 747–752.

Amyotte, P. (2013a). *An introduction to dust explosions. Understanding the myths and realities of dust explosions for a safer work place.* Waltham, MA: Elsevier/Butterworth-Heinemann.

Amyotte, P. (2013b). Process safety educational determinants. *Process Safety Progress, 32*, 126–130.

Amyotte, P. R., Berger, S., Edwards, D. W., Gupta, J. P., Hendershot, D. C., Khan, F. I., et al. (2016). Why major accidents are still occurring. *Current Opinion in Chemical Engineering, 14*, 1–8.

Amyotte, P. R., Goraya, A. U., Hendershot, D. C., & Khan, F. I. (2007). Incorporation of inherent safety principles in process safety management. *Process Safety Progress, 26*, 333–346.

Amyotte, P. R., & Khan, F. I. (2016). What do gas blows, iron dust accumulations and sulfidation corrosion have in common? In *15th International Symposium on Loss Prevention and Safety Promotion in the Process Industries, Chemical Engineering Transactions, 48, Freiburg, Germany (June 5–8, 2916)* (pp. 739–744).

Amyotte, P., Khan, F., & Lupien, C. (2016). Different hazards, similar causes, same results. *Loss Prevention Bulletin*, (253), 14–18.

Amyotte, P. R., MacDonald, D. K., & Khan, F. I. (2011). An analysis of CSB investigation reports concerning the hierarchy of controls. *Process Safety Progress, 30*, 261–265.

Amyotte, P., Margeson, A., Chiasson, A., Khan, F., & Ahmed, S. (2014). There is no such thing as a black swan process incident. In *Proceedings of Hazards 24, IChemE, Edinburgh, UK (May 7–9, 2014).*

Antonsen, S., Skarholt, K., & Ringstad, A. J. (2012). The role of standardization in safety management—A case study of a major oil & gas company. *Safety Science, 50*, 2001–2009.

API. (2010). *Process safety performance indicators for the refining and petrochemical industries. ANSI/API Recommended Practice 754* (1st ed.). Washington, DC: American Petroleum Institute.

Arendt, S. (2006). Continuously improving PSM effectiveness—A practical roadmap. *Process Safety Progress, 25*, 86–93.

Arntz-Gray, J. (2016). Plan, do, check, act: The need for independent audit of the internal responsibility system in occupational health and safety. *Safety Science, 84*, 12–23.

Asfahl, C. R., & Rieske, D. W. (2010). *Industrial safety and health management* (6th ed.). Upper Saddle River, NJ: Prentice Hall.

Aziz, H. A., Shariff, A. M., & Rusli, R. (2014). Managing process safety information based on process safety management requirements. *Process Safety Progress*, *33*, 41–48.

Aziz, H. A., Shariff, A. M., & Rusli, R. (2016). Interrelations between process safety management elements. *Process Safety Progress*. http://dx.doi.org/10.1002/prs.11824.

Azizi, W. (2016). Predict incidents with process safety performance indicators. *Chemical Engineering Progress*, *112*(2), 22–25.

Battaglia, M., Passetti, E., & Frey, M. (2015). Occupational health and safety management in municipal waste companies: A note on the Italian sector. *Safety Science*, *72*, 55–65.

Baybutt, P. (2015). Competency requirements for process safety auditors. *Process Safety Progress*, *34*, 179–182.

Baybutt, P. (2016a). The meaning and importance of process safety competence. *Process Safety Progress*, *35*, 171–175.

Baybutt, P. (2016b). Insights into process safety incidents from an analysis of CSB investigations. *Journal of Loss Prevention in the Process Industries*, *43*, 537–548.

Bianchini, A., Donini, F., Pellegrini, M., & Saccani, C. (2017). An innovative methodology for measuring the effective implementation of an occupational health and safety management system in the European Union. *Safety Science*, *92*, 26–33.

Bird, F. E., Jr., & Germain, G. L. (1996). *Practical loss control leadership*. Loganville, GA: Det Norske Veritas (USA), Inc. revised ed.

Bloch, K. P., & Wurst, D. M. (2010). Process safety management lessons learned from a petroleum refinery spent caustic tank explosion. *Process Safety Progress*, *29*, 332–339.

Bottani, E., Monica, L., & Vignali, G. (2009). Safety management systems: Performance differences between adopters and non-adopters. *Safety Science*, *47*, 155–162.

Boustras, G., Hadjimanolis, A., Economides, A., Yiannaki, A., & Nicolaides, L. (2015). Management of health and safety in micro-firms in Cyprus—Results from a nationwide survey. *Safety Science*, *79*, 305–313.

Brackey, A. C. (2013). Process safety management: 21 years plus or minus. What I wish I'd known then and what we can't afford to forget now! *Process Safety Progress*, *32*, 260–263.

Bragatto, P. A., Ansaldi, S. M., & Agnello, P. (2015). Small enterprises and major hazards: How to develop an appropriate safety management system. *Journal of Loss Prevention in the Process Industries*, *33*, 232–244.

Brauer, R. L. (2006). *Safety and health for engineers* (2nd ed.). Hoboken, NJ: John Wiley and Sons, Inc.

Bresland, J. (2016). 10 rules to succeed at process safety management. *Chemical Processing*. September 9. Available at: http://www.chemicalprocessing.com/articles/2016/10-rules-to-succeed-at-process-safety-management/. last accessed November 7, 2016.

Cacciabue, P. C., Cassani, M., Licata, V., Oddone, I., & Ottomaniello, A. (2015). A practical approach to assess risk in aviation domains for safety management systems. *Cognition, Technology & Work*, *17*, 249–267.

CAPP (Canadian Association of Petroleum Producers). (2014). *Process safety management: Regulatory scan*. Calgary, AB: Canadian Association of Petroleum Producers. Publication Number 2014-0026.

CASA (Civil Aviation Safety Authority). (2014). *SMS 1. SMS for aviation—A practical guide. Safety management system basics* (2nd ed.). Canberra, Australia: Civil Aviation Safety Authority.

Cazabon, M. D., & Erickson, K. (2010). An oven explosion: Lessons learned on PSM applications. *Process Safety Progress*, *29*, 87–93.

CCPS (Center for Chemical Process Safety). (1992). *Plant guidelines for technical management of chemical process safety*. New York, NY: American Institute of Chemical Engineers.

CCPS (Center for Chemical Process Safety). (2007). *Risk based process safety*. Hoboken, NJ: John Wiley & Sons, Inc.

CCPS (Center for Chemical Process Safety). (2009). *Inherently safer chemical processes. A life cycle approach* (2nd ed.). Hoboken, NJ: John Wiley & Sons, Inc.
CCPS (Center for Chemical Process Safety). (2011a). *Conduct of operations and operational discipline for improving process safety in industry.* Hoboken, NJ: John Wiley & Sons, Inc.
CCPS (Center for Chemical Process Safety). (2011b). *Process safety leading and lagging metrics…you don't improve what you don't measure.* New York, NY: Center for Chemical Process Safety.
CCPS (Center for Chemical Process Safety). (2012). *Recognizing catastrophic incident warning signs in the process industries.* Hoboken, NJ: John Wiley & Sons, Inc.
CCPS (Center for Chemical Process Safety). (2014). *Risk based process safety overview.* New York, NY: American Institute of Chemical Engineers.
CCPS (Center for Chemical Process Safety). (2015). *Guidelines for defining process safety competency requirements.* Hoboken, NJ: John Wiley & Sons, Inc.
CCPS (Center for Chemical Process Safety). (2016a). *Introduction to process safety for undergraduates and engineers.* Hoboken, NJ: John Wiley & Sons, Inc.
CCPS (Center for Chemical Process Safety). (2016b). *Guidelines for integrating management systems and metrics to improve process safety performance.* Hoboken, NJ: John Wiley & Sons, Inc.
CCPS (Center for Chemical Process Safety). (2016c). *Guidelines for implementing process safety management* (2nd ed.). Hoboken, NJ: John Wiley & Sons, Inc.
Celik, M. (2009). Designing of integrated quality and safety management system (IQSMS) for shipping operations. *Safety Science, 47*, 569–577.
Chang, J. I., & Liang, C.-L. (2009). Performance evaluation of process safety management systems of paint manufacturing facilities. *Journal of Loss Prevention in the Process Industries, 22*, 398–402.
Chosnek, J., & Clifton, R. (2008). Improved process safety management and simple metrics. *Process Safety Progress, 27*, 284–288.
Coulter, P. D. (1995). Applying total quality management (TQM) to health, safety, and environmental auditing at Union Carbide. In L. Harrison (Ed.), *Environmental, health, and safety auditing handbook* (2nd ed., pp. 113–130). New York, NY: McGraw-Hill, Inc.
CSB (Chemical Safety Board). (2010a). *Case study. Texas Tech University. Laboratory explosion. No. 2010-05-I-TX.* Washington, DC: US Chemical Safety and Hazard Investigation Board.
CSB (Chemical Safety Board). (2010b). *Investigation report. Xcel Energy hydroelectric plant. Penstock fire. No. 2008-01-I-CO.* Washington, DC: US Chemical Safety and Hazard Investigation Board.
CSB (Chemical Safety Board). (2010c). *Urgent recommendations from Kleen Energy investigation.* Washington, DC: US Chemical Safety and Hazard Investigation Board.
CSB (Chemical Safety Board). (2011). *Case study. Hoeganaes Corporation: Gallatin, TN. Metal dust flash fires and hydrogen explosion. No. 2011-04-I-TN.* Washington, DC: US Chemical Safety and Hazard Investigation Board.
CSB (Chemical Safety Board). (2013). *Interim investigation report. Chevron Richmond refinery fire.* Washington, DC: US Chemical Safety and Hazard Investigation Board.
CSB (Chemical Safety Board). (2014a). *Regulatory report. Chevron Richmond refinery. Pipe rupture and fire. Report No. 2012-03-I-CA.* Washington, DC: US Chemical Safety and Hazard Investigation Board.
CSB (Chemical Safety Board). (2014b). *Investigation report. Catastrophic rupture of heat exchanger. Tesoro Anacortes Refinery. Report No. 2010-01-I-WA.* Washington, DC: US Chemical Safety and Hazard Investigation Board.
CSB (Chemical Safety Board). (2014c). *Case study. AL Solutions, Inc., New Cumberland, WV. Metal dust explosion and fire. Report No. 2011-03-I-WV.* Washington, DC: US Chemical Safety and Hazard Investigation Board.

CSB (Chemical Safety Board). (2016a). *Case study. Williams Geismar olefins plant. Reboiler rupture and fire. No. 2013-03-I-LA.* Washington, DC: US Chemical Safety and Hazard Investigation Board.
CSB (Chemical Safety Board). (2016b). *Investigation report. West Fertilizer Company fire and explosion. Report No. 2013-02-I-TX.* Washington, DC: US Chemical Safety and Hazard Investigation Board.
CSChE (Canadian Society for Chemical Engineering). (2012a). *Process safety management guide* (4th ed.). Ottawa, ON: Canadian Society for Chemical Engineering.
CSChE (Canadian Society for Chemical Engineering). (2012b). *Process safety management standard* (1st ed.). Ottawa, ON: Canadian Society for Chemical Engineering.
Cummings, D. E. (2009). The evolution and current status of process safety management metrics. *Process Safety Progress, 28*, 147–155.
De Rademaeker, E., Suter, G., Pasman, H. J., & Fabiano, B. (2014). A review of the past, present and future of the European loss prevention and safety promotion in the process industries. *Process Safety and Environmental Protection, 92*, 280–291.
Di Menna, J. (2012). Safety haven. *Canadian Chemical News*, 24–27. October 2012.
Donaldson, A., Borys, D., & Finch, C. F. (2013). Understanding safety management system applicability in community sport. *Safety Science, 60*, 95–104.
Dougherty, T. M. (1999). How to conduct an accident investigation. In L. J. DiBerardinis (Ed.), *Handbook of occupational health and safety* (pp. 103–126). New York, NY: John Wiley & Sons, Inc.
Edwards, D. (2014). Trevor kletz: 1922–2013. *The Chemical Engineer*, (870/871), 60–61. December 2013/January 2014.
Energy Institute. (2010). *High level framework for process safety management.* London, UK: Energy Institute.
EPA (Environmental Protection Agency). (2004). *Chemical accident prevention provisions. 40 CFR 68.* Washington, DC: US Environmental Protection Agency. Available at: https://www.epa.gov/rmp/general-rmp-guidance-appendix-40-cfr-68. last accessed November 16, 2016.
Ferguson, A., & Moritz, M. (2015). *Leading. Learning from life and my years at Manchester United.* New York, NY: Hachette Books.
Fernandez-Muniz, B., Montes-Peon, J. M., & Vazquez-Ordas, C. J. (2007). Safety management system: Development and validation of a multidimensional scale. *Journal of Loss Prevention in the Process Industries, 20*, 52–68.
Forest, J. J. (2012). Management discipline. *Process Safety Progress, 31*, 334–336.
Forest, J. J. (2014). Management discipline: Defining a process safety strategy. *Process Safety Progress, 33*, 162–165.
Frank, W. L. (2007). Process safety culture in the CCPS risk based process safety model. *Process Safety Progress, 26*, 203–208.
Gerbec, M. (2016). Safety change management—A new method for integrated management of organizational and technical changes. *Safety Science.* http://dx.doi.org/10.1016/j.ssci.2016.07.006.
Gholami, P. S., Nassiri, P., Yarahmadi, R., Hamidi, A., & Mirkazemi, R. (2015). Assessment of health safety and environment management system function in contracting companies of one of the petro-chemistry industries in Iran, a case study. *Safety Science*, 77, 42–47.
Gill, S. (2015). PSM auditing: Thinking beyond compliance. *Process Safety Progress, 35*, 295–299.
Gnoni, M. G., & Saleh, J. H. (2017). Near-miss management systems and observability-in-depth: Handling safety incidents and accident precursors in light of safety principles. *Safety Science, 91*, 154–167.
Goddard, S. (2012). Preparing for process safety management. *Process Safety Progress, 31*, 355–358.

Goetsch, D. L. (2010). *Establishing a safety-first corporate culture in your organization. An integrated approach for safety professionals and safety committees.* Upper Saddle River, NJ: Pearson Educational, Inc.

Goetsch, D. L. (2015). *The basics of occupational safety* (2nd ed.). Boston, MA: Pearson.

Goh, Y. M., Tan, S., & Lai, K. C. (2015). Learning from the Bhopal disaster to improve process safety management in Singapore. *Process Safety and Environmental Protection, 97*, 102–108.

Goraya, A., Amyotte, P. R., & Khan, F. I. (2004). An inherent safety-based incident investigation methodology. *Process Safety Progress, 23*, 197–205.

Grote, G. (2012). Safety management in different high-risk domains—All the same? *Safety Science, 50*, 1983–1992.

Guss, D. (2015). Implementing PSM—Where is the finish line? CSChE PSM award presentation. In *65th Canadian Chemical Engineering Conference, Calgary, AB (October 4–7, 2015).*

Haesle, J., Devlin, C., & McCavit, J. (2009). Improving process safety by addressing the human element. *Process Safety Progress, 28*, 325–330.

Hale, A., Borys, D., & Adams, M. (2015). Safety regulation: The lessons of workplace safety rule management for managing the regulatory burden. *Safety Science, 71*, 112–122.

Hammer, W., & Price, D. (2001). *Occupational safety management and engineering* (5th ed.). Upper Saddle River, NJ: Prentice Hall.

Havold, J. I. (2010). Safety culture and safety management aboard tankers. *Reliability Engineering and System Safety, 95*, 511–519.

Hayes, J. (2015). Lessons for effective integrity management from the San Bruno pipeline rupture. *Process Safety Progress, 34*, 202–206.

Hendershot, D. C. (2012). Process safety management—You can't get it right without a good safety culture. *Process Safety Progress, 31*, 2–5.

Hendershot, D. C. (2015). Guest perspective on Bhopal: Why can't we do better? Thoughts on the 30th anniversary of the Bhopal tragedy. *Journal of Loss Prevention in the Process Industries, 36*, 183–184.

Hendershot, D. (2016). Process safety: More to process safety than regulations. *Journal of Chemical Health and Safety, 23*(2), 37.

Herber, J. W. (2012). Smaller companies struggle with process safety. *Process Safety Progress, 31*, 346–349.

Hopkins, A. (2000). *Lessons from Longford. The Esso gas plant explosion.* Sydney, Australia: CCH Australia Limited.

Hopkins, A. (2005). *Safety, culture and risk.* Sydney, Australia: CCH Australia Limited.

Hopkins, A. (2009a). *Failure to learn. The BP Texas City refinery disaster.* Sydney, Australia: CCH Australia Limited.

Hopkins, A. (Ed.), (2009b). *Learning from high reliability organisations.* Sydney, Australia: CCH Australia Limited.

Howell, P. P. (2010). Plant explosion emphasizes importance of implementing PSM. *Process Safety Progress, 29*, 144–149.

HSE (Health and Safety Executive). (2006). *Developing process safety indicators. A step-by-step guide for chemical and major hazard industries. HSG254.* London, UK: Health and Safety Executive.

HSE (Health and Safety Executive). (2015). *The control of major accident hazards regulations 2015* (3rd ed.). London, UK: Health and Safety Executive.

Huat, G. H. (2012). Singapore's approach to process safety. In *Process Safety Management Seminar, 14th Asia Pacific Confederation of Chemical Engineering Congress (APCChE 2012). Singapore (February 21–24, 2012).*

Irvine, Y., Amyotte, P., & Khan, F. (2016). An analysis of CSB investigation reports. In *66th Canadian Chemical Engineering Conference, Quebec City, QC (October 16–19, 2016).*

ISO (International Organization for Standardization). (2016). *Management system standards.* International Organization for Standardization. Available at: http://www.iso.org/iso/home/standards/management-standards.htm. last accessed November 9, 2016.

Jeon, J., Lee, J., Shin, D., & Park, H. (2009). Development of dam safety management system. *Advances in Engineering Software*, *40*, 554–563.

Kadri, S., Peters, G., VanOmmeren, J., Fegley, K., Dennehy, M., & Mateo, A. (2014). So we have all been implementing process safety metrics—What next? *Process Safety Progress*, *33*, 172–178.

Kelly, B. D. (2011). Why process safety programs sometimes fail. *Process Safety Progress*, *30*, 307–309.

Kelly, B. D. (2013). Management of change—Enabler or inhibitor? *Process Safety Progress*, *32*, 140–141.

Kenan, S., & Kadri, S. (2014). Process safety leading indicators survey—February 2013. Center for Chemical Process Safety—White Paper. *Process Safety Progress*, *33*, 247–258.

Kerin, T. (2016). Bolster your lead process safety metrics. *Chemical Processing*. November 7. Available at: http://www.chemicalprocessing.com/articles/2016/bolster-your-lead-process-safety-metrics/. last accessed November 7, 2016.

Khan, F., Abunada, H., John, D., & Benmosbah, T. (2010). Development of risk-based process safety indicators. *Process Safety Progress*, *29*, 133–143.

Khan, F., Hashemi, S. J., Paltrinieri, N., Amyotte, P., Cozzani, V., & Reniers, G. (2016). Dynamic risk management: A contemporary approach to process safety management. *Current Opinion in Chemical Engineering*, *14*, 9–17.

Khan, F., Rathnayaka, S., & Ahmed, S. (2015). Methods and models in process safety and risk management: Past, present and future. *Process Safety and Environmental Protection*, *98*, 116–147.

King, C. (2013). The importance of leadership and management in process safety. *Process Safety Progress*, *32*, 179–184.

Klein, J. A., & Dharmavaram, S. (2012). Improving the performance of established PSM programs. *Process Safety Progress*, *31*, 261–265.

Kletz, T. (2012). The history of process safety. *Journal of Loss Prevention in the Process Industries*, *25*, 763–765.

Kletz, T., & Amyotte, P. (2010). *Process plants. A handbook for inherently safer design* (2nd ed.). Boca Raton, FL: CRC Press (Taylor & Francis Group).

Knegtering, B., & Pasman, H. J. (2009). Safety of the process industries in the 21st century: A changing need of process safety management for a changing industry. *Journal of Loss Prevention in the Process Industries*, *22*, 162–168.

Knijff, P., Allford, L., & Schmeizer, P. (2013). Process safety leading indicators—A perspective from Europe. *Process Safety Progress*, *32*, 332–336.

Koivupalo, M., Sulasalmi, M., Rodrigo, P., & Vayrynen, S. (2015). Health and safety management in a changing organisation: Case study global steel company. *Safety Science*, *74*, 128–139.

Kontogiannis, T., Leva, M. C., & Balfe, N. (2016). Total safety management: Principles, processes and methods. *Safety Science*. http://dx.doi.org/10.1016/j.ssci.2016.09.015.

Kwon, H.-M., Lee, C.-J., Seo, D., & Moon, I. (2016). Korean experience of process safety management (PSM) regulation for chemical industry. *Journal of Loss Prevention in the Process Industries*, *42*, 2–5.

Lee, K., Kwon, H.-M., Cho, S., Kim, J., & Moon, I. (2016). Improvements of safety management system in Korean chemical industry after a large chemical accident. *Journal of Loss Prevention in the Process Industries*, *42*, 6–13.

Leveson, N. (2015). A systems approach to risk management through leading safety indicators. *Reliability Engineering and System Safety*, *136*, 17–34.

Liao, M.-Y. (2015). Safety culture in commercial aviation: Differences in perspective between Chinese and Western pilots. *Safety Science*, *79*, 193–205.

Lingard, H., Hallowell, M., Salas, R., & Pirzadeh, P. (2017). Leading or lagging? Temporal analysis of safety indicators on a large infrastructure construction project. *Safety Science*, *91*, 206–220.

Louvar, J. F. (2008). Improving the effectiveness of process safety management in small companies. *Process Safety Progress*, *27*, 280–283.
Luo, H. (2010). The effectiveness of U.S. OSHA process safety management inspection—A preliminary quantitative evaluation. *Journal of Loss Prevention in the Process Industries*, *23*, 455–461.
Lutchman, C., Maharaj, R., & Ghanem, W. (2012). *Safety management. A comprehensive approach to developing a sustainable system.* Boca Raton, FL: CRC Press (Taylor & Francis Group).
Macza, M. (2008). A Canadian perspective of the history of process safety legislation. pp. 12/1–12/22. *Available at: In 8th Internationale Symposium. Programmable Electronic System in Safety-Related Applications, Cologne, Germany (September 2–3, 2008).* http://wishlist.vogue.com/sites/default/files/news-article-pdfs/A%20Canadian%20Perspective%20of%20the%20History%20of%20PSM%20Legislation.pdf. last accessed November 13, 2016.
Maher, S. T., Long, G. D., Cromartie, R. S., Sutton, I. S., & Steinhilber, M. R. (2016). Paradigm shift in the regulatory application of safety management systems to offshore facilities. *Process Safety Progress*, *35*, 317–329. http://dx.doi.org/10.1002/prs.11558.
Majid, N. D. A., Shariff, A. M., & Loqman, S. M. (2016a). Ensuring emergency planning & response meet the minimum Process Safety Management (PSM) standards requirements. *Journal of Loss Prevention in the Process Industries*, *40*, 248–258.
Majid, N. D. A., Shariff, A. M., & Rusli, R. (2015). Process Safety Management (PSM) for managing contractors in process plant. *Journal of Loss Prevention in the Process Industries*, *37*, 82–90.
Majid, N. D. A., Shariff, A. M., Rusli, R., & Azman, K. I. (2016b). Trade secret model based on OSHA process safety management requirement. *Procedia Engineering*, *148*, 1089–1095.
Mannan, M. S. (2015). Guest perspective on Bhopal. *Journal of Loss Prevention in the Process Industries*, *38*, 298–299.
Mason, E. (2001a). Elements of process safety management: Part 1. *Chemical Health & Safety*, *8*(4), 22–24.
Mason, E. (2001b). Elements of process safety management: Part 2. *Chemical Health & Safety*, *8*(5), 23–26.
McCavit, J., Berger, S., Grounds, C., & Nara, L. (2015). A call to action—Next steps for Vision 20/20. *Process Safety Progress*, *34*, 111–118.
McCavit, J., Berger, S., & Nara, L. (2014). Characteristics of companies with great process safety performance. *Process Safety Progress*, *33*, 131–135.
Mendeloff, J., Han, B., Fleishman-Mayer, L. A., & Vesely, J. V. (2013). Evaluation of process safety indicators collected in conformance with ANSI/API recommended practice 754. *Journal of Loss Prevention in the Process Industries*, *26*, 1008–1014.
Meyer, T., & Reniers, G. (2016). *Engineering risk management* (2nd ed.). Berlin, Germany: De Gruyter.
MKOPSC (Mary Kay O'Connor Process Safety Center). (2012). Texas A&M University System, A Frontiers of Research Workshop. In *Process safety research agenda for the 21st century. A policy document developed by a representation of the global process safety academia. October 21–22, 2011, College Station, TX.* College Station, TX: Mary Kay O'Connor Process Safety Center, Texas A&M University System.
Mooren, L., Grzebieta, R., Williamson, A., Olivier, J., & Friswell, R. (2014). Safety management for heavy vehicle transport: A review of the literature. *Safety Science*, *62*, 79–89.
Moorkamp, M., Kramer, E.-H., van Gulijk, C., & Ale, B. (2014). Safety management theory and the expeditionary organization: A critical theoretical reflection. *Safety Science*, *69*, 71–81.
Nascimento, F. A. C., Majumdar, A., Ochieng, W. Y., Schuster, W., & Studic, M. (2016). Fundamentals of safety management: The offshore helicopter transportation system model. *Safety Science*, *85*, 194–204.

Newslow, D. L. (2014). Food safety and quality management systems. In Y. Motarjemi, G. Moy, & E. Todd (Eds.), *Food safety management: Vol. 4. Encyclopedia of food safety* (pp. 149–158). San Diego, CA: Academic Press.
OECD (Organisation for Economic Co-operation and Development). (2008). Guidance on developing safety performance indicators related to chemical accident prevention, preparedness and response for public authorities and communities/public *OECD environment, health and safety. Chemical accidents programme* (2nd ed.). Paris, France: Organisation for Economic Co-operation and Development.
OECD (Organisation for Economic Co-operation and Development). (2012). Corporate governance for process safety. OECD guidance for senior leaders in high hazard industries. *OECD environment, health and safety. Chemical accidents programme*. Paris, France: Organisation for Economic Co-operation and Development.
OGP (Oil and Gas Producers). (2011). *Process safety—Recommended practice on key performance indicators. Report no. 456*. London, UK: The International Association of Oil and Gas Producers.
Olewski, T., Ahammad, M., Quraishy, S., Gan, N., & Vechot, L. (2016). Building process safety culture at Texas A&M University at Qatar: A case study on experimental research. *Journal of Loss Prevention in the Process Industries*, *44*, 642–652. http://dx.doi.org/10.1016/j.jlp.2016.08.022.
OSHA (Occupational Safety acbend Health Administration). (2000). *Process safety management. OSHA 3132*. Washington, DC: US Department of Labor, Occupational Safety and Health Administration.
OSHA (Occupational Safety and Health Administration). (2013). *Process safety management of highly hazardous chemicals. 29 CFR 1910.119*. Washington, DC: US Department of Labor, Occupational Safety and Health Administration. Available at https://www.osha.gov/pls/oshaweb/owadisp.show_document?p_table=STANDARDS&p_id=9760. last accessed November 16, 2016.
Paradies, M. (2011). Has process safety management missed the boat? *Process Safety Progress*, *30*, 310–314.
Pasman, H. J., Knegtering, B., & Rogers, W. J. (2013). A holistic approach to control process safety risks: Possible ways forward. *Reliability Engineering and System Safety*, *117*, 21–29.
Pasman, H., & Rogers, W. (2014). How can we use the information provided by process safety performance indicators? Possibilities and limitations. *Journal of Loss Prevention in the Process Industries*, *30*, 197–206.
Payne, S. C., Bergman, M. E., Rodriguez, J. M., Beus, J. M., & Henning, J. B. (2010). Leading and lagging: Process safety climate–incident relationships at one year. *Journal of Loss Prevention in the Process Industries*, *23*, 806–812.
Philley, J. (2002). Potential impacts to process safety management from mergers, acquisitions, downsizing, and re-engineering. *Process Safety Progress*, *21*, 151–160.
Piette, R. (2012). Culture club: Process safety experts bring Canada up to snuff. *Canadian Chemical News, October 2012*, 7.
Pilbeam, C., Doherty, N., Davidson, R., & Denyer, D. (2016). Safety leadership practices for organizational safety compliance: Developing a research agenda from a review of the literature. *Safety Science*, *86*, 110–121.
Pillay, M. (2015). Accident causation, prevention and safety management: A review of the state-of-the-art. *Procedia Manufacturing*, *3*, 1838–1845.
PSV (Process Safety Visions). (2016). Committed culture. *Chemical Engineering Progress*, *112*(11), 22.
Puyosa, H. D. (2012). Knowledge management applied to chemical process safety. In *Proceedings of the 14th IFAC Symposium on Information Control Problems in Manufacturing, Bucharest, Romania (May 23–25, 2012)*.

Qian, Q., & Lin, P. (2016). Safety risk management of underground engineering in China: Progress, challenges and strategies. *Journal of Rock Mechanics and Geotechnical Engineering*, *8*, 423–442.

Rashid, M. I., Ramzan, N., Iqbal, T., Yasin, S., & Yousaf, S. (2013). Implementation issues of PSM in a fertilizer plant: An operations engineer's point of view. *Process Safety Progress*, *32*, 59–65.

Reason, J. (1990). *Human error*. Cambridge, UK: Cambridge University Press.

Remawi, H., Bates, P., & Dix, I. (2011). The relationship between the implementation of a safety management system and the attitudes of employees towards unsafe acts in aviation. *Safety Science*, *49*, 625–632.

Rezvani, Z., & Hudson, P. (2016). Breaking the clay layer: The role of middle management in the management of safety. *Journal of Loss Prevention in the Process Industries*, *44*, 241–246.

Rosen, R. (2015). The ten commandments of risk based process safety. *Process Safety Progress*, *34*, 212–213.

Rosenthal, I., Kleindorfer, P. R., & Elliott, M. R. (2006). Predicting and confirming the effectiveness of systems for managing low-probability chemical process risks. *Process Safety Progress*, *25*, 135–155.

Rowe, S., & Francois, J.-M. (2016). Process safety data—The cornerstone of PSM and often its undermining. *Journal of Loss Prevention in the Process Industries*, *43*, 736–740.

Roy, N., Eljack, F., Jiminez-Gutierrez, A., Zhang, B., Thiruvenkataswamy, P., El-Halwagi, M., et al. (2016). A review of safety indices for process design. *Current Opinion in Chemical Engineering*, *14*, 42–48.

Schein, E. H. (2010). *Organizational culture and leadership* (4th ed.). San Francisco, CA: Jossey-Bass.

Scholtz, C. R., & Maher, S. T. (2014). Tips for the creation and application of effective operating procedures. *Process Safety Progress*, *33*, 350–354.

Seveso III. (2012). Directive 2012/18/EU of the European Parliament and of the Council of 4 July 2012 on the control of major-accident hazards involving dangerous substances, amending and subsequently repealing Council Directive 96/82/EC. *Official Journal of the European Union*, *L 197*, 1–37.

Shariff, A. M., Aziz, H. A., & Majid, N. D. A. (2016). Way forward in Process Safety Management (PSM) for effective implementation in process industries. *Current Opinion in Chemical Engineering*, *14*, 56–60.

Sheehan, C., Donohue, R., Shea, T., Cooper, B., & De Cieri, H. (2016). Leading and lagging indicators of occupational health and safety: The moderating role of safety leadership. *Accident Analysis and Prevention*, *92*, 130–138.

Shin, I. J. (2013). The effective control of major industrial accidents by the Major Accident Prevention Centers (MAPC) through the process safety management (PSM) grading system in Korea. *Journal of Loss Prevention in the Process Industries*, *26*, 803–814.

Sousa, V., Almeida, N. M., & Dias, L. A. (2014). Risk-based management of occupational safety and health in the construction industry—Part 1: Background knowledge. *Safety Science*, *66*, 75–86.

Sousa, V., Almeida, N. M., & Dias, L. A. (2015). Risk-based management of occupational safety and health in the construction industry—Part 2: Quantitative model. *Safety Science*, *74*, 184–194.

Sreenevasan, R. (2015). *The effect of regulations in improving process safety. In Risk Engineering Society WA Technical Event. Engineers Australia, Perth, Australia (June 11, 2015)*. Available at: https://www.engineersaustralia.org.au/sites/default/files/shado/Learned%20Groups/Technical%20Societies/Risk%20Engineering%20Society/australian_regulations_res_wa_paper.pdf. last accessed November 13, 2016.

Stroeve, S. H., Som, P., van Doorn, B. A., & Bakker, G. J. (2016). Strengthening air traffic safety movement by moving from outcome-based towards risk-based evaluation of runway incursions. *Reliability Engineering and System Safety, 147*, 93–108.

Sutton, I. (2014). *Offshore safety management. Implementing a SEMS program* (2nd ed.). Waltham, MA: Elsevier. Chapter 8—Safety Cases.

SWA (Safe Work Australia). (2012). *Guide for major hazard facilities. Safety management systems.* Canberra, Australia: Safe Work Australia.

Swuste, P., Theunissen, J., Schmitz, P., Reniers, G., & Blokland, P. (2016). Process safety indicators, a review of the literature. *Journal of Loss Prevention in the Process Industries, 40*, 162–173.

Swuste, P., van Gulijk, C., Zwaard, W., Lemkowitz, S., Oostendorp, Y., & Groeneweg, J. (2016). Developments in the safety science domain, in the fields of general and safety management between 1970 and 1979, the year of the near disaster on Three Mile Island, a literature review. *Safety Science, 86*, 10–26.

Vaughen, B. K. (2015). Three decades after Bhopal: What we have learned about effectively managing process safety risks. *Process Safety Progress, 34*, 345–354.

Vaughen, B. K., Downes, A., Fox, J., & Belonger, D. (2015). Guidelines for integrating management systems and metrics to improve process safety performance. *Process Safety Progress, 34*, 259–266.

Vaughen, B. K., & Kletz, T. A. (2012). Continuing our process safety management journey. *Process Safety Progress, 31*, 337–342.

Vinodkumar, M. N., & Bhasi, M. (2010). Safety management practices and safety behaviour: Assessing the mediating role of safety knowledge and motivation. *Accident Analysis and Prevention, 42*, 2082–2093.

Wahlstrom, B., & Rollenhagen, C. (2014). Safety management—A multi-level control problem. *Safety Science, 69*, 3–17.

Wang, M., Mentzer, R. A., Gao, X., Richardson, J., & Mannan, M. S. (2013). Normalization of process safety metrics. *Process Safety Progress, 32*, 337–345.

Warmerdam, A., Newnam, S., Sheppard, D., Griffin, M., & Stevenson, M. (2017). Workplace road safety risk management: An investigation into Australian practices. *Accident Analysis and Prevention, 98*, 64–73.

Wilson, L., & McCutcheon, D. (2003). *Industrial safety and risk management.* Edmonton, AB: The University of Alberta Press.

Wincek, J., Sousa, L. S., Myers, M. R., & Ozog, H. (2015). Organizational change management for process safety. *Process Safety Progress, 34*, 89–93.

Wold, T., & Laumann, K. (2015). Safety management systems as communication in an oil and gas producing company. *Safety Science, 72*, 23–30.

WSHCouncil (Workplace Safety and Health Council). (2012). *Workplace safety and health guidelines. Process safety performance indicators.* Singapore: Workplace Safety and Health Council.

Wu, B., Xu, Z., Zhou, Y., Peng, Y., & Yu, Z. (2014). Study on coal mine safety management system based on "hazard, latent danger and emergency responses". *Procedia Engineering, 84*, 172–177.

Young, C. W., & Hodges, K. J. (2012). Process safety management mentoring: Subjects to convey and the methods for conveying. *Process Safety Progress, 31*, 350–354.

Zhang, L., Wu, X., Chen, Q., Skibniewski, M. J., & Hsu, S.-C. (2014). Towards a safety management approach for adjacent buildings in tunneling environments: Case study in China. *Building and Environment, 75*, 222–235.

Zhao, J., Suikkanen, J., & Wood, M. (2014). Lessons learned for process safety management in China. *Journal of Loss Prevention in the Process Industries, 29*, 170–176.

Zhou, Z., Goh, Y. M., & Li, Q. (2015). Overview and analysis of safety management studies in the construction industry. *Safety Science, 72*, 337–350.

CHAPTER FOUR

Human Factors in the Chemical Process Industries

Kathryn Mearns[1]
Human Factors Consultant
[1]Corresponding author: e-mail address: k.j.mearns@gmail.com

Contents

1. INTRODUCTION

The chemical process industries encompass a number of different hazardous substances and processes, ranging from the manufacture of pharmaceuticals and industrial chemicals through to processing hydrocarbons into petrochemicals. Nevertheless, some common principles are necessary to prevent the major accident hazard (MAH) risks that are inherent in these

Methods in Chemical Process Safety, Volume 1
ISSN 2468-6514
http://dx.doi.org/10.1016/bs.mcps.2017.01.002

industries from being realized. These principles have been covered elsewhere in this volume, and this chapter focuses on how the human element can contribute to major accidents and how contributory "performance influencing factors" (PIFs) can be managed and mitigated.

Of course, the chemical process industries are not just susceptible to MAH risks, i.e., loss of containment, fires, etc. The likelihood of an MAH risk being realized is fortunately very small, and the vast majority of incidents in the chemical process industries tend to be occupational injuries arising from slips, trips, and falls. What is relevant about human factors is that the same personal and system failings can lead to occupational injuries as well as to major accident harm. While it is unacceptable for people to suffer any accident or injury at work, the main focus for most human factors work in the chemical process industries is to prevent major accidents. These accidents have not only the potential to affect the people working directly at the plant but also local populations in residence close by and the physical environment.

Major accidents such as Seveso, Flixborough, Bhopal, and Texas City serve to highlight the costs both in human life and financially. For example, 15 people died and 180 were injured at Texas City and the financial losses were in the order of $1.5 billion. The immediate and underlying causes of this accident are well documented in the US Chemical Safety Board Report (2007). In relation to the "Key Issues" identified as being underlying causes of this accident, it is worth noting that safety culture and human factors are two of the four, the other two being process safety metrics and regulatory oversight. The panel reports that simply focusing on the actions of the front-line operators at the plant and their immediate supervisors misses the point. Tackling the underlying causes of the accident such as a poor safety culture; overfocus on occupational safety indicators rather than process safety indicators; and a cost-cutting regime and lack of incident reporting will have a greater impact on preventing future major accidents. This recommendation is supported by the incident investigations of other major accidents over the decades in a number of safety critical industries.

This chapter is organized in the following way. First, there is an outline of the human factors definitions and terminology, to ensure that readers have a consistent understanding of the terms used throughout the chapter. Second, there is a section on human error and violations (also known as noncompliances/nonconformances), which are often the final triggering actions that can lead to an adverse event. Third, there is a description of the factors that can influence human performance (so-called PIFs), which need to be managed in order to keep the risk of human error as low as reasonably practicable. Finally, techniques used to measure and manage human error and PIFs are discussed throughout the chapter.

2. HUMAN FACTORS DEFINITIONS AND TERMINOLOGY

Human Factors (HF) or Human and Organizational Factors (HOF), as it is sometimes called, is a wide-ranging discipline incorporating elements of psychology, physiology, ergonomics, sociology, engineering, and management science. The discipline is often labeled Ergonomics or Human Factors Engineering, but in reality the scope is much broader than that, with an increasing emphasis on the knowledge, skills, and techniques that characterize the social sciences such as sociology and management studies, i.e., qualitative techniques such as interviews, focus groups, workshops, as well as the traditional quantitative techniques, e.g., Human Reliability Analysis (HRA), Human Error Quantification, and questionnaire survey design and deployment, which have traditionally been associated with human error and human factors measurement and management.

For the purposes of this chapter, the terminology used in research and assessment of human factors is clarified below since many of these terms are used interchangeably and inconsistently in the literature:

1. "Ergonomics" literally means "laws of work" and is the term mainly used in Europe. The term *Human Factors* originated in the United States but is now used more widely. Ergonomics tends to be associated more with physical workplace assessment, which can also be referred to as *Human Factors Engineering*.
2. *Cognitive Ergonomics or Engineering Psychology* emphasizes the study of cognitive or mental aspects of work, especially where there is a high level of human–machine interaction, automation, decision making, and mental workload.
3. *Human–Machine Interaction or Human–Computer Interaction* is the applied study of how people interact with technology.
4. *Working Environment* emphasizes the environmental and task factors that influence human performance.
5. *Human and Organizational Factors (HOF)* emphasizes the organizational aspects that influence human performance such as leadership style, management systems, safety culture and climate, training and competency arrangements, incident reporting systems, behavioral safety, and human resource practices. As such, HOF mainly concerns itself with PIFs, which will be covered later in the chapter.

Members of the human factors community will often specialize in more than one of these subdisciplines and will be familiar with a range of different techniques to measure and manage both human error and PIFs.

The UK Health and Safety Executive (i.e., UK Regulator) considers Human Factors to encompass "environmental, organizational and job factors, and human and individual factors, which influence behavior at work in a way which can affect health and safety" (HSE, 1999). Using the above definition as a basis, the UK HSE has grouped the following human factors issues under each theme (see Table 1).

Table 1 provides a useful framework for considering the topic of human factors, although it is often difficult (and indeed inadvisable) to consider these factors in isolation, since they often are interrelated and interact with each other. For example, "competence" could be considered as an individual responsibility, in the sense that the individual should ensure that they are trained and competent for the tasks they are involved. It could also be considered an organizational responsibility, in the sense that the organization should ensure that arrangements are in place for their staff to be properly trained and competent for their roles and responsibilities.

It is not possible to discuss all the issues in Table 1 in great detail; therefore, this chapter will only focus on a few issues under each heading. For *individual factors*, the focus will be on skills and personality, thereby covering competence and attributes. For *job factors* the focus will be on displays and controls and procedures, thereby including task and environment. For *organizational factors* the focus will be on culture and leadership, thereby encompassing resources and communication, both of which are key aspects of leadership.

One of the first points to be addressed is the definition of human error (human failure), which can be caused by many of the PIFs outlined in Table 1.

Table 1 Human Factors Issues According to HSG48 (HSE, 1999)

Individual	**Job**	**Organization**
Competence	Task	Culture
Skills	Workload	Leadership
Personality	Environment	Work patterns
Attributes	Displays and controls	Resources
Risk Perception	Procedures	Communications

2.1 Human Error

Professor James Reason defines Human Error as "*a generic term to encompass all the occasions in which a planned sequence of mental or physical activities fails to achieve its intended outcome and when these failures cannot be attributed to the intervention of some chance agency*" (Reason, 1991, p. 9).

Reason further distinguishes between *slips, lapses, and mistakes*. *Slips* are associated with faulty actions, where actions do not proceed as intended, e.g., misreading a display. *Lapses* are failures of memory, e.g., forgetting to press a switch. Slips and lapses tend to occur during routine tasks in familiar surroundings where the operator may be on "automatic pilot" or attention is captured by something other than the task in hand, e.g., a work colleague distracting the operator with a question. They are categorized as "skill-based" errors and are relatively easy to recover from because the operator will receive direct feedback that their actions have not led to the anticipated outcome.

Mistakes occur when the intended plans for action are wrong in the first place. In other words, intended actions may proceed as planned, but fail to achieve their intended outcome. There are two types of mistake "*rule-based mistakes*" and "*knowledge-based mistakes.*" Rule-based mistakes are where an incorrect diagnostic rule is applied. An example of this is where an experienced operator on a batch reactor may have learned diagnostic rules, which are inappropriate for continuous process operations. If an attempt is made to apply these rules to work out the cause of a continuous process upset, a misdiagnosis can occur, leading to an inappropriate action. There is also a tendency for people to apply strong but successful rules that they have applied in the past, even if they are not appropriate for the situation.

Knowledge-based mistakes occur when the information-processing capabilities of the operator are being tested by an unfamiliar situation that has to be worked out from first principles and there are no apparent rules or procedures to deal with the situation. Quite often, only information that is readily available will be used to evaluate the situation or people will depend on a "gut feel" that their course of action is correct, perhaps based on similar but unrelated incidents in the past. Sometimes "attention capture" occurs where the operators become focused on one small part of an overall problem, e.g., at Three-Mile Island. In other situations, operators might switch their attention between one task and another, thereby not really solving the problem.

Mistakes are difficult to detect and are more subtle, complex, and dangerous than slips. Detection may depend on someone else intervening or unwanted consequences becoming apparent.

The final type of "error" is a *violation* or nonconformance/noncompliance. These are considered "intentional" because the operator deliberately carries out actions that contravene the organization's policies, rules, or safe operating procedures. However, violations often tend to be well intentioned, either because their objective is to complete a task or simplify it. Fortunately, violations are very rarely conducted in order to sabotage a process or a plant.

It is important to note that violations can occur due to both internal and external pressures on the operator (i.e., pressure to get a task done on time, which can come from the individual or it can come from perceived pressures from supervisors and managers). Violations can be routine, situational, optimizing, or exceptional.

Routine violations refer to when violations become the norm, i.e., what is normally done in the workplace, learned by others therefore becoming embedded as the way of doing things. They are usually shortcuts taken to get the job done more quickly, more efficiently, or more easily. Unless these types of violation are monitored and controlled, the organization can develop a culture that tolerates violations. Ways of counteracting routine violations are through good supervision, proper training (with explanations of *why* certain procedures are in place), good procedures and work practices, and, as a final measure, behavioral safety programs, which coach to reinforce the correct behavior and challenge and change incorrect behavior.

Situational violations tend to occur when there is a gap between what the rules or procedures say should be done, and what is actually available in order to get the job done. For example, a lack of trained and competent staff to conduct a task or lack of procedural clarity can lead to a situational violation. This can often occur when management are unaware of what resources are required, when procedures are out of date, or when there has been a cost-cutting program and staff have been made redundant, thus leading to reduced manning levels. Obviously, the same measures that are applied to manage routine violations can be used to manage situational violations.

Optimizing violations refer to when individuals carry out an activity for personal gain or simply for "kicks," e.g., seeing how far they can go by testing the system to its limits. However, organizations often provide incentives such as bonuses for meeting production targets, which can encourage "organizational" optimizing violations. If brought out into the open through incident/near miss reporting programs and good communication up and down the organizational hierarchy, these types of violation can help to identify measures that can be taken to improve both production and safety within the organization.

Exceptional violations occur when there is an unusual or unanticipated situation where no rules or procedures apply or where the rules/procedures cannot be applied. Perhaps the most poignant example of an exceptional violation can be seen during the Piper Alpha disaster, where personnel had been told that jumping off the platform into the sea was not survivable. In reality, many of those who jumped into the sea that night did survive unlike those who stayed on the platform and made their way to the muster point in the accommodation block as the procedures and their training dictated where they perished.

This example serves to remind us that human beings do not always follow rules and procedures blindly but are capable of interpreting and adapting to situations and solving problems in situ. This inherent behavioral flexibility and adaptability is what keeps us safe in a complex and constantly changing environment, an issue that we will return to later in the chapter.

Human error can be seen as a *consequence* or *outcome* of our performance limitations, and we regularly make errors on an everyday basis. Even highly trained and competent individuals working in control rooms and on maintenance tasks within the chemical process industries are error-prone, but fortunately, most errors are captured before they develop into something more serious. There are a number of challenges facing the human factors specialist. One challenge is to design plant, equipment, and systems, which make the chance of human error very low or As Low As Reasonably Practicable (ALARP), i.e., at a point where the costs are not so prohibitive that the benefits are seriously undermined. Another challenge is to identify where human error is most likely to occur, what type of error might occur, and what is the likelihood of it occurring. A third challenge is to measure human error and identify what factors make it more likely to occur (the so-called PIFs). Finally, once error has occurred, it would be of interest how these errors are identified and recovered from.

The remainder of this chapter covers how these many challenges can be identified and overcome. Due to the diverse and extensive nature of the human factors subject matter, not all of the factors identified by the HSE or other international bodies will be reported here. Those areas in which the author has relevant knowledge and expertise will be the main focus of attention. As a result, most of the discussion will be centered on human performance issues and will cover individual and organizational factors, since these are the areas in which the author has conducted most of her research. Ergonomics and Human Factors Engineering, particularly regarding workplace design, human–machine interaction, and characteristics of the working environment, are not areas that the author is familiar with and therefore will only be discussed in passing. Nevertheless, it is worth noting that "Designing for Humans" (Noyes, 2001) is an important exercise and can potentially prevent the development of many of the human performance issues that will be discussed in this chapter.

2.2 Measuring and Managing Human Error

In a paper prepared for presentation at the Institute of Chemical Engineers (IChemE) XX Hazards symposium in April 2008, Visscher of the US Chemical Safety Board presented summaries of investigations of some of the 50 chemical-related incidents in the United States since 1998. In his conclusions, he notes that many of the incidents occurred in facilities where chemicals are stored and used for other purposes rather than at chemical processing companies per se. He also observes that the controls and safeguards that rely on *human judgment and reliability* are revealed as a particular area of vulnerability and that management should focus on these issues in their operations. Furthermore, Visscher reports that the major accidents that have occurred at the larger companies have highlighted the important role of *corporate leadership and oversight* in assuring process safety integrity. Corporate leadership and oversight and the important role they play in maintaining safety are explored later on in this chapter.

The chemical process industries use a number of techniques for measuring and managing human error/human failure. The Center for Chemical Process Safety (CCPS) (1994) has produced a set of guidelines for preventing human error in process safety, to which the interested reader is referred.

This section will only deal with a few of these techniques, but it is also to remember that it is difficult to estimate when and where human failure is likely to occur and the most effective way to prevent errors is to focus on managing the PIFs in general. Thorough and detailed incident investigation, with a strong focus on human and organisational factors, is important to identify which PIFs should be focused on in order to prevent further incidents. In the author's experience the following PIFs are the most commonly identified contributory factors in major accidents: poor design, poor procedures, lack of supervision, fatigue, poor safety climate/culture and lack of safety leadership, and underinvestment by senior management in safety improvements.

2.3 Human Reliability Analysis

Human factors specialists attempt to measure the likelihood of human error in predetermined situations through HRA and Human Error Probability (HEP). HRA techniques are used to support the minimization of risks associated with human failure. They are both quantitative (e.g., HEP) and qualitative (e.g., safety critical task analysis (SCTA)) in nature; however the application of quantitative techniques can be difficult due to the fact that HEPs in particular are often used without sufficient justification. In particular, new processes and new technologies will not have sufficient data available to generate HEPs. SCTA (see Energy Institute, 2011) considers the impact that PIFs can have on the likelihood of error, which will be affected by job and organizational factors such as design of plant and equipment, the quality of procedures, and the time available to get the job done. Using HEPs without task context can therefore lead to inaccuracies in analysis.

Henderson and Embrey (2012) produced guidance for the Energy Institute on Quantified Human Reliability Analysis (QHRA) (see Energy Institute, 2012), in order to reduce instances of poorly developed or executed analyses. These authors recommend an eight-stage Generic HRA process as follows:

1. Preparation and problem definition
2. Task analysis
3. Failure identification
4. Modeling
5. Quantification
6. Impact assessment
7. Failure reduction
8. Review

Bow-tie diagrams have become a popular way of illustrating how initiating and response failures can occur. For example, Henderson and Embrey (2012) use the following figure to show how different human failures can affect the initiation and mitigation and escalation of a hypothetical event (see Fig. 1).

From a practical point of view, a number of factors can undermine the validity of an HRA. As a starting point, the analyst will need a thorough understanding of the task and the environment it is conducted in. Therefore, the input of skilled and experienced operators will be required. A walk-through of the tasks and subtasks involved in the activity at the location and/or a detailed talk-through of the tasks and subtasks in a task analysis

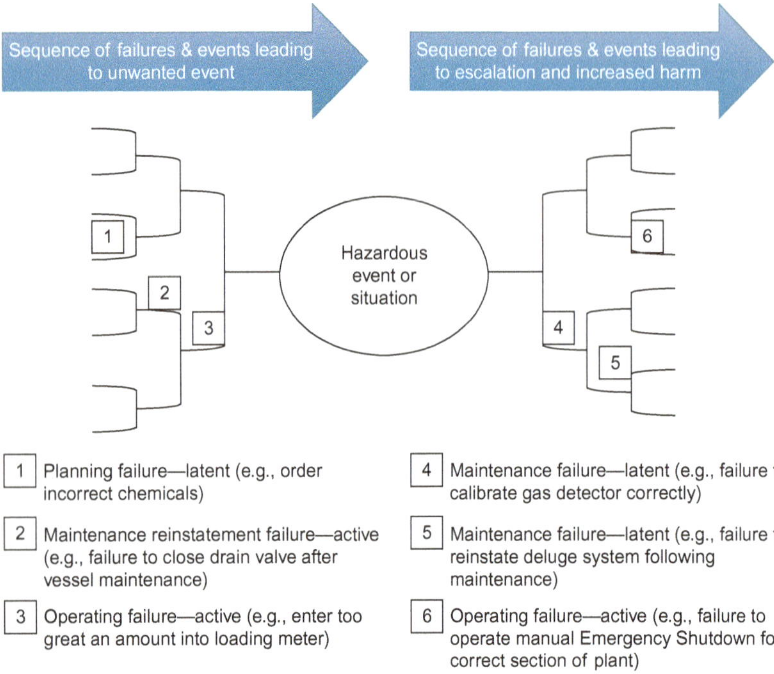

Fig. 1 Examples of the potential impact of human failures on an event sequence. *From Henderson, J., & Embrey, D. (2012). Quantifying human reliability in risk assessments.* Petroleum Review.

workshop is necessary. It is important to understand which PIFs might be exerting an influence in the actual working situation. Also if any HEPs have been imported (usually from a general database if such a database exists), the rationale from including these HEPs must be clearly articulated. It is normal practice to use a set of guidewords to identify the potential failures.

There are a number of guides to HRA (see Health and Safety Executive, 2009; Kirwan, 1994). The HSE (2009) report RR679 provides a review of human reliability assessment methods. Out of a total of 72 human reliability tools identified, the report authors considered 35 to be potentially relevant to hazardous industries such as the chemical process industries. This list was then reduced to 17 tools, most of which had only been applied in the nuclear industry. Only five of these tools were considered to have "Generic" capability, although some of the nuclear tools were considered to have wider application.

The lack of space precludes detailed discussion of all 17 HRA tools covered in RR679, and the interested reader is advised to access the report (which is available on the Health and Safety Executive website, for further details). Two of the tools are discussed here based on the fact that they are generic or have the potential to be applied more widely than in the nuclear sector: Human Error Assessment and Reliability Technique (HEART) and Technique for Human Error Rate Prediction (THERP). In addition, HEART and THERP are two of the few HRA methods that have been empirically validated (see Kirwan, 1996; Kirwan, Kennedy, Taylor-Adams, & Lambert, 1997).

2.3.1 Human Error Assessment and Reliability Technique

Williams (1985, cited in HSE, 2009) is attributed as being the first to refer to HEART in a series of conference papers. According to the review of HRA methods in RR679, HEART has been applied across a number of high-hazard industries where human reliability is critical, including the chemical process industry. It is designed to be a relatively quick to apply and is easily understood by both human factors specialists and engineers. There are nine Generic Task Types (GTTs) described in HEART, each with an associated HEP and 38 Error Producing Conditions (EPCs). EPCs affect task reliability, each with a maximum amount by which the nominal HEP can be multiplied. The key stages of HEART are the following:

- classify the task for analysis into one of the nine GTTs;
- assign the nominal HEP to the task;
- identify which EPCs may affect task reliability;
- consider the proportion of effect for each EPC;
- calculate the task HEP.

There are a number of premises that have to be taken into consideration when the technique is applied: (1) human reliability will be dependent upon the task to be performed; (2) this level of human reliability will tend to be achieved with a given likelihood within probabilistic limits in perfect conditions; (3) since perfect conditions rarely exist, human reliability will degrade as a function of the extent to which EPCs apply. It should be noted that the total probability of failure should never be more than 1.00, so if the multiplication of factors goes above 1.00, the probability of failure can only ever be assumed to be 1.00.

2.3.2 Technique for Human Error Rate Prediction

Swain and Guttman (1983) developed THERP for the US Nuclear Regulatory Commission (NRC). According to Kirwan (1994), THERP is a total methodology for assessing human reliability and, like HEART, it has also been validated by Kirwan et al. (1997). The THERP handbook prepared by Swain and Guttman (1983) for the NRC presents methods, models, and estimated HEPs to allow analysts to make either quantitative or qualitative assessments of human errors in nuclear power plants. It includes task analyses, error identification, and quantification of HEPs. Although THERP was developed for and has been used extensively in the nuclear industry, it has also been applied to the offshore and medical sectors and would no doubt also have applications in the chemical process industries. RR679 (HSE, 2009) outlines the key steps for applying THERP as:

- Decomposition of tasks into elements
- Assignment of nominal HEPs to each element
- Determination of effects of PSF on each element
- Calculation of the effects of dependence between tasks
- Modeling in an HRA event tree
- Quantification of total task HEP

Estimating the overall probability of failure involves summing the probabilities of all failure paths in the event tree. Summing only the primary failure paths and ignoring all the success limbs, when all the HEPs are 0.01 or smaller, can give an approximation of failure, expressed as an equation.

In conclusion, the human factors community has developed a range of different techniques to assess human reliability across a range of high hazard industries. The HSE report RR679 presents a review of 17 of these methods out of the total 72 identified. Most of the HRA tools have been developed for the nuclear industry; however, five of these tools were considered to have "Generic" application having been used in other industries, including the chemical process industry. It is worth noting that there have been three generations of HRA tools and the selection of an appropriate tool may be dependent upon the maturity of the site where it is being applied. For sites that are just attempting to quantify the risk of human error for the first time, first-generation tools may be the most useful. They may not be able to give insight into dependency or errors of commission, but they will be able to give a fundamental insight into the issue of human error. HEART and THERP are examples of these first-generation tools. Second-generation tools are more appropriate for more "mature" sites, with a long tradition of applying the first-generation tools but that want to understand context and errors of commission when predicting human error. CREAM, ATHEANA, and MERMOS are examples of second-generation tools, but they do not seem to be widely applied and they have yet to be empirically validated.

Third-generation tools are now being developed using first-generation tools such as HEART as a basis. It is the current author's belief that if an organization does not have a full understanding of the context under which tasks are executed, there is little chance of accurately assessing human reliability and therefore the impact of PIFs should be the foremost consideration in the application of any tool.

2.4 Safety Critical Task Analysis

SCTA describes a process whereby the impact of potential human error on MAHs can be assessed. SCTA is an extension of Task Analysis, which is the study of the actions and mental processes an employee is expected to carry out in order to achieve a goal. Task Analysis can be used for activities such as assessing staff levels and improving training programs; however, it is not discussed in any detail in this chapter. For the interested reader, there are some excellent resources available, e.g., Kirwan and Ainsworth (1992) and Shepherd (2001). The purpose of SCTA is to identify where human failure can contribute to MAHs and can include initiating events, prevention and detection, control and mitigation, maintenance tasks, and emergency response. The process involves identifying which tasks on a site are safety critical in relation to MAHs, understanding whether human error or violations might contribute to initiating an adverse event, and understanding what preventative measures or layers of protection could be put in place to reduce the likelihood or mitigate the consequences of human failure.

A number of publications exist to support the implementation of SCTA. The Energy Institute (2011) has developed a clear set of guidance on SCTA and the UK Health and Safety Executive has produced a guidance paper for its inspectors (*identifying human failures, HSE Core Topic 3*), describing a seven-step approach for SCTA. This seven-step approach consists of the following:

1. Identify the main hazards—e.g., from HAZIDs, Safety Reports, or Risk Assessments.
2. Identify safety critical tasks associated with those main hazards and prioritize those tasks where there are many MAHs. Procedures and discussions with staff are the main techniques recommended.
3. Understand the safety critical tasks, i.e., who does what, when and in what sequence? Again this can be derived from procedures, checklists, interviews with staff (walkthroughs/talkthroughs), and observations of staff conducting the tasks.

4. Represent the safety critical tasks, i.e., through breakdown of the tasks in tables or diagrams in sufficient detail for further analysis.
5. List all the potential human failures and their consequences, through representing the safety critical tasks from step 4. Also, list the potential PIFs that could influence human performance for each task and consider what safety measures are already in place. The different types of human failure/error and PIFs used in SCTA are listed in Tables 2 and 3.
6. Identify any additional safety measures that could be implemented to further mitigate PIFs and the risk of human failure.
7. Review the effectiveness of the process contributes to a wider understanding of SCTA and improvements to the technique.

A common problem is that all tasks are considered to be "safety critical." Usually, it is the operational tasks that are the focus on the SCTA,

Table 2 Human Error Guidewords for Use in SCTA (Energy Institute, 2011)

Action failures	Checking failures
Operation omitted Operation incomplete Operation mistimed Operation in wrong direction Operation too long/short Operation too little/much Operation too fast/slow Operation too early/late Operation in wrong order Right operation on wrong object Wrong operation on right object Misalignment Misplacement	Check omitted Check incomplete Right check on wrong object Wrong check on right object Wrong check on wrong object Check mistimed
Retrieval failures	**Selection failures**
Information not obtained Wrong information obtained Information retrieval incomplete Information incorrectly interpreted	Selection omitted Wrong selection made
Communication failures	**Planning failures**
Information not communicated Wrong information communicated Information communication incomplete Information communication unclear	Plan omitted Plan incorrect

Table 3 PIFs for SCTA (Energy Institute, 2011)

Individual factors
Physical capability
Fatigue
Stress/morale
Work overload/underload
Competence
Motivation
Job factors
Clarity of signs, signals, instructions
System/equipment interface (labeling, alarms)
Difficulty/complexity of task
Routine or unusual task
Divided attention
Inadequate procedures
Task preparation (PTW, risk assessment, checklists)
Time available/required
Tools appropriate for the task
Working environment (noise, heat, space, lighting)
Organizational factors
Work pressure
Supervision/leadership
Communication
Staffing levels
Peer pressure
Clarity of roles and responsibilities
Consequences of failure to follow rules/procedures
Effectiveness of learning from incidents
Safety culture
Change management

Adapted from Health and Safety Executive (1999). *Reducing error and influencing behaviour (HSG48)*. Suffolk: HSE Books.

e.g., chemical offloading operations, control room operations, or blending chemicals; however, checking tasks, emergency response, and maintenance tasks might also be included.

The UK Health and Safety Executive has made SCTA a requirement for acceptance of its Control of Major Accident Hazards (COMAH) Safety Reports. Table 2 lists the human error guidewords suggested by the HSE.

3. PERFORMANCE INFLUENCING FACTORS

The management of human failure is dependent upon understanding and responding appropriately to PIFs. The list of factors identified by the HSE (1999) shown in Table 1 outlines some PIFs; however for the purposes of SCTA, more comprehensive lists should be used (see Table 3).

Having outlined techniques used to identify different types of human error, the remainder of the chapter discusses some of the PIFs identified under the HSE, HSG48 framework shown in Table 1.

4. INDIVIDUAL FACTORS

The HSE (1999) lists the individual factors that can affect performance such as competence, skills, personality, and various person attributes such as attitudes to risk and safety. Competence can be defined as "*A cluster of related abilities, commitments, knowledge, and skills that enable a person (or an organization) to act effectively in a job or situation*" (Business Dictionary, accessed October 2016). The same source defines "skills" as "*An ability or capacity acquired through deliberate, systematic and sustained effort to carry out complex activities or job functions involving ideas (cognitive skills), things (technical skills), and/or people (interpersonal skills).*" Personality refers to a "*relatively stable, consistent and distinctive set of mental and emotional characteristics a person exhibits when alone, or when interacting with people and his or her external environment.*" Within the context of the chemical process industry, attitudes to risk and safety would refer to a tendency to respond positively or negatively toward hazards, their likelihood, their consequences, and the measures in place to mitigate the realization of the risks inherent in those hazards (author's own definition). Hence, someone could hold a positive attitude toward risk and a negative attitude to safety or a negative attitude toward risk and a positive attitude toward safety. This section will now expand on each of these individual factors.

4.1 Competence and Skills

Competence and skills are clearly of prime importance; however, there can be some debate about who is responsible for developing skills and competencies and keeping them up to date. Depending on the nature of the job/tasks and the skills and competencies associated with it, initial responsibility may lie with the individual or with the schools, colleges, and universities that are responsible for ensuring the people receive the necessary education. No matter what the job or tasks entail, it is safe to assume that *basic numeracy and literacy skills* will be required. It is up to the organization to provide an accurate *job description* and outline the relevant *person characteristics* required to carry out the job so that the right people can be selected from the outset. In the author's own experience, these job descriptions and person characteristics are sometimes inadequate and the suspicion is that the Human Resources Department has been left to write it, without any input from skilled practitioners; therefore, there is a lack of understanding of the true nature of the job and the type of person required for it. In order to recruit and retain the right people for the job roles and tasks in the organization, it is crucial that the right people are involved in writing the job description and person characteristics.

If one assumes that the right person is selected for the right job, then the organization may need to deliver some "on-the-job" training so that the new recruit can learn the tasks that he or she will be required to carry out in context. Again, in the author's experience, a school-leaver or graduate can very rarely come straight into a job and conduct their activities with the requisite level of competency; development "in role" under the tutelage of an experienced supervisor or mentor will be critical.

Once the basic skills and knowledge are acquired, there is a need for a period of on-the-job supervision to determine whether the individual is competent to carry out the assigned tasks. However, competencies should be kept up to date for example, when there are changes to the job or when there are new developments in systems, technology, legislation, or equipment. Refresher training may also be required for some tasks. Employees in the chemical industry will be trained and competent in the *technical skills* they require for their job, e.g., control room operations, chemical engineering, or maintenance; however, there are a set of *nontechnical skills*, which are widely used in other high-hazard, high-reliability industries, but do not appear to be prevalent in the chemical process industries.

4.2 Nontechnical Skills

Nontechnical skills training began in the aviation industry as far back as the 1970s, after the industry identified that many aviation accidents did not occur due to technical problems with the aircraft or the pilots' lack of technical flying ability. Instead, flight data recorders and cockpit voice recorders identified nontechnical skills such as poor situation awareness, communication, teamwork, decision making, leadership, fatigue, and stress, as major contributors to aviation accidents. The industry therefore started to develop nontechnical skills training as part of Crew Resource Management (CRM), which has now become mandatory for all pilots and cabin crew. CRM refers to the flight crew's use of *all necessary resources* (systems, technology, equipment, and human) to ensure safe and efficient operation of the aircraft. Nontechnical skills are a critical part of CRM, particularly in the identification and recovery from errors (threat and error management). Nontechnical skills are generally divided into two subgroups: (1) cognitive skills (decision making, situational awareness) and (2) social skills (leadership, teamwork, and communication).

The use of simulators and Line-Oriented Flight Training (LOFT), which involves testing the crew's nontechnical skills, i.e., situation awareness, team working, decision making, in abnormal situations which have not been prebriefed, has facilitated this type of training over the decades. Test scenarios can be developed from various sources, but accident reports with a full emphasis on human factors issues are most often used. CRM and nontechnical skills are examined both in the simulator and in normal flight operations, and pilots have to keep these skills up to date in order to fly commercial aircraft. Pilots and cabin crew are also required to undertake regular refresher training.

Other sectors such as aircraft maintenance (Sian, Robertson, & Watson, 2016), maritime (STCW, 2010), nuclear (INPO, 1993), offshore production platforms (O'Connor & Flin, 2003), hospital operating theatres (Flin & Maran, 2004; Mitchell & Flin, 2008; Yule, Flin, Paterson-Brown, & Maran, 2006), and offshore well operations (Energy Institute, 2014; International Association of Oil and Gas Producers (OGP), 2014) have also adopted the principles of CRM and nontechnical skills training. It is worth noting that the development of guidance for CRM/nontechnical skills training in well operations has arisen directly as a result of the findings of the inquiries into the Deepwater Horizon disaster.

The content of CRM and associated nontechnical skills training will, out of necessity, reflect the industry in which the training is being designed and implemented (see Flin, O'Connor, & Mearns, 2002 for a review). Nonetheless, reference to the research papers and training development around these programs indicates that the subject areas of situation awareness, decision making, team working, leadership, and managing stress and fatigue are key components. Communication is considered to be the cornerstone of CRM and nontechnical skills training.

The author is not aware of CRM and nontechnical skills training in the onshore chemical process industries; however, it is likely that the components outlined earlier, particularly team working, leadership, and communication, are covered in some training programs for the industry. The importance and utility of CRM and nontechnical skills training is that it takes incidents and accidents as its starting point, and based on the detailed human factors analysis of those incidents, trains the skills that have been found to be lacking and have led to the incident occurring. It is therefore a way of "closing the loop" (Flin et al., 2002) to prevent the likelihood of human failure in the future. Through a focus on CRM and nontechnical skills, personnel can also be trained to recognize and trap errors before they develop into more series adverse events. The other interesting point about this type of training is that the aviation industry has developed and persisted with it, whereas other industries "chop and change" their training regimes to adopt whatever is fashionable at the time (the offshore oil and gas industry is a case in point). This consistency of approach means that these programs have had time to be integrated into the systems and processes of the industry and deliver the performance improvements they were designed to deliver.

Proper development of nontechnical skills training also requires sets of "behavioral markers" that can be used to identify whether employees are exhibiting the trained behaviors or not. Obviously, the development of behavioral markers will be industry specific and will require the combined skills of operational personnel (e.g., operators, engineers, maintenance staff) and human factors specialists. Performance can be assessed in training simulators if they are available or online, when the personnel will be assessed while actually conducting their tasks. Studies into the development of behavioral markers for nontechnical skills training in the aviation sector and hospital operating theaters include Flin and Martin (2001), Flin (2003), Fletcher et al. (2003), and Yule et al. (2008). CRM/nontechnical skills training courses will tend to consist of both theoretical and applied teaching.

In conclusion, nontechnical skills training provides a means by which human performance can be managed and improved. A number of safety critical industries have adopted this type of training, and it could be considered as a form of training for the chemical process industries.

4.3 The Role of Personality

The role of personality in determining safety performance has been well researched starting back in the 1950s with the idea of the "accident-prone" personality. Subsequent research and development has shown that so-called accident proneness may be a misnomer and again we find that a focus on PIFs is a more productive route to follow. Nonetheless, there are clearly individual differences in intelligence, aptitude, dexterity, skills, education, and motivation, and potential employees will tend to be selected for these characteristics rather than their personality, per se.

There is a vast literature on personality and a wide range of personality tests developed by psychologists over many decades. According to the American Psychological Association (APA, 2016), "personality" is defined as "*individual differences in characteristic patterns of thinking, feeling and behaving.*" The assumption is that "personality" is relatively stable and resistance to change and can be measured along a number of measurable "dimensions," "factors," or "traits." The biological definition of "traits" is characteristics or attributes of an organism as expressed by genes and/or by the environment. When it comes to personality, a combination of genetics and the environment will lead to a particular trait developing in an individual. Many of the personality tests developed by psychologists are designed to identify individuals, who are suffering from clinical psychological conditions such as anxiety, depression, dementia, obsessive–compulsive disorders, and schizophrenia. From an occupational or industrial/organizational perspective, a number of personality tests exist for selection and development purposes. These tests are designed to not just measure personality but also assess aptitudes such as problem solving, social interaction, and situation judgment. One of the best-known occupational personality tests is the Myers Briggs, which assesses employees' tendencies toward Introversion/Extraversion; Thinking/Feeling; Judging/Perceiving; and Intuition/Sensing. Results from these tests place people into one of a range of 16 different personality types, each of which has its own strengths and weaknesses. Although the Myers Briggs claims to identify 16 different personality types, most psychologists working in the area of personality testing nowadays recognize five main dimensions (i.e., the so-called Big 5, also known by the acronyms OCEAN or CANOE) with people scoring "high" to "low" on each trait:

- Openness to Experience—This personality trait reflects preferring variety in life, being attentive to inner feelings and having an active imagination, aesthetic sensitivity, and intellectual curiosity.
- Conscientiousness—Individuals who are conscientious tend to have high levels of self-discipline and are good at planning and striving to achieve long-term goals. They are often perceived as being responsible and reliable, but they can also be perfectionists and workaholics.

- Extraversion—Extraverts score highly on scales measuring sociability, assertiveness, energy, and talkativeness. By contrast, those scoring on the other end of the scale are Introverts, who enjoy spending time alone with their own thoughts and prefer solitary work and hobbies.
- Agreeableness—This trait speaks for itself. People who score highly tend to be warm, friendly, and tactful, with a positive opinion of others and are able to get along well with other people.
- Neuroticism—Neurotic individuals display the characteristics of being anxious, worried, moody, frustrated, or afraid. These characteristics are generally not considered to be conducive to good work performance; however, when it comes to risk and safety, neuroticism may have an important role to play.

The research evidence for the role of personality in industrial safety and accident involvement tends to be quite limited, with much of the literature focusing on driver personality and accident involvement rather than how personality traits affect attitudes to risk and safety in the workplace, particularly the chemical process industries. It is important to remember that any personality trait will not be exerting its influence in isolation but will also be subject to influence from the working environment (job factors) and the social environment (organizational factors) that the individual is exposed to.

4.3.1 Personality and Safety in the Workplace

A study conducted by Hansen (1989) investigated the relationship between accidents, biodata (e.g., age, job experience), cognitive factors, and personality in a sample of 362 chemical industry workers in the United States. The only personality trait investigated appears to have been neuroticism, in particular the social maladjustment and distractibility components of neuroses, both of which were found to have a relationship with accident involvement. Thus, individuals who were more socially maladjusted and prone to distractibility were more likely to have experienced a work accident.

Another study by Cellar, Yorke, Nelson, and Carroll (2004) examined the relationships between the Big 5 and workplace accidents in a sample of 202 undergraduate volunteers (134 women and 68 men). Clearly, this is not the best sample of "industrial workers" to focus on and the validity of this study could be challenged since it can be assumed that these students would probably not be working full time and may only have temporary jobs while completing their studies. The results showed that more Agreeable and Conscientious individuals were less likely to have been involved in accidents.

Clarke (2006a) investigated the role of perceptions of safety (known as "safety climate"; see later in this chapter), attitudes to safety (a personal expression of favor or disfavor toward risk and safety), and personality characteristics of traits on accident involvement in a wide variety of industries. The study took a "meta-analytic" approach, where relevant studies were identified from a literature search and the correlations between the variables of interest, i.e., safety perceptions, attitudes to safety, personality traits, and accident involvement, are analyzed. After an initially screening exercise against a set of criteria, only 19 studies out of total of 51 originally identified were included in the meta-analysis.

The results showed that negative safety perceptions were correlated most highly with accident involvement, followed by negative attitudes to safety and last of all, the personality characteristics with relatively low correlations. However, the personality trait "*Agreeableness*" was a better predictor of accident involvement than either safety attitudes or safety climate. Basically, people who demonstrate lower levels of agreeableness are more likely to be involved in accidents. This could be due to the fact that such individuals are less likely to be socialized into the norms of the organization and may also be less compliant with regulations, rules, procedures, and work instructions. Such individuals may be more likely to violate the rules and procedures that are in place to keep plant, people, and the environment safe. However, this is only conjecture on my part.

Clarke and Robertson (2008) demonstrated the role of agreeableness in accident involvement in another meta-analytic study. They identified 24 studies that had investigated the relationship between the "Big 5" Personality dimensions and self-reported accidents or personal injuries. The studies covered a wide range of occupations and nationalities, including Hansen's study on US chemical processing workers (production and maintenance) and UK personnel on offshore drilling rigs and production platforms. It should be noted that several of the studies were conducted on taxi or bus drivers from India, Turkey, South Africa, and the United States, thus covering the area of driver behavior rather than industrial safety. Although other personality dimensions were measured in some of the 24 studies, Clarke and Robertson (2008) only focused on the measurements of Openness, Conscientiousness, Extraversion, Agreeableness, and Neuroticism. Their meta-analytic study found that low Agreeableness, low Conscientiousness, and high levels of Openness and Neuroticism were all positively correlated with accident involvement, with little evidence of a relationship between Extraversion and accidents. In other words, people who were disagreeable and lacked conscientiousness were more likely to have been involved in accidents as were people who were more open to experience. This would seem to make sense because, as noted earlier, such individuals would have less tendency to comply with social norms and rules and regulations and a higher tendency to explore new ways of working, which again may not be compliant with the organization's way of doing things. However, somewhat surprisingly, people who were more neurotic also showed higher levels of accident involvement. This seems counterintuitive since one would have expected highly neurotic individuals to be risk averse and safety conscious. However, all personality dimensions except Agreeableness demonstrated evidence of being situation specific according to certain statistical controls used in the meta-analysis. This means that the effects of the personality characteristics were only evident in specific contexts, for example, a particular industry or occupation. For Agreeableness, the relationship with workplace accidents appears to be consistent across occupations.

There is one last study worth mentioning in the context of personality and accident involvement in a chemical process industry. This study was conducted in the UK offshore oil and gas industry (Sutherland & Cooper, 1991) and did not focus on the big 5 but instead took measurements of the Type A/Type B personality types, along with neuroticism and extraversion/introversion. Type As typically display behavior patterns such as competitiveness, hostility, and time pressure. These individuals are often categorized as "workaholics" and appear to live with high levels of stress in their lives. Type Bs, on the other hand, display more relaxed behavior, less competitiveness, and less hostility. Apart from questionnaire measurements of Type A/Type B personality, Sutherland and Cooper (1991) also used the questionnaire to measure neuroticism, extraversion/introversion, job satisfaction, self-reported stress, and accident involvement. The study involved 360 personnel working in the European offshore oil and gas industry on both production platforms and drilling rigs. These personnel had been surveyed previously, and this was a follow-up study 1 year later. The results showed that both the Type A behavior patterns and neuroticism were associated with increased accident involvement, lower job satisfaction, and higher levels of stress. Extraverts actually seemed to report more accidents than introverts, but Sutherland and Cooper found that introverts had also been involved in accidents, leading to personal injury; however, they were less likely to report them. The propensity to report may therefore be a characteristic of the extraverts. Type As were more prevalent in the sample of offshore workers used in this study and they also seem to be more characteristic of offshore workers in general (author's own observations and results from unpublished research), but Type Bs do exist offshore and there may be aspects of self-selection into offshore work and certain job roles, according to personality type. Not everyone can cope with the rigors of offshore life and long periods of isolation from family and friends. The role of stress was also examined in this study, and this topic will be covered in the next section.

In conclusion, the meta-analytic studies mentioned earlier indicate that only the personality trait of "Agreeableness" seems to have a consistent relationship with accident involvement across a wide range of industries, professions, and occupations with disagreeable personalities being more likely to have been involved in self-reported accidents. The other personality traits appear to exert their influence via associations with other factors such as safety climate, attitudes to safety, stress, and job satisfaction, which are factors that also exert an influence on the likelihood of being involved in an accident. It is therefore difficult to disentangle cause and effect here.

From the perspective of making interventions to improve safety, it is the author's personal belief that there is little an organization can do to change someone's personality; however, safety climate, levels of stress, job satisfaction, and attitudes to risk and safety can be managed and modified. Of course, organizations can select for particular personality types; for example, one could argue that only highly agreeable people should be selected to work for high-risk industries such as the chemical process industry. However, selecting for one particular personality trait may mean selecting out other personality traits that are beneficial for the organization in other ways. For example, the organization may also want personnel who are conscientious or competitive, or are quiet, thoughtful, analytic, and introverted. Therefore, it is suggested that personality is not an area an organization should focus on if it wants a viable, flexible, and competitive business. Nevertheless, it may be a factor to take into consideration when candidates are closely matched on a person specification for a job description and recruiters are having difficulty in making a recruitment selection. This may be particularly relevant when selecting candidates for a safety critical position in the chemical process industry.

4.4 Occupational Stress

No discussion of PIFs would be complete without reference to occupational stress, although where it fits within the individual, job, and organizational factors framework is a point of debate, since all these factors can be associated with the experience of occupational stress. There is an extensive literature on the subject and it is one factor that has been consistently related to accident involvement in a number of industries. Sutherland and Cooper's (1991) study on stress has already been mentioned, but there are many other studies, some of which have been conducted in the chemical process industries. The word "stress" is widely used, but occupational or job stress is specific to the workplace and arises from the conditions experienced there. Like so many human factors issues, stress will not exist in isolation from other PIFs and it can be debated as to whether occupational stress comes under the category of an individual, job, or organizational factors. In reality, it can span all three. For example, personality, lack of skills and competence, excessive workload, inadequate supervision, badly written procedures, and a poor safety climate impact on stress and therefore on the performance of both the individual and the organization. Of course, factors external to the work environment will also play a role in creating stress, for example, problems with work–life balance due to shift work or excessive overtime. *Fatigue* is often treated as a separate issue, but when considered within an occupational setting the author considers fatigue to be closely related to job stress, creating a vicious circle where the stress leads to tension, worry and lack of sleep, and the increase in fatigue makes it less easy to cope and therefore creates more stress. The techniques to manage both stress and fatigue are very similar and will be covered in the discussion later.

Occupational stress has been defined as the physical and psychological states that arise when an individual no longer has the resources to cope with the demands and pressures of the situation (see Michie, 2002 for a comprehensive review). It is considered to be the result of the interaction between the person and the environment. This is important to remember, since one person's "stress" may be another's "challenge"; therefore the feeling of being stressed and unable to cope is very much an individual-level phenomenon.

The causes of stress are manifold. They include factors that are intrinsic to the job such as work overload, time pressure, and poor working environment, e.g., noise, cramped conditions, lack of adequate lighting. The person's job role may also be an issue, such as role ambiguity and conflict; i.e., goals and expectations are not clear. Employees might have a poor relationship with their boss or with colleagues or they may feel that they are not being promoted quickly enough or are insecure about their job. Finally, the climate, culture, and structure of the organization may be a problem with a lack of participation in decision making, communication, and consultation.

We have already covered some of the factors that the individual will bring to the situation that contribute to feelings of being stressed. These include personality characteristics such as anxiety, neuroticism, and the Type A behavioral pattern. Factors external to the job such as family problems or life crises can be stressors, but research has shown that the main contributor to job stress tends to the organization itself and how people are being managed within that organization. Well-designed workplaces with clear roles, responsibilities, and expectations for their staff will tend to experience less work-related stress. Adequate training and competency programs, well-written procedures, an engaged workforce, and supportive supervision and management will also contribute to a reduction in stress. As is so often the case, addressing PIFs is the key to good safety performance.

The consequences of stress for the individual include physiological responses such as increased blood pressure and heart rate and behavioral responses such as increased smoking and drinking and either over- or under-eating. Needless to say, prolonged exposure to stress can have a long-term impact on both physical and mental health, e.g., coronary heart disease and depression. Researchers and medical professionals distinguish between acute and chronic stress. Acute stress is the immediate "flight or flight" response to a perceived threat, which leads to physiological changes in the body such as the release of adrenalin (associated with "butterflies in the stomach" and increased breathing rate). The body is then activated to deal with the threat either by fighting it or by running away, but in modern-day society, this is not necessarily a practical solution and so the individual starts to suffer from chronic stress, which is the prolonged exposure to stressors from the environment. It is this prolonged exposure that is considered to be dangerous and can lead to mental and physical consequences.

The consequences for the organization of chronic stress can include reduced quantity and quality of work, increased absenteeism and high turnover of staff, and reduced job satisfaction and morale. Ultimately, if employees are increasingly stressed and end up leaving, the organization's reputation might be damaged and it might find it harder to recruit personnel in the future. Of course, other potential negative consequences for both the individual and the organization are reduced levels of safety and increased accident involvement. There are many studies reporting the impact of stress on safety performance in the industrial context, although very few appear to have been conducted in the onshore chemical process industries. However as way of example, some of the studies conducted in the offshore oil and gas industry will be discussed here.

4.4.1 Occupational Stress in the Offshore Oil and Gas Industry

The offshore environment is an inhospitable place; installations are usually located far from land and are effectively chemical processing plants with accommodation built on top. For anyone who has ever visited an offshore installation, it will immediately become apparent that they are not built with humans in mind. Space is at a premium and stairways, doors, and gangways can be small and narrow. Parts of the plant can be difficult to access, and there appears to have been minimal ergonomic input into the design of installations. Due to their remote nature, workers are often required to spend up to 3 weeks offshore providing an added stressor of being remote from family and friends. Apart from being exposed to the risks arising from loss of containment and from flying in a helicopter to reach the installation, offshore workers are exposed to other psychosocial stressors such as noise, vibration, weather (cold, heat, wind, rain, snow), 12-h shifts, and sometimes dull monotonous work with little opportunity for developing new skills.

Sutherland and Cooperoper (1986, 1991, 1996) conducted some of the earliest studies investigating the relationship between occupational stress, mental health, and accidents in the offshore industry. Their studies included two samples of offshore personnel from the European offshore industry (mostly British): a sample of 190 in 1986 and a sample of 310 in 1991. A questionnaire for "stress auditing" was developed from interviews with offshore personnel, asking about aspects of their job, lifestyle, and accident involvement. Job satisfaction, psychological health, and social support were also measured. Analysis of the "stressor" questionnaire indicated 12 factors: career prospects and reward; safety and insecurity at work; home/work interface; under stimulation, i.e., low demand; physical conditions—working and living; unpredictability of work pattern; living conditions; physical climate and work; organization structure and climate; physical well-being; work overload; and transportation (e.g., flying in a helicopter). The top three causes of stress were pay and conditions, i.e., rate of pay, lack of paid holidays, and pay differentials between operating and contracting staff but relationships at home and at work were also of significance and these measures were correlated with job satisfaction and mental health. Accident victims reported less job satisfaction and poorer mental health, but it is difficult to ascertain cause and effect here since in relation to this chapter, the key issue is whether stress impacts on safety performance and leads to accidents.

Sutherland and Cooper (1986, 1991) raised the possibility of this relationship, and Rundmo (1992) suggested that stress could play an indirect role to play in accident causation among Norwegian offshore workers. However, in an extensive study of health and well-being in the UK offshore industry sponsored by the Health and Safety Executive, Mearns and Hope (2005) used the "Health Offshore" questionnaire to survey 1928 offshore workers on 31 installations to evaluate their perceptions of Health Climate and the measures that had been put in place to manage their health and well-being. In one section, respondents were asked to rate the extent to which they felt they were able to cope with any pressures experienced at work. Most respondents' felt they coped well and there did not seem to be a significant relationship between self-reported levels of stress and accident involvement. Overall, 21% of respondents indicated that they had received support (in the form of advice, information, guidance) to help them cope with the stress they experienced in the workplace. When the effectiveness of this support was rated, there was considerable variation between installations, but the general consensus was that current forms of support offered were only moderately useful. This is somewhat surprising given that 50% of the offshore medical staff who responded to the questionnaire reported that they were trained in stress management. Furthermore, 37% of the 31 installations involved in the study offered training courses on stress management for their workforce.

In conclusion, the research evidence suggests that it is the consequences of work-related stress, such as psychological and physical ill-health, fatigue, and the way that people are thinking, feeling that can cause human error or violations and this is what causes accidents rather than stress in itself. It is also worth noting that people suffering from the symptoms of stress may be self-medicating or receiving assistance from medical practitioners to alleviate those symptoms; thus the medication may impact on performance causing drowsiness, disturbed vision, and so on.

4.4.2 Managing Occupational Stress

The best way to manage occupational stress is to address the organizational issues that affect it such as improving employee awareness of stress; implementing regular stress assessments; developing a stress management policy and procedures; mitigating the impact of organizational change and job uncertainty on the workforce by good communications; and developing a positive health and safety culture which includes the reporting of psychological ill-health and considers the impact of stress and its associated symptoms when investigating accidents and incidents. Furthermore, as the next section will demonstrate, attending to job factors such as good design, procedures, and working environment will further mitigate the causes and consequences of occupational stress.

5. JOB FACTORS

Job factors include the nature of the task and the environment it is conducted in. This covers factors such as equipment, workload, procedures, and displays and controls. As for all the other sections of this chapter, it is not possible to focus in any detail on the entire list of job factors, so this section will only discuss a few. As a starting point, it is important to know that a raft of measures exist to ensure that equipment, control rooms, plant, and processes are designed in accordance with key ergonomics standards (see the Engineering Equipment & Material Users Association—EEMUA). For example, equipment should be designed in accordance with EN614 Parts 1 and 2 and control rooms should be designed in accordance with BS EN11064, EEMUA 191, and EEMUA 201. The reader is referred to the EEMUA website and these standards for further information. Other recommendations are that consideration should be given to the operators' body size, strength, and mental capability, and both plant and process should be designed to facilitate operation and maintenance. The design should take account of all phases of the plant life cycle, including decommissioning and all foreseeable operating conditions such as plant upsets and emergencies. Finally, consideration should be given to the interface between the end user and the system and one way to ensure this is to involve users in the design process. Users should include plant operators, control room operators (CROs), maintenance staff, and systems support personnel. Unfortunately, this involvement often seems to be overlooked in the design process. Cost also seems to play a role in the suboptimum design; however, it is important to note that the proper design with humans in mind can prevent accidents and incidents further down the line in the life history of the plant and equipment.

5.1 Design of Control Rooms

Control rooms provide an important safety critical barrier to MAHs in the chemical process industries; however, CROs can have a number of challenges to deal with. These challenges include having to deal with too many alarms simultaneously (alarm flooding), several safety critical tasks that have to be performed simultaneously (workload), communications equipment and display equipment positioned apart in the control room even when they need to be used together (design), and uneven workloads, with long periods of monitoring tasks interspersed with periods of high intensity when dealing with abnormal situations.

It is therefore recommended that due care is put into designing control rooms for normal, abnormal, and emergency situations. There are six main areas that require attention:

1. Layout
2. Working Environment
3. Control and Safety Systems
4. Job Organization
5. Procedures and Work Descriptions
6. Training and Competence

The Safety and Reliability Group at SINTEF (2004) developed the CRIOP methodology to contribute to "*verification and validation of the ability of a control center to safely and efficiently handle all modes of operation including start-up, normal operations, maintenance and revision maintenance, process disturbances, safety critical situations and shut down*" (p. 2). CRIOP is a registered trademark and stands for a scenario method for Crisis Intervention and Operability Analysis. Organizations such as Statoil and Norsk Hydro were involved in its development for the Norwegian offshore sector, and the methodology was based on interviews, user discussions, workshops, and contributions from experts.

e-Operations have been added as a seventh area in the 2004 version of CRIOP because remote control and remote operations have been a recent development for the Norwegian offshore industry. This is partly to reduce risk to personnel, since fewer people will be required to travel offshore to conduct the work. It also can be considered a cost-efficiency measure.

CRIOP works on the basis of checklists and scenarios. The checklists used in design and operations are based on a "best practice," including standards and guidelines such as ISO11064, EEMUA 191, and IEC 61508. The Norwegian Petroleum Directorate regulations have also been taken into account. The checklists have been laid out in a particular way for clear and easy usage. There is a checklist for each of the six areas, which comprises numbered points, e.g., 1, 1.1, 1.2, etc.; a Description, e.g., Are breaks planned/coordinated with control center tasks? Yes/No/Not Applicable; Reference to Documentation, e.g., ISO 11064-1; Comments and Recommendations and Responses.

The final stage of the CRIOP process is Scenario Analysis, which consists of Introduction, Planning, Participants, Duration, Group discussions, Documentation, Number of Scenarios, Organizational Learning, and

Framework. This is followed by the identification of actors who will be involved in the events making up the scenario. Observations are made of how the events of the scenario are enacted and interpreted including planning and decision making and action/execution. A checklist of PIFs (referred to as Sociotechnical factors in CRIOP) is used in the scenario analysis. Actions plans are developed from the outcomes of the scenario analysis with the implication that management implements these action plans.

In conclusion, CRIOP provides a method whereby the full complement of human factors issues can be addressed through a systemic, applied enactment of events within process industry-specific scenarios, particularly involving the all important control room at the center of operations.

5.2 Procedures

In the author's personal experience in over 25 years as an academic and consultant, frontline workers in the chemical process industries constantly raise the usability of procedures as an issue. This is particularly the case for maintenance personnel, who seem to be largely ignored when it comes to procedural usability. One of the main complaints raised in questionnaires and focus groups with frontline workers is that "engineers," who never visit the worksite to understand the context in which those procedures are to be used, write the procedures. Furthermore, work as "planned" rarely fits exactly with work as "executed," with the exception of the simplest of tasks. This means that frontline workers constantly have to adapt and adjust the procedures to use in order to achieve their work objectives. When this happens, the new way of working can gradually become the "norm" and drift away from the original intent of the procedure. New recruits observe, copy, and conduct their work in the new way, thereby being "normalized" by the more experienced workers, who may have forgotten what the correct procedure is. Another problem with procedures is that they can grow "arms and legs" and become lengthy and unwieldy. This often occurs in response to accidents and incidents, where a new section (or sections) is incorporated to prevent such an incident happening again.

It is important to clarify what is meant by the word "procedures." This term can mean different things to different people, depending on their role in the organization. The Business dictionary (http://www.businessdictionary.com) refers to a *procedure* as "*A fixed, step-by-step sequence of activities or course of action (with definite start and end points) that must be followed in the same order to correctly perform a task.*" Repetitive procedures are called routines, but there are also *method statements/specifications* and *work instructions*. A method specification is described as a "*Statement of requirements that prescribes a method of achieving a desired standard, instead of prescribing the standard itself.*" Method statements are likely to be used by designers, engineers, or managers. A work instruction is "*A description of the specific tasks and activities within an organization. A work instruction in a business will generally outline all of the different jobs needed for the operation of the firm in great detail and is a key element to running a business smoothly.*" In the author's experience, frontline workers are usually operating according to work instructions but often with reference to higher order procedures. In addition, they have to comply with *rules* and *regulations*. A rule is defined as "*an authoritative statement of what to do or not to do in a specific situation, issued by an appropriate person or body. It clarifies, demarcates, or interprets a law or policy.*" A regulation is "*a principle or law (with or without the coercive power of law) employed in controlling, directing or managing an activity, organization or system.*" It goes without saying that rules and regulations are the province of managers and leaders, but the expectation is that they are understood and followed throughout the whole organization.

Frontline workers will be expected to work according to a set of procedures or work instructions. The procedures might be generalized to a number of different tasks and situations, but the work instructions will be specific to the tasks being conducted on a day-to-day basis. Work instructions tend to be relatively clear and straightforward, but procedures can often be seen as lengthy and ambiguous. In order to help with the process of writing clear and concise procedures the UK Health and Safety Executive has produced guidance to preparing procedures in their Human Factors Briefing Note No. 4. The Major Accident Prevention Policy (MAPP) should describe how procedures should be developed, reviewed, revised, and publicized. In particular, COMAH sites should have a procedure for writing procedures, which should cover which tasks require procedures, how detailed those procedures should be, how to keep them up to date, and how to ensure compliance with procedures. It is also important that procedures are reviewed on a regular basis through consultation with users, walkthroughs of the procedures during actual work tasks, and identification of "informal" procedures or "workarounds." Analysis of incidents in which noncompliance with procedures has been identified as a causal factor is also recommended. Table 4 outlines a procedural checklist developed by the HSE.

The HSE guidance recommends using task analysis to fully understand how a job is actually done, including the identification of hazards and

Table 4 Procedures Checklist from HSE Human Factors Briefing Note 4

Procedures should be	Examples
Always easy to find, particularly for:	Operational, commissioning, maintenance, abnormal/emergency tasks
Are completely up to date	Involving users should help to ensure this
Set out in logical steps	Starts with general instructions and works down to specifics
Very easy to read	Use words people understand Use diagrams, pictures, flowcharts, checklists Size, color, style of lettering, and illustrations are clear
Are accurate	There are no inconsistencies and inaccuracies in the content
Highlight steps where care is required	For example, when a particular hazard is present
Describes items of special equipment	For example, tools and clothing
In good condition	Not dirty, torn or with pieces missing
Used to train people to do the job	This ensures compliance with the correct procedure
Changed quickly if the job changes	Should be considered under management of change
Consistent with other information	For example, verbal instructions from supervisors
Supplemented by other job aids	Pocket-sized checklists, reference materials

whether a procedure is the best way of controlling the hazard. As noted earlier, it is important to involve the people who actually do the task in writing the procedure. They will have the most realistic view of how the task is done (as compared to managers who may have a view on how the task should be done). Those who use the procedures can also advise on how and why the procedure might not be complied with (i.e., anticipate potential violations) and can also advise on the best wording and style of layout to use. Finally, involving the frontline workforce in developing procedures ensures ownership of the procedure and therefore more motivation and commitment to actually using the procedure in the way it is intended.

Other advice covers training in the procedure, including the training of contractors. This will ensure workers are familiar with the content and can point out any errors or impracticalities. It is important that contractors are familiar with the terminology used and that the needs of novice users are taken into account. Finally, make sure that procedures can be found quickly and easily on the organizational system.

Proper management of procedures is also important. Keep checking that they are being used properly and if not, find out why. Workers may have found an easier way to complete the task, but there new method might entail some inherent risk. On the other hand, the new way of doing things might be both more efficient and safer. Make sure that there is a system for capturing problems and that any problems are dealt with as quickly as possible. If a problem cannot be dealt with efficiently, then an explanation is required as to why.

Managers should plan for any changes to the task due to changes in equipment or the methods used. If it is not possible to change the procedure quickly and get people trained up in it, then temporary working instructions or extra supervision may be required. Procedures should be controlled. This often occurs by an instruction that the procedure is uncontrolled if printed (for example, only procedures accessed from the company website are controlled). Finally, a log should be kept of who is responsible for the procedure and any out-of-date procedures should be disposed of.

6. ORGANIZATIONAL FACTORS

At the organizational level culture, leadership, shift and rotation patterns, staffing levels, and communications are key factors. It could be argued that many individual and job factors originate at the organizational level, and if human factors is not integrated into the safety management system (SMS), it is unlikely that the SMS will achieve its objectives. The integration of HF into the SMS will lead to a positive safety culture, where the processes and practices as outlined in the SMS become recognized good practice.

6.1 Safety Climate and Safety Culture

Safety climate and safety culture deserve a lengthy section in this chapter, reflecting both their complexity and the academic and practitioner interest shown in these concepts. They are particularly important because public inquiries into most major accidents involving highly hazardous materials have found a distinct lack of safety culture (or climate) to be one of the underlying factors contributing to the accident.

There are numerous research articles and book chapters available, describing how to measure both safety culture and safety climate and demonstrating their relationship with safety performance. Although used interchangeably, the history and etiology of the two concepts is very different and ultimately the concepts reflect two different but overlapping aspects of organizational safety in high-hazard industries such as the chemical industry.

6.1.1 Safety Climate

Professor Dov Zohar is credited with publishing the first study on safety climate in *Journal of Applied Psychology* in 1980. According to Zohar (1980), safety climate represents a summary of the molar perceptions that workers have about how safety is managed in their working environment. He developed a 40-item questionnaire measuring 8 safety climate factors in the Israeli manufacturing industry. These factors were:

- importance of safety training;
- effects of required work pace on safety;
- status of the safety committee;
- status of the safety officer;
- perceived level of risk at the work place;
- management attitudes toward safety;
- effect of safe conduct on promotion; and
- effect of safe conduct on social status.

The technique Zohar used to analyze the responses to his questionnaire is referred to as Exploratory Factor Analysis (EFA), which uses a mathematical approach to reduce a number of measurable and observable variables, i.e., responses on a 40-item questionnaire measuring perceptions of safety, to fewer number of "latent variables," i.e., 8 factors, which are unobservable. These 8 factors are therefore hypothetical constructs used to represent the measurable variables. Effectively, the EFA attempts to find out which items on the questionnaire "go together" to form the fewest number of "factors" possible. If questionnaire items share a "commonality" (determined largely by correlational techniques), it makes sense to progress research and development on a few factors rather than on 40 questionnaire items.

In order to determine whether the "factors" measure what they are supposed to measure, they are often validated against another measure of safety, e.g., accidents or near misses. In Zohar's study, the safety climate measurement was validated against inspectors' evaluations of the safety performance in the various factories, where the climate survey was deployed.

Brown and Holmes (1986) tried to replicate this factor structure using confirmatory factors analysis in a US sample (this statistical technique sets up a predetermined theoretical "model," which the questionnaire data are tested against using similar statistical principles to the EFA—in this case an eight-factor model), but found a three-factor model instead which they labeled; *employees' perceptions of management concern about their well-being, management activity in responding to problems with well-being, and their own physical risk.* Further research on measuring safety climate appeared to become conflated with the measurement of "safety culture" (probably due to the emergence of the concept of safety culture following the Chernobyl nuclear disaster in 1986), until the early 1990s when Dedobbeleer and Beland (1991) investigated safety climate in the construction sector and tried to replicate Brown and Holmes' factor structure. Instead, they found that construction safety climate was best represented by two factors: *management commitment and workforce involvement.* The literature then became studded with a plethora of studies on safety climate in various high-hazard industries such as offshore oil and gas (Cox & Cheyne, 2000; Mearns, Flin, Gordon, & Fleming, 1998—although they refer to safety culture in offshore environments), construction (Poussette, Larsson, & Törner, 2008), road administration (Niskanen, 1994), the chemical industry (Berg, Shahriari, & Kines, 2013; Donald & Canter, 1994; Vinodkumar & Bhasi, 2009), and in nuclear reprocessing (Lee, 1998) and nuclear power plants (Lee & Harrison, 2000)—again these two nuclear studies refer to "safety culture" no doubt reflecting the introduction of the term following the Chernobyl disaster. Each study found that the statements used in their questionnaires loaded on slightly different factor structures, possibly reflecting the management structure and arrangements and underlying "professional culture" of the organizations and industries being targeted. It is also the case that questionnaires used in the various studies are developed from first principles with the researchers developing their own question sets which will differ from study to study. It is therefore not surprising that different factor structures emerge, since different "measurable" questionnaire items are used to measure the underlying, latent constructs.

In many ways, one would not expect "safety climate" to remain a static phenomenon as lessons are learned from incidents and there are new developments in legislation, technology, equipment, workforce training and competency, leadership and management for safety, etc. Therefore, over time, different factors may develop or become more relevant. Flin, Mearns, O'Connor, and Bryden (2000) reviewed a number of safety climate studies and concluded that the main themes in safety climate measurement tools, appearing in two-thirds of questionnaires available at that time, were related to *management, safety systems, and risk.* A more recent cross-validation of safety climate scales using confirmatory factor analysis (Seo, Torabi, Blair, & Ellis, 2004) found that safety climate grouped around five key themes: *management commitment to safety, supervisor support for safety, coworker support for safety, employee participation, and training and competence.* It is the current author's belief that this makes empirical sense, since the safety climate is supposed to measure how safety is managed by the organization, site or plant, depending on the level being targeted by the study.

One aspect that is strongly supported by safety climate research is the notion that climate is a "group phenomenon," i.e., the perceptions are shared among members of the workforce (see Zohar, 2010). This means that members of the workforce from the same location or within the same work group, e.g., operations, maintenance, show significant consensus about their safety perceptions, compared to their attitudes to safety (Poussette et al., 2008). Indeed, the prevailing *atmosphere* with regard to risk and at the place of work is critical in defining the characteristics of a safety climate. This is demonstrated most clearly by the work of Mearns et al. (1998), who demonstrated that offshore installations operated by the same oil and gas company, e.g., BP, Total, Conoco, have very different "safety climates" as reflected by workforce perceptions of supervisor and management commitment to safety.

6.1.2 Safety Climate in the Chemical Industries

From the perspective of the chemical process industries, each study seems to use its own safety climate measurement tool, making direct comparisons between studies difficult. Donald and Canter (1994) developed a questionnaire to measure safety climate and safety attitudes in UK chemical plants. Their study established the validity and reliability of their questionnaire and found a relationship between the safety climate and occupational injuries. It has not been possible for the current author to access the full paper and therefore the findings reported here only reflect the content of the abstract.

Vinodkumar and Bhasi (2009) developed a new questionnaire to measure safety climate in the chemical industry in India. This questionnaire was based on existing measures but was designed to better reflect the Indian "national culture." The questionnaire as then tested on 2536 employees (workers, firstline supervisors, and managers) across 8 chemical industrial units in Kerala, India. The results indicated eight factors:

- management commitment and actions for safety;
- workers knowledge and compliance with safety measures;
- workers attitudes toward safety;
- workers participation and commitment to safety;
- safeness of work environment;
- emergency preparedness;
- priority for safety over production; and
- risk justification.

These eight factors are similar to those identified in other industries and in other countries. Also, the safety climate scores differed between the chemical companies and across the different levels of hierarchy with workers being the least positive in their safety climate perceptions and management being the most positive. Finally, the more positive safety climates were correlated with lower self-reported accidents, a finding which has also been corroborated in other safety climate studies.

Berg et al. (2013) used the Nordic Occupational Safety Climate Questionnaire (NOSACQ-50) to survey workers at two chemical plants in Sweden. The NOSACQ-50 consists of 50 items measuring perceptions of: management commitment and priority to safety; worker commitment and priority to safety; safety empowerment for the workforce, safety justice, safety communication and trust in safety systems. In common with other safety climate studies in other industries, the workforce had more negative perceptions of safety climate than supervisors and management, although perceptions of the climate were generally positive. Furthermore, on both plants, shift workers had significantly lower scores on all the safety climate scales than daytime workers.

In conclusion, these safety climate studies indicate a certain level of consistency with regard to the emerging factors. Management commitment and the priority given to safety seem to predominate, with worker engagement and commitment, supervisor support, understanding the risks and safety systems also prevalent. The importance of management and leadership cannot be underestimated and this chapter devotes a separate section to the subject.

The next section deals with the concept of safety culture and how it differs from safety climate.

6.1.3 Safety Culture

The term "safety culture" seems to have first been used in relation to the 1986 Chernobyl nuclear disaster. Both the International Atomic Energy Authority and the OECD Nuclear Agency identified a "poor safety culture" in the former Soviet Union nuclear industry as a contributory factor in the accident. In the wake of Chernobyl, the UK Advisory Committee on the Safety of Nuclear Installations (HSC, 1993) developed what has become one of the most cited definitions of the concept, i.e.:

> *The safety culture of an organization is the product of individual and group values, attitudes, perceptions, competencies and patterns of behaviour that determine the commitment to, and the style and proficiency of, an organization's health and safety management.*
>
> ***HSC (1993, p. 23)***

Despite its widespread use, on first appearance this definition appears to be too general and not theoretically or empirically grounded; however; Cooper (2000) defends this wide-ranging definition by pointing out that the reciprocal nature of interactions between individuals, groups, situations; and behavior is the essence of a safety culture. Moreover, the HSC report makes the valid point that a safety culture is "more than a sum of its parts" and is therefore challenging to both measure and manage. Ultimately, the goal must be to demonstrate a focus on safety through both management commitment and workforce engagement, both reflecting a "top-down" and a "bottom-up" approach, rather than reflecting one or the other group's perspective.

It is interesting to note that the concept of "safety culture," invoked in response to this disaster, was then mentioned in Public Inquiries into other major accidents throughout the 1980s and 1990s, e.g., the Herald of Free Enterprise ferry sinking (1987), the Ladbroke Grove rail crash (1999), and the destruction of the Piper Alpha offshore installation (1988). The term has continued to be applied throughout the 2000s and 2010s, e.g., Columbia Space Shuttle explosion (2003), Texas City refinery explosion (2005), the loss of the RAF Nimrod MR2 Aircraft in Afghanistan (2006), Deepwater Horizon (2010), and the Fukushima Daichii nuclear disaster (2011). As Mearns, Whitaker, and Flin (2003) note, safety culture was a concept originally used to describe the inadequacies of safety management that resulted in major disasters; however, the concept is now being applied to explain accidents at the individual level, despite the fact that the validity of safety culture with regard to individual accidents is yet to be established. However, the validity of safety climate in relation to occupational accidents and injuries has been clearly demonstrated in a number of studies (e.g., Christian, Bradley, Wallace, & Burke, 2009; Clarke, 2006b; Mearns et al., 2003).

6.1.4 Theoretical Approaches to Safety Culture

Reason (1998), Pidgeon and O'Leary (2000), and Guldenmund (2000), among others, have provided theoretical perspectives on safety culture. For example, Reason (1998) proposed that safety culture consisted of *being informed* about the organization's state of safety, which is dependent in turn on having a *reporting culture* and learning from these reports. However, people will only report errors and incidents if they are dealt with in a *just and fair way*, i.e., the organization operates a *just culture*. Finally, Reason believed that *being flexible* was also important. This means having the ability to reconfigure in the face of high tempo operations or certain kinds of danger, usually by moving from a typical organizational hierarchical structure to a flatter structure where people with the right level of expertise make the assessments and take the decisions.

Reason believed that it should be possible to "engineer a safety culture" by identifying and fabricating these essential components and building them into a working whole; however, a perusal of some of the other theoretical musings about the nature of culture seems to indicate that this may be easier said than done. The simple fact is that an organization does not just have one single "safety culture," and it is made up of "subcultures" of personnel from different occupational and professional disciplines. Measuring and managing a safety culture involves understanding those subcultures and the interaction between them.

Pidgeon and O'Leary (2000) argued a "good" safety culture might be promoted by four factors: senior management commitment to safety; realistic and flexible customs and practices for handling both well-defined and ill-defined hazards; continuous organizational learning through practices such as feedback systems, monitoring, and analysis; and shared care and concern for hazards across the entire workforce. This closely resembles both definitions of safety climate and culture, but, in the author's opinion, reflects more deeply seated beliefs and assumptions about the nature of risk and safety and how they are being managed.

An increasing number of organizations are starting to understand the importance of measuring safety culture as "leading indicators" of the organization's state of safety; however, my own personal conclusion is that the lack of a culture is more likely to be associated with major process incidents rather than occupational injuries such as slips, trips, and falls.

Seo et al. (2004) suggest that climate is more of a state (or mood), provides a "snapshot" of the organization's safety, and is measured "quantitatively" via a questionnaire. They suggest that culture is a stable, deeply held trait, more akin to personality and is measured via "qualitative" methods such as interviews and focus groups. This is the definition and the measurement methods that the current author subscribes to although other definitions and methods for measuring safety culture exist.

6.1.5 Conclusion

In conclusion, both the theoretical and empirical evidence suggest that safety climate and safety culture are distinct but overlapping concepts. Safety climate refers to the shared perceptions of the "state of safety" at a place of work, largely inferred from how supervisors and management behave in relation to the management of risk and safety, i.e., their demonstration of safety commitment. Safety climate can therefore be considered to reflect a particular "atmosphere" at the workplace and has mostly been studied in relation to its impact on occupational accidents and injuries. Given its location-specific focus, it is possible that the key to good safety climate is heavily dependent on how supervisors' commitment to safety is perceived by the workforce. Nevertheless, proper supervision can only be achieved if more senior management commits time and resources to the supervisors in order to meet their safety commitments.

Safety culture is considered to be a more deeply rooted concept based on the values, beliefs, attitudes, and behavior of organizational members and how this impacts on the safety performance of the organization and the safety systems developed to protect that organization. The notion of a "lack of safety culture" tends to be invoked after major accidents, where government organizations and the general public at large consider it almost inconceivable that any high-hazard industry should be so remiss in its approach to safety, as to allow such adverse events to occur. Increasingly, the research evidence seems to suggest that senior management oversight and commitment to safety is the key factor influencing safety culture. This very much subscribes to a "top-down" approach to safety culture largely influenced by the psychology and management disciplines. Another body of thought, mostly derived from the sociology and social anthropology traditions, considers safety culture from the "bottom up," i.e., that the workforce and the groups that make up the workforce develop their own "cultures," which may keep them "safe" despite poor management oversight and inadequate of safety management processes and procedures. The interested reader is referred to a Special Issue of Safety Science published in 2000 for further discussion.

Workforce engagement and involvement is an important component of safety culture, but again, the true value of that engagement can only come about if management is willing to relinquish some power and control in order for the workforce to contribute to SMSs and processes. For many years, the focus in the chemical process and other high-hazard industries has been on "human error" and the mistaken belief that frontline workers are to "blame" for those errors. However, Reason (1997) has pointed out that "*Rather than being the main instigators of an accident, operators tend to be the inheritors of system defects created by poor design, incorrect installation, faulty maintenance and bad management decisions. Their part is usually that of adding the final garnish to a lethal brew whose ingredients have already been long in the cooking.*" This therefore leads on to the final section of this chapter, which will focus on leadership and management for safety.

6.2 Leadership and Management for Safety

There has long been a focus on what constitutes effective leadership and management in organizational settings, with a multitude of books, chapters, research articles, consultancy reports, training programs, and theories espousing and promoting the value of good leadership and management skills to ensure the success of an organization. While the general area of leadership and management has been well researched and reported, there has been less focus on the skills, knowledge, and attitudes that leaders and managers require to "lead and manage" for safety. The argument could apply that it would be same set of attributes that are required for good general leadership and management; however, there is something different about the motivational drivers for safety performance compared to other organizational performance metric such as balance sheets, turnover, and productivity. The one major difference (in this author's opinion) is that with safety "nothing happens." Safe performance is the expectation and the norm. "Unsafe" performance is usually presented as occupational accident statistics that have an impact on individual lives but less of an impact on the overall organization and many lives. Major process incidents can lead to multiple fatalities of both employees and the general public and considerable damage to the plant and environment, e.g., Bhopal, Flixborough, Texas City. As has been clearly demonstrated for accidents such as Texas City and Deepwater Horizon, reductions in lost-time injuries are not a reliable indicator of how well major hazards are being managed. The organization must be set up to prevent all forms of harm and the driving force for this state of affairs must be senior management, who set the overall direction and ethos for the organization. The key role of senior management in setting and promoting the safety culture of the organization has been demonstrated in most of the research on the subject (Health and Safety Executive, 2003; International Association of Oil and gas producers, 2013; Prior, 2003).

The following sections consider the role of safety leadership at all levels of the management chain from supervisors and team leaders, through middle managers to CEO and Board level. Leadership and management are closely related, with managers planning, organizing, and maintaining the structure and systems of the organization and leaders inspiring and motivating the workforce to achieve organizational goals (of which safety is one of many). A successful business depends on having people with both qualities in the ranks of its management teams. Much of the research on leadership and management for safety has been focused at the supervisor level, but Public Inquiries into major accidents have identified that culpability ultimately lies at senior management level and this accountability is now recognized in legislation, for example, in the UK HSE's Corporate Manslaughter and Corporate Homicide Act 2007. The UK HSE has also published guidance in support of this legislation: "Leading health and safety at work, Actions for directors, board members and organizations of all sizes" (HSE, 2013). Other guidance exists in the form of a Best Practice Guide from the Chemical Industries Association's "Process safety leadership in the chemical industry" and the International Association of Oil and Gas Producers "Shaping safety culture through safety leadership" (OGP, 2013). Finally, the UK HSE has published an extensive review of the literature on effective leadership behaviors for safety (Health and Safety Executive, 2012).

In a review paper for leadership in healthcare, Flin and Yule (2004) present an excellent overview of some of the main findings generated by research into safety leadership behavior at the supervisor, middle management, and senior management levels, with a focus on Multifactor Leadership theory otherwise known as Transactional/Transformational Leadership theory (Bass, 1998; Bass & Avolio, 1990). Transactional leadership refers to the "exchange" type of relationship that normally occurs between an individual and their superior. In other words, the supervisor/manager set goals or targets, which are then rewarded when they are achieved (and perhaps punished when they are not). The most common styles of Transactional leadership are "Management by exception" (i.e., only intervening when goals/targets are not being achieved) and "Contingent reward" (reinforcement for achieving goals). Transformational leadership is believed to "augment" Transactional leadership and consists of four components:

1. Idealized influence—Transformational leaders act as role models for their followers, embodying vision into action.
2. Individualized consideration—Transformational leaders attend to each follower's needs, attending to individual strengths and development.
3. Inspirational motivation—Transformational leaders inspire followers to go beyond their level of comfort by linking purpose and meaning and driving people forward.
4. Intellectual stimulation—Transformational leaders encourage innovation and creativity as well as critical thinking and problem-solving skills among followers.

Bass and Avolio (1990) have demonstrated that Transformational leadership style improves employee job satisfaction, organizational commitment, and workplace performance, and recent research has demonstrated that it has also has a positive impact on the safety behavior of subordinates. Flin and Yule (2004) point out that the style of leadership for safety will vary according to management level within the organization, but ultimately both styles of leadership will be apparent depending on the situation. The following sections discuss safety leadership at the supervisory, middle management, and senior management levels.

6.2.1 Supervisors

Supervisors are in a difficult position as the individual "in the middle" facing more senior management and requirements to meet the strategic objectives of the organization on the one side and deploying their work teams and resources to achieve frontline goals on the other. As a result, supervisors face considerable challenges in the trade-off between "production and safety." In one of the earliest studies of its kind, Andriessen (1978) concluded that supervisors have a direct influence on the safety motivation and safety behavior of the workforce, but they themselves will be influenced by their more senior managers; thus, senior management objectives and priorities will filter down through supervisors to the workforce level. Nonetheless, some research evidence suggests that supervisors can act as a "buffer" between senior management and their subordinates, protecting them from unreasonable or unachievable goals (Fleming, 2000). They do this by effective safety communication; valuing their subordinates and their contribution; making frequent visits to the worksites to encourage and support their subordinates rather than trying to "catch them out"; and allowing the work group to participate in decision making.

Experimental studies using worksite observations in the manufacturing sector have demonstrated how training in Transactional and Transformational leadership practices in supervisors leads to improved safety behavior in subordinates (Zohar, 2002; Zohar & Luria, 2003), a finding echoed in studies in the construction sector (Conchie, Moon, & Duncan, 2013; Mattila, Hyttinen, & Rantanen, 1994). Other studies have indicated how supervisory safety practices influence safety in the military (Hofmann, Morgeson, & Gerras, 2003) and hospitality sectors (Barling, Loughlin, & Kelloway, 2002).

Fuller and Vassie (2005) conducted a benchmarking study into the role of the supervisor in relation to health and safety performance in the chemical process industries. Their objective was to benchmark supervisory styles for both low- and high-risk activities in chemical sites of different sizes and functions. Data from a questionnaire completed by 84 sites in the UK chemical sector provided the basis for the study. The findings showed no causal relationship between supervisory methods and health and safety performance, and most supervisors were focused on compliance with health and safety legislation and risk control measures. The study did indicate however that operational responsibilities lay with management and organizational support was important for supervisory methods, particularly in the larger chemical organizations. This leads us on to the role of more senior managers in maintaining high safety performance in the chemical industries.

6.2.2 Middle Managers

There is surprisingly little research on the role of departmental or site leaders in safety (Flin & Yule, 2004). Part of the problem is that studies often do not clarify what level of "management" is being targeted, with the general term being used, which could mean supervisors, site managers, or senior managers. Most studies of management commitment to safety are related to safety climate and safety culture (see earlier) and therefore will not be covered in any further detail here.

O'Dea and Flin (2001) carried out one of the few studies of site level safety leadership in the UK offshore oil and gas industry. They surveyed 200 offshore installation managers (OIMs) across 157 offshore production platforms and drilling rigs to investigate their leadership style. O'Dea and Flin found that these OIMs reported a more "telling and selling" style, rather than a "participative, inspirational" style, even though the managers were aware that the latter style was probably preferable when it came to influencing safety performance. In another paper reported by O'Dea and Flin (2000) in the offshore industry, the relationship between workforce safety compliance and involvement perceived OIM commitment to safety and Transactional/Transformational Leadership style was reported. Transformational leadership was associated with more safety initiative among the platform workforce, whereas Transactional Leadership showed no effect.

In conclusion, it would appear that the role of "middle management" in leadership for safety has not been well delineated in research studies. Their influence has largely been captured through studies of safety climate, where workforce perceptions of management commitment to safety are the key influence on workplace safety climate. Since members of the workforce rarely have any contact with senior managers, i.e., CEOs and Board members, it can be assumed that any reference to *management* as opposed to *supervisors* in safety climate studies will be targeting site or departmental managers.

6.2.3 Senior Managers

Flin (2003) referred to senior managers as a "neglected species," but recently, they have come very much under the spotlight. The legal requirements of senior managers are articulated at the start of this section, and as a result of the outcomes of major accident Public Inquiries, their key role in determining and reinforcing the safety culture of the organization and associated standards and expectations has become very apparent, e.g., Chernobyl (International Atomic Energy Authority, 1992), NASA (NASA, 2003), and Deepwater Horizon (National Commission on the BP Deepwater Horizon Oil Spill and Offshore Drilling, 2011). Although generally far removed from front-line operations, senior management attitudes, behaviors, decision making, and deployment of resources can have an impact on safety. They are effectively the "controlling mind" and subordinates will look to them for guidance on what is recognized and rewarded. It is acknowledged that senior managers have to balance a number of organizational goals, e.g., budgets, productivity, safety, but where MAHs can have devastating consequences if realized, senior management focus should be on safety. A study of

70 European chemical and petrochemical companies (Keller & Huwaishel, 1993) showed that only 23% reported safety as the top priority for management. As demonstrated by Fuller and Vassie's (2005) benchmarking study, management focus tends to be on compliance with health and safety legislation in order to prevent legal action being taken by the authorities.

Up until relatively recently, there has been very little research on senior management attitudes and behavior in relation to safety. Rundmo and Hale (2003) investigated attitudes toward safety and accident prevention in 210 presidents, vice presidents, and managers in a Norwegian petrochemical company. They found that high management commitment, low fatalism, high-risk awareness, and high safety priority were key attitudes. Research into senior leadership and safety in the UK petrochemical industry (Roger, 2013; Roger, Flin, & Mearns, 2011) led to the development of a six-category framework for key senior leadership functions:

1. Establishing safety as an organizational priority.
2. Establishing clear and open communication for safety.
3. Active involvement in safety activities.
4. Setting and maintaining safety standards.
5. Maintaining risk awareness.
6. Motivating and supporting the workforce.

In conclusion, senior managers should adopt a Transactional approach by ensuring compliance with regulatory requirements and providing resources for a comprehensive SMS. The research evidence also indicates the importance of a Transformational Leadership Style at senior management level demonstrated through visible and consistent commitment to safety, showing concern for people, encouraging participatory styles in their subordinates, and ensuring allocation of their time to safety issues (Flin & Yule, 2004).

7. GENERAL CONCLUSION

This chapter has covered a cross-section of the subdisciplines and areas that make up the discipline of Human Factors. The main role of Human Factors is to design, develop, and maintain systems of work that improve human efficiency but above all keep people, plant, and the environment healthy and safe. This applies to both occupational health and safety and process safety. Application of Human Factors begins at the design stage with decades of research and practitioner experience providing examples of standards, legislation, and good practice. It is generally accepted that good design and engineering are the best way to prevent process incidents; however, sometimes the costs associated with designing for humans are considered to be prohibitive for many companies and therefore human factors engineering is sometimes not implemented to a high enough standard. If designing for humans is not achieved, a second barrier to major incidents involves administrative controls such as procedures. Clearly written and usable procedures are a perquisite to a safe workplace, as is adequate training and regular assessment of competency. Understanding the organization and how people are managed provides another important barrier to preventing both process and occupational incidents. This can be achieved through safety climate and safety culture assessment, depending on whether the objective is to address underlying attitudes and beliefs or perceptions of how safety is being managed at the workplace. Ultimately, although everyone is responsible for safety, it is management who are accountable and that accountability has to be understood and complied with.

Apart from management understanding and executing their accountabilities, Human Factors should be integrated into the SMS. An SMS is unlikely to achieve its full potential to improve safety performance unless staff have a full understanding of Human Factors principles and are able to apply this understanding to support a positive safety culture. Human Factors should be considered as routinely as other important SMS activities such as risk assessment, cost–benefit analyses, and access to and deployment of resources. Human Factors principles can be incorporated into hazard identification and reducing risks to As Low As Reasonably Practicable (ALARP); designing systems, equipment, jobs, and tasks; staff training and competency assessment; and the management of change. At the end of the day, the SMS is only as good as the people that operate within it. To coin an idiom, understanding and applying good human factors practice is "where the rubber meets the road." It is the moment of truth for many process industry organizations when it comes preventing major incidents.

REFERENCES

American Psychological Association (APA, 2016). www.apa.org (accessed 23rd October, 2016).

Andriessen, J. (1978). Safe behaviour and safety motivation. *Journal of Occupational Accidents*, *1*, 363–376 [now Safety Science].

Barling, J., Loughlin, C., & Kelloway, E. K. (2002). Development and test of a model linking safety-specific transformational leadership and occupational safety. *Journal of Applied Psychology*, *87*(3), 488–496.

Bass, B. (1998). *Transformational leadership*. Mahwah, NJ: LEA.

Bass, B., & Avolio, B. (1990). The implications of transactional and transformational leadership for individual, team and organizational development. *Research in Organizational Change and Development*, *4*, 231–272.

Berg, M., Shahriari, M., & Kines, P. (2013). Occupational safety climate and shift work. *Chemical Engineering Transactions, 31*, 403–408.
Brown, R. L., & Holmes, H. (1986). The use of a factor analytic procedure for assessing the validity of an employee safety climate model. *Accident Analysis and Prevention, 18*(6), 445–470.
BS EN ISO 11064 (Parts 1 to 6). (2000–2007). *Ergonomic design of control centres.*
Business Dictionary (2016). http://www.businessdictionary.com/definition/competence.html.
Cellar, D. F., Yorke, C. M., Nelson, Z. C., & Carroll, K. A. (2004). Relationships between five factor personality variables, workplace accidents and self-efficacy. *Psychological Reports, 94*(3 Pt. 2), 1437–1441.
Center for Chemical Process Safety (CCPS). (1994). *Guidelines for preventing human error in process safety.* New York: American Institute of Chemical Engineers. Inter Science.
Chemical Industries Association. (2008). *Process safety leadership in the chemicals industry—Best practice.* London: CIA.
Christian, M. S., Bradley, J. C., Wallace, J. C., & Burke, M. J. (2009). Workplace safety: A meta-analysis of the roles of person and situation factors. *Journal of Applied Psychology, 29*(5), 1103–1127.
Clarke, S. (2006a). Contrasting perceptual, attitudinal and dispositional approaches to accident involvement. *Safety Science, 44*, 537–550.
Clarke, S. (2006b). The relationship between safety climate and safety performance: A meta-analytic review. *Journal of Occupational Health Psychology, 11*, 315–327.
Clarke, S., & Robertson, I. (2008). An examination of the role of personality in work accidents using meta-analysis. *Applied Psychology. An International Review, 57*(1), 94–108.
Conchie, S., Moon, S., & Duncan, M. (2013). Supervisors' engagement in safety leadership: Factors that help and hinder. *Safety Science, 51*(1), 109–117.
Cooper, D. (2000). Towards a model of safety culture. *Safety Science, 36*, 111–136.
Cox, S., & Cheyne, A. (2000). Assessing safety culture in offshore environments. *Safety Science, 34*, 111–119.
Dedobbeleer, N., & Beland, F. (1991). A safety climate measure for construction sites. *Journal of Safety Research, 22*(2), 97–103.
Donald, I., & Canter, D. (1994). Employee attitudes and safety in the chemical industry. *Journal of Loss Prevention in the Process Industries*, 7(3), 203–208. Abstract only accessed.
Energy Institute. (2011). *Guidance on human factors safety critical task analysis.* London: Energy Institute.
Energy Institute. (2012). *Guidance on quantified human reliability analysis (QHRA).* London: Energy Institute.
Energy Institute. (2014). *Guidance on crew resource management (CRM) and non-technical skills training* (1st ed.). London: Energy Institute.
Engineering Equipment and Material Users Association (EEMUA). See https://www.eemua.org/tni/About-EEMUA/What-we-do/Standards-guides.aspx.
EEMUA 191. (2007). *Alarm systems. A guide to design, management and procurement: Engineering equipment and materials users association publication 191* (2nd ed.). EEMUA.
EEMUA 201. (2009). *Process plant control desks utilising human-computer interfaces: A guide to design, operational and human-computer interface issues: Engineering equipment and materials users association publication no. 201* (2nd ed.). EEMUA.
Fleming, M. (2000). Effective supervisory leadership behaviour in the offshore oil industry. In *Institute of chemical engineers (IChemE), symposium series 147, paper 29.*
Fletcher, G., Flin, R., McGeorge, P., Glavin, R., Maran, N., & Patey, R. (2003). Anaesthetists' non-technical skills (ANTS): Evaluation of a behavioural marker system. *British Journal of Anaesthesia, 90*(5), 580–588.
Flin, R. (2003). Danger—Men at work. *Human Factors and Ergonomics in Manufacturing, 13*, 261–268.

Flin, R., & Maran, N. (2004). Identifying and training non-technical skills for teams in acute medicine. *Quality & Safety in Health Care, 13*, 180–184.
Flin, R., & Martin, L. (2001). Behavioural markers for crew resource management: A review of current practice. *The International Journal of Aviation Psychology, 11*(1), 95–118.
Flin, R., Mearns, K., O'Connor, P., & Bryden, R. (2000). Measuring safety climate: Identifying the common features. *Safety Science, 34*, 177–193.
Flin, R., O'Connor, P., & Mearns, K. (2002). Crew resource management: Improving teamwork in high reliability industries. *Team Performance Management: An International Journal, 8*(3/4), 68–78.
Flin, R., & Yule, S. (2004). Leadership for safety: Industrial experience. *Quality and Safety in Healthcare, 13*(Suppl. II), ii45–ii51.
Fuller, C., & Vassie, L. (2005). *Benchmarking employee supervisory processes in the chemical industry: Research report 312.* Norwich: HSE Books.
Guldenmund, F. W. (2000). The nature if safety culture: A review of theory and research. *Safety Science, 34*(1–3), 215–257.
Hansen, C. P. (1989). A causal model of the relationship among accidents, biodata, personality and cognitive factors. *Journal of Applied Psychology, 74*(1), 81–90.
Health and Safety Commission (HSC). (1993). Advisory committee on the safety of nuclear installations. *Organising for safety.* London: HSE Books.
Health and Safety Executive. (1999). *Reducing error and influencing behaviour (HSG48).* Suffolk: HSE Books.
Health and Safety Executive. (2003). *The role of managerial leadership in determining safety outcomes: Research report 044.* HM Stationary Office: Norwich.
Health and Safety Executive. (2009). *Research Report RR679: Review of human reliability assessment methods.* Norwich, UK: HSE Books.
Health and Safety Executive. (2012). *A review of the literature on effective leadership behaviours for safety: Research report 952.* Norwich, UK: HSE Books.
Health and Safety Executive (2013). Leading health and safety at work. Actions for directors, board members, business owners and organisations of all sizes. Industry Guidance 417 (rev 1). Leaflet available at www.hse.gove.uk/pubns/indg417.htm.
Health and Safety Executive (HSE). (2005). Core topic 3. Identifying human failures. Accessed from http://www.hse.gov.uk/humanfactors/topics/core3.pdf.
Health and Safety Executive (HSE). (2005). Core topic 4. Procedures. Briefing note 4. Accessed from http://www.hse.gov.uk/humanfactors/topics/core4.pdf.
Henderson, J., & Embrey, D. (2012). Quantifying human reliability in risk assessments. *Petroleum Review*, 30–34.
Hofmann, D., Morgeson, F., & Gerras, S. (2003). Climate as a moderator of the relationship between leader-member exchange and content specific citizenship: Safety climate as an exemplar. *Journal of Applied Psychology, 88*, 170–178.
Institute of Nuclear Power Operations. (1993). *Control of room teamwork development training: Course administration and facilitation guide.* Atlanta, GA: National Academy for Nuclear Training.
International Association of Oil and gas producers. (2013). *Shaping safety culture through safety leadership: OGP report 45.* October 2013.
International Association of Oil and Gas Producers. (2014). *Crew resource management for well operations teams: Report 501.* April 2014.
International Atomic Energy Authority. (1992). The Chernobyl accident: Updating of INSAG-1. A report by the international nuclear safety advisory group (INSAG), Safety Series No. 75-INSAG-7.24.
Keller, A. Z., & Huwaishel, A. M. (1993). Top management attitude toward safety in the western European chemical and petrochemical industries. *Disaster Prevention and Management, 2*, 48–57.
Kirwan, B. (1994). *A guide to practical human reliability assessment.* London: Taylor & Francis.

Kirwan, B. (1996). The validation of three human reliability quantification techniques THERP, HEART and JHEDI: Part 1. Technique descriptions and validation issues. *Applied Ergonomics*, *27*(6), 359–373.
Kirwan, B., & Ainsworth, K. (1992). *A guide to task analysis*. London, UK: Taylor and Francis.
Kirwan, B., Kennedy, R., Taylor-Adams, S., & Lambert, B. (1997). The validation of three human reliability quantification techniques, THERP, HEART and JHEDI: Part II—Results of validation exercise. *Applied Ergonomics*, *28*(1), 17–25.
Lee, T. R. (1998). Assessment of safety culture in a nuclear reprocessing plant. *Work and Stress*, *2*, 217–237.
Lee, T. R., & Harrison, K. (2000). Assessing safety culture in nuclear power stations. *Safety Science*, *34*, 61–97.
Mattila, M., Hyttinen, M., & Rantanen, E. (1994). Effective supervisory behaviour and safety on a building site. *International Journal of Industrial Engineers*, *13*, 85–93.
Mearns, K., Flin, R., Gordon, R., & Fleming, M. (1998). Measuring safety climate on offshore installations. *Work and Stress*, *12*, 238–254.
Mearns, K., & Hope, L. (2005). *Health and well-being in the offshore environment: The management of personal health*. HSE Research Report 305. Norwich, UK.
Mearns, K., Whitaker, S. M., & Flin, R. (2003). Safety climate, safety management practice and safety performance in offshore environments. *Safety Science*, *41*, 641–680.
Michie, S. (2002). Causes and management of stress at work. *Occupational and Environmental Medicine*, *59*, 67–72.
Mitchell, L., & Flin, R. (2008). Non-technical skills of the operating theatre scrub nurse: Literature review. *Journal of Advanced Nursing*, *63*, 15–24.
NASA. (2003). *Columbia accident investigation board: Vol. 1*. United States: NASA [chapter 7].
National Commission on the BP Deepwater Horizon Oil Spill and Offshore Drilling. (2011). *Deepwater: The Gulf oil disaster and the future of offshore drilling*. Report to the President [chapter 8].
Niskanen, T. (1994). Safety climate in road administration. *Safety Science*, *17*, 237–255.
Noyes, J. (2001). Designing for humans. In *Psychology at work series*. Hove, East Sussex, UK: Psychology Press; Taylor and Francis.
O'Connor, P., & Flin, R. (2003). Crew Resource Management training for offshore oil production teams. *Safety Science*, *41*, 591–609.
O'Dea, A., & Flin, R. (2000). Site managers and safety leadership in the offshore oil and gas industry. In *Paper presented at the academy of management conference, August* [cited in Flin & Yule, 2004].
O'Dea, A., & Flin, R. (2001). Site managers and safety leadership in the offshore oil and gas industry. *Safety Science*, *37*, 39–57.
Oil and Gas Producers (OGP). (2013). *Shaping safety culture through safety leadership: Report no. 452*. London: International Association of Oil and Gas Producers.
Poussette, A., Larsson, S., & Törner, M. (2008). Safety climate cross-validation, strength and prediction of safety behaviour. *Safety Science*, *46*, 398–404.
Pidgeon, N., & O'Leary, M. (2000). Man-made disasters: Why technology and organizations (sometimes) fail. *Safety Science*, *34*, 15–30.
Prior, R. (2003). Top management behaviours—The determining role in changing safety culture. In *Institute of chemical engineers (IChemE), symposium series no. 149* (pp. 733–744).
Reason, J. (1991). *Human error*. Aldershot: Ashgate Books.
Reason, J. (1997). *Managing the risks of organizational accidents*. Aldershot: Ashgate books.
Reason, J. (1998). Achieving a safe culture: Theory and practice. *Work and Stress*, *12*, 293–306.
Roger, I. (2013). *Safety leadership in the energy industry: The development and testing of a framework outlining Key behaviours of senior managers*. Doctoral Thesis School of Psychology, University of Aberdeen.

Roger, I., Flin, R., & Mearns, K. (2011). Safety leadership from the top: Identifying the key behaviours. In *Proceedings of the human factors and ergonomics society 55th annual meeting. Las Vegas, USA*.
Rundmo, T. (1992). Risk perception and safety on offshore petroleum platforms—Part II: Perceived risk, job stress and accidents. *Safety Science*, *15*(1), 53–68.
Rundmo, T., & Hale, A. (2003). Managers' attitudes towards safety and accident prevention. *Safety Science*, *41*, 557–574.
Seo, D. C., Torabi, M. R., Blair, E. H., & Ellis, N. T. (2004). A cross-validation of safety climate scales using a confirmatory factor analytic approach. *Journal of Safety Research*, *35*(4), 427–445.
Shepherd, A. (2001). *Hierarchical task analysis*. London and New York: Taylor and Francis.
Sian, I. B., Robertson, M., & Watson, J. (2016). *Maintenance resource management handbook*. Washington DC: Federal Aviation Authority.
SINTEF. (2004). Industrial Management Safety and Reliability. *CRIOP®: A scenario method for Crisis Intervention and Operability Analysis*. Trondheim, Norway.
STCW. (June 2010). International convention on standards of training, certification and watchkeeping for seafarers (the STCW Convention), and its associated code. In *Diplomatic conference in Manila, the Philippines, 21–25*.2011.
Sutherland, V. J., & Cooper, C. L. (1986). *Man and accidents offshore: The costs of stress among workers on oil and gas rigs*. London: Lloyd's List/Dietsmann.
Sutherland, V. J., & Cooper, C. L. (1991). Personality, stress and accidents in the offshore oil and gas industry. *Personality and Individual Differences*, *12*, 195–204.
Sutherland, V. J., & Cooper, C. L. (1996). Stress in the offshore oil and gas exploration and production industries: An organizational approach to stress control. *Stress Medicine*, *12*, 27–34.
Swain, A. D., & Guttman, H. E. (1983). *Handbook of human reliability analysis with emphasis on nuclear power plant operations*. Washington, DC: US Nuclear Regulatory Commission. NUREG/CR-1278.
US Chemical Safety Board Report. (2007). *Refinery explosion and fire. BP Texas City, Texas: Report no. 205-04-I-TX*.
Vinodkumar, M. N., & Bhasi, M. (2009). Safety climate factors and its relationship with accidents and personal attributes in the chemical industry. *Safety Science*, *47*, 659–667.
Visscher, G. (2008). Some observations about major chemical accidents from recent CBI Investigations. In *Paper presented at the IChemE XX Hazards symposium, symposium series 154*.
Williams, J. (1985). HEART—A proposed method for achieving high reliability in process operations by means of human factors engineering technology. In *Proceedings of a symposium on the achievement of reliability in operating plant, safety and reliability society, 16th September 1985, Southport* [cited in HSE RR679, 2009].
Yule, S., Flin, R., Paterson-Brown, S., & Maran, N. (2006). Non-technical skills for surgeons. A review of the literature. *Surgery*, *139*, 140–149.
Yule, S., Flin, R., Maran, N., Youngson, G., Rowley, D., & Paterson-Brown, S. (2008). Surgeons' non-technical skills in the operating room: Reliability testing of the NOTSS behaviour rating system. *World Journal of Surgery*, *32*, 548–556.
Zohar, D. (1980). Safety climate in industrial organizations: Theoretical and applied implications. *Journal of Applied Psychology*, *65*, 96–102.
Zohar, D. (2002). Modifying supervisory practices to improve sun-unit safety: A leadership-based intervention model. *Journal of Applied Psychology*, *87*, 156–163.
Zohar, D. (2010). Thirty years of safety climate research: Reflections and future directions. *Accident; Analysis and Prevention*, *42*, 1517–1522.
Zohar, D., & Luria, G. (2003). The use of supervisory practices as leverage to improve safety behaviour: A cross-level intervention model. *Journal of Safety Research*, *34*, 567–577.

CHAPTER FIVE

Introduction to Dynamic Risk Analyses

Warren D. Seider*,[1], Ankur Pariyani[†], Ulku G. Oktem*,[†], Ian Moskowitz[‡], Jeffrey E. Arbogast[‡], Masoud Soroush[§]

*University of Pennsylvania, Philadelphia, PA, United States
[†]Near-Miss Management LLC, Philadelphia, PA, United States
[‡]American Air Liquide, Newark, DE, United States
[§]Drexel University, Philadelphia, PA, United States
[1]Corresponding author: e-mail address: seider@seas.upenn.edu

Contents

Methods in Chemical Process Safety, Volume 1
ISSN 2468-6514
http://dx.doi.org/10.1016/bs.mcps.2017.02.001

1. INTRODUCTORY CONCEPTS AND CHAPTER OBJECTIVES

Automated control and safety systems are prevalent in modern chemical plants, as they help plants return to normal operating conditions when abnormal events occur. The databases associated with these systems contain a wealth of information about near miss occurrences. Frequent statistical analyses of the information can help identify problems and prevent accidents and expensive shutdowns. Such analyses are referred to as dynamic risk analyses.

Predictive maintenance is an important evolution in the effective and efficient management of industrial chemical processes. Often predictive maintenance is focused upon individual equipment within a process (e.g., compressors). Lately, in addition, proactive risk management is recognized as a key concept for the evolution to the next level of safety, operability, and reliability performance in chemical operations involving plant-wide analyses toward prescriptive maintenance. Dynamic risk analysis refers to a methodology that utilizes various tools to help collectively to achieve proactive risk management. Hence, its application has been gaining significant importance in the chemical and process industries.

This chapter discusses dynamic risk analysis of alarm data. It provides a general overview of what these analyses are, how they can be used in chemical processing to improve safety, and challenges that must be addressed over the next 5–10 years. It also highlights current research in this area and offers perspectives on methodologies most likely to succeed.

1.1 Alarms, Near Misses, and Accidents

Fig. 1 is a generic control chart for a primary process variable (P), which is divided into three zones: green-belt, yellow-belt, and red-belt. An *abnormal event* occurs when a process variable leaves its normal operating range (green-belt zone), which triggers an alarm indicating transition into the yellow-belt zone. If the variable continues to move away from its normal range, the variable may transition into its red-belt zone, indicated by second-level alarm (e.g., LL, HH) activation. Once a variable remains in its red-belt zone for a prespecified length of time (typically on the order of seconds), an interlock activates and an automatic shutdown occurs. If the automatic shutdown is unsuccessful, an accident would occur

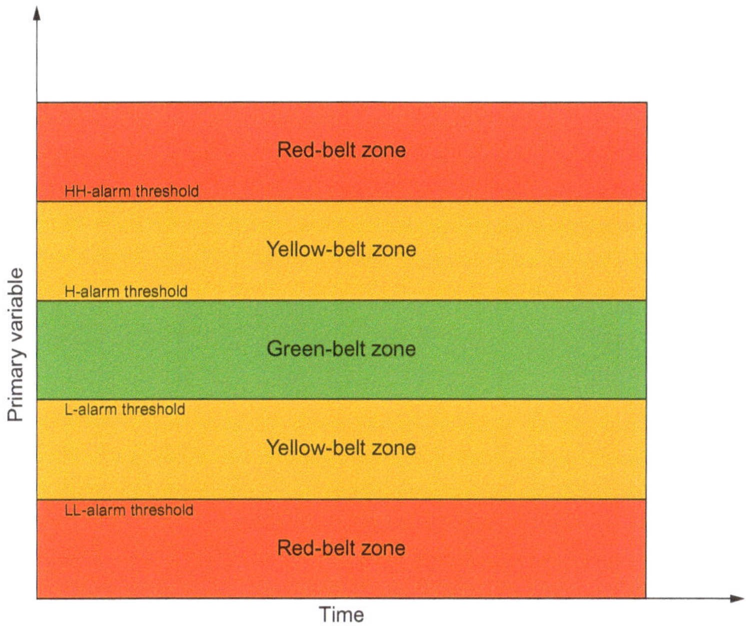

Fig. 1 A control chart for a primary process variable (*P*). An abnormal event occurs when the variable moves outside of the green-belt zone. From Oktem, U. G., Seider, W. D., Soroush, M., & Pariyani, A. (2013). Improve process safety with near-miss analysis. *Chemical Engineering Progress, 109,* 20–27. Used with permission.

if not prevented by physical protections (e.g., pressure relief devices, containment).

The accidents are rare, high-consequence (e.g., significant human health, environmental, and/or economic impact) events. The costs of unplanned shutdowns—which happen infrequently, but not rarely—are also quite significant. Of course, the layers of protection in place are usually successful, and therefore, the vast majority of abnormal events are arrested before accidents occur. When an abnormal event is stopped and the variable returns to its green-belt zone, this is considered a process near miss event (which is simply referred to as a near miss in this chapter). Accidents are typically preceded by several near misses, as the latter are higher-probability, lower-consequence events. As a result, vast amounts of near miss data can be extracted from alarm activations for dynamic risk analyses. Depending on their criticality, abnormal events can be classified into different categories (Pariyani, Seider, Oktem, & Soroush, 2010). In this chapter, the following two categories are used: *least-critical* abnormal events that cross the high/low alarm thresholds, but do not cross the high–high/low–low alarm thresholds, and *most-critical* abnormal events that cross the high–high/low–low thresholds, often associated with interlock systems or emergency shutdown (ESD) systems.

Many companies record these alarm occurrences in databases and operators, engineers, and managers seek guidance from these databases by evaluating various *key performance indicators* (see alarm management standards published by ANSI/ISA 18.2 (2009) and EEMUA (2007)) and paying attention when special events such as alarm flooding and chattering occur (Kondaveeti, Izadi, Shah, Shook, & Kadali, 2010). Most of the time, further analysis is done after process upsets, unanticipated trips, and accidents occur.

Companies are becoming increasingly aware that these alarm databases are rich in information related to near misses. In recent years, researchers have been developing key performance indicators, or metrics, associated with potential trips (shutdowns with no associated personal injury, equipment damage, or significant environmental problem) and accidents; leading indicators (i.e., events or trends indicating the times these trips and accidents are likely to occur); and probabilities of failure of the individual safety systems and the occurrence of trips and accidents (ANSI/ISA 18.2, 2009; EEMUA, 2007; Khan, Rathnayaka, & Ahmed, 2015; Pariyani et al., 2010; Villa, Paltrinieri, Khan, & Cozzani, 2016; Ye, Liu, Fei, & Liang, 2009). When conducted at frequent intervals, the analyses that are associated with these performance indicators are often referred to as dynamic risk analyses or, simply, near miss analyses.

1.2 Conventional Risk Analyses

Risk assessment is an important component of the US Occupational Safety and Health Administration's (OSHA) process safety management (PSM) standard, which includes (among other elements) inherently safer design, hazard identification, risk assessment, consequence modeling and evaluation, auditing, and inspection. PSM has become a popular and effective approach to maintain and improve the safety, operability, and productivity of plant operations. As part of this, several risk assessment methods have been developed.

The use of quantitative risk analysis (QRA), which was pioneered in the nuclear industry in the 1960s, was extended to the chemical industry in the late 1970s and early 1980s after major accidents such as the 1974 Flixborough explosion in the UK and the 1984 Bhopal incident in India. Chemical process quantitative risk analysis (CPQRA) was introduced as a safety assessment tool by the AIChE's Center for Chemical Process Safety (CCPS) in the 1990s as a means to evaluate potential risks when qualitative methods are inadequate. CPQRA is used to identify incident scenarios and evaluate their risk by defining the probability of failure, the various consequences, and the potential impacts of those consequences. This method typically relies on historical data, including chemical process and equipment data, and human reliability data to identify hazards and risk-reduction strategies. A recent review article by Khan et al. (2015) examines qualitative, semi-quantitative, and quantitative methods developed in the past 10–15 years. Much emphasis is shifting toward quantitative techniques involving fault trees, event trees, and bowtie networks to understand better safety events. This quantification empowers plant managers and operators to make decisions based upon their current risk assessment. In offshore drilling, bowtie models have been used to represent the potential accident scenarios, their causes, and the associated consequences (Abimbola, Khan, & Khakzad, 2014; Khakzad, Khan, & Amyotte, 2013a, 2013b). These involve dynamic risk analysis using Bayesian statistics.

Other risk assessment methods were subsequently developed to analyze industry-wide incident databases (Anand et al., 2006; CCPS, 1999; Elliott, Wang, Lowe, & Kleindorfer, 2004; Kleindorfer et al., 2003; Meel et al., 2007). These databases include: CCPS's Process Safety Incident Database (CCPS, 1999, 2007) which tracks, pools, and shares process safety incident information among participating companies; the Risk Management Plan database, RMP*Info (Kleindorfer et al., 2003), developed by the US Environmental Protection Agency (EPA); the National Response Center (NRC) database, an online tool set-up by NRC to allow users to submit and share incident reports; and the Major Accident-Reporting System (MARS), which is maintained by the Major Accident Hazards Bureau (MAHB). Recent risk analyses associated with chemical plant safety and operability have used Bayesian statistics to incorporate expert opinion (Meel & Seider, 2006; Valle, 2009; Yang, Rogers, & Mannan, 2010), and fuzzy logic to account for knowledge uncertainty and data imprecision (Ferdous, Khan, Veitch, & Amyotte, 2009; Markowski, Mannan, & Bigowszewska, 2009). Such methods have significantly improved quantitative risk assessment.

While these methods have been important in quantifying safety performance, a large amount of precursor information pointing to unsafe conditions has been overlooked and unutilized, because it resides in large alarm databases associated with the distributed control system (DCS) and ESD system (also referred to as "alarm data"). The alarms help plant operators assess and control plant performance, especially in the face of potential safety and product-quality problems. The alarm data, therefore, contain information on the progression of disturbances and the performance of regulating and protection systems. However, despite advances in alarm management standards and procedures, alarm data analysis methods reported in the literature quantify only performance, not risk. Risk is qualitatively inferred from these qualitative performance metrics (e.g., alarm frequency).

Several comprehensive algorithms and software packages to evaluate process safety risks with an eye toward developing and implementing appropriate protective measures have been developed over the last 2 decades (Risk World, 2013; Vinnem, 2010). Most of these systems rely on accident and failure databases mentioned above, which provide information such as accident frequencies, consequences, and associated economic losses, to perform QRAs. Other tools, discussed later, utilize a quantitative methodology for the risk analysis either in real time or on-demand, but they do not focus on estimating the likelihood of incidents or the failure of safety systems. The analyses that involve accidents and failures only, and exclude day-to-day alarm information and associated near miss data, are not highly predictive. They overlook the progression of events leading up to near misses—information that can only be found by analyzing data found in alarm databases.

A study of an ammonia storage facility conducted by the Joint Research Center and Denmark Risk National Laboratory of the European Commission (Lauridsen, Kozine, Markert, & Amendola, 2002) found that risk estimates based on generic databases of reliability and failure data for commonly used equipment and instruments are prone to biases, providing widely varying results depending on data sources.

For these reasons, the importance of utilizing process-specific databases for risk analyses has been gaining recognition. The next section discusses dynamic risk analysis methods and how they extract valuable risk information not available from conventional risk analyses.

1.3 Dynamic Risk Analysis

Accidents are rare events, occurrence of which is often described using the popular Swiss-cheese model (Reason, 1990). In this model, each layer of protection (safety system) is considered as a layer (slice) of Swiss cheese, with the holes in the slice (varying in size and placement) corresponding to

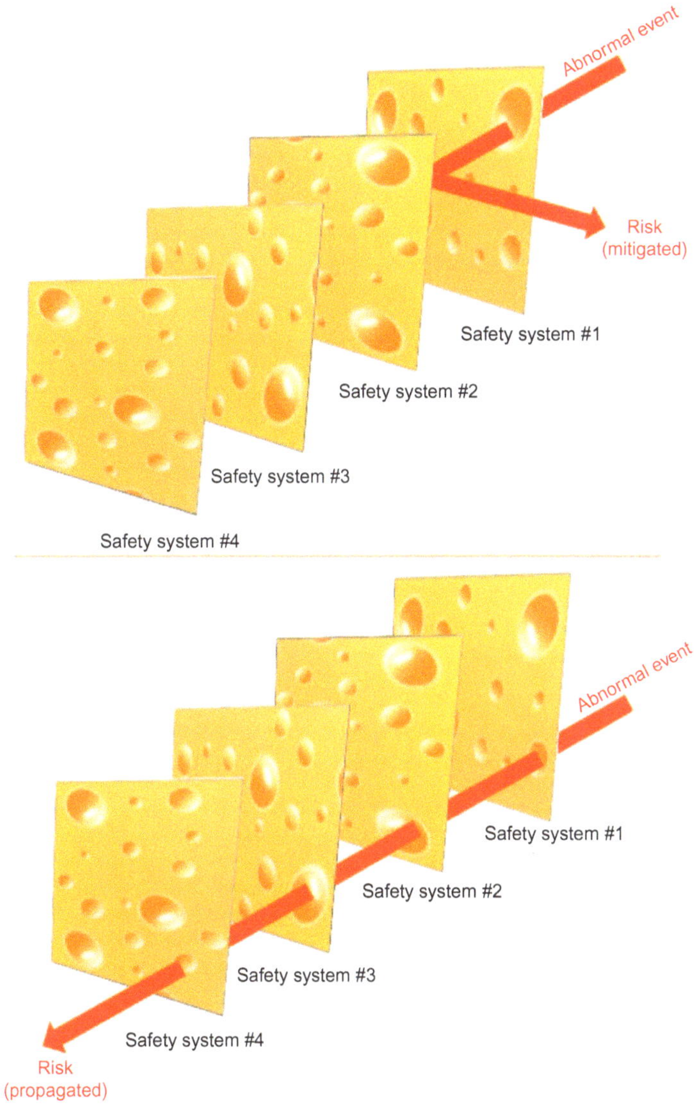

Fig. 2 Swiss-cheese model depicts the relationship between layers of protection and accidents From Oktem, U. G., Seider, W. D., Soroush, M., & Pariyani, A. (2013). Improve process safety with near-miss analysis. *Chemical Engineering Progress, 109,* 20–27. Used with permission.

weaknesses in the layer of protection (Fig. 2), and safety systems of a process are represented by several layers (slices) of Swiss cheese lined up in a row. According to this model, failures occur when the holes in the individual slices line up, creating trajectories of accident opportunities. This model accounts for the element of chance that is involved in the occurrence of failures.

Reports of major accident investigations often list several observable near misses—i.e., less-severe events, conditions, and consequences that occur before the accident. Regular, thorough analysis of process near misses could prevent the emergence of risky conditions from being overlooked.

On the basis of the sensitivity and importance, certain plant variables are labeled as *primary* variables—denoted by pP, p for primary and P for *process* variable (typically temperature, pressure, flow rate, ...). These variables are closely related to process safety and are associated with the *interlock* system. When these variables move beyond their interlock limits, ESDs or "trips" are triggered, often after a small time delay. The remaining variables in the plant are referred to as *secondary* variables, which are not associated with the interlock system—and denoted by sP, s for primary and P for process variable. For large-scale processes, typically thousands of variables are monitored; however, only a small percentage (less than 1%–10%) are chosen as primary variables. The primary variables are selected during the design and commissioning of plants by carrying out analyses of tradeoffs between the safety and profitability of the plant. Note that the control chart described in Fig. 1 is for a primary variable. For secondary variables, similar control charts exist, however, they do not have red-belt zones. As a result, they can indicate only least-critical abnormal events. Also note that, for many processes, control charts are provided for *quality* variables (typically, viscosity, density, average molecular weight, ...)—denoted by pQ, p for primary and Q for quality variable, and sQ, for secondary quality variables. See Pariyani, Seider, Oktem, and Soroush (2012a) for more complete coverage involving quality variables.

Although the layers of protection are designed to keep process (and quality) variables within acceptable limits, the chemical processes frequently encounter special causes (i.e., sudden or unexpected causes of variations in process conditions due to unexpected phenomena), resulting in *abnormal events*. An event-tree corresponding to a *primary* variable's transition between belt zones (in response to an abnormal event) is shown in Fig. 3. The first-level (e.g., L, H) alarm system activates safety system 1 (SS_1), which is typically an operator action. When SS_1 is successful, with probability $1-x_1$, continued operation (CO) is achieved, as indicated by path P^1. The second-level (e.g., LL, HH) alarm system activates SS_2, which is typically a more aggressive operator action. When successful, with probability

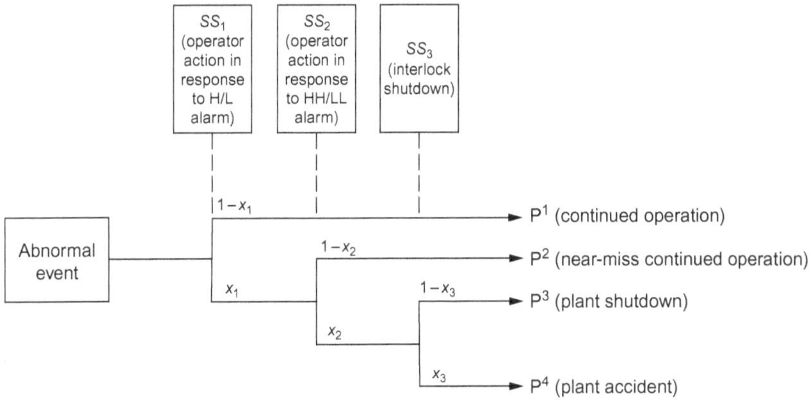

Fig. 3 Event tree for an abnormal event.

$1-x_2$, near miss CO is achieved, indicated by path P^2. If the primary variable occupies the red-belt zone for a predetermined length of time (on the order of seconds), SS_3, the automatic interlock system for plant shutdown, will become activated. The interlock system is designed to be independent of alarm systems, and the activation of SS_3 is determined by an independent set of sensors. It should be noted that if the interlock system is designed to have no delay time the probability of SS_2 success is equal to zero ($x_2=1$) since the operator has precisely no time to respond for such a design. If SS_3 succeeds, with probability $1-x_3$, the interlock shutdown occurs and an accident is avoided, represented by path P^3. If the interlock shutdown is unsuccessful, an accident occurs at the plant, represented by path P^4. With a proper design, x_3 should be very small consistent with the specified safety integrity level (SIL). Since the interlock system is independent of the alarm system, the success of SS_3 does not depend on factors such as operator skill and alarm sensor faults. However, it can be concluded that if either SS_1 or SS_2 is successful in arresting the special-cause event (SCE), the activation of the interlock system is avoided altogether. In some cases, alarms are officially considered as a layer of protection and contribute to the SIL rating of the overall safety system, composed of SS_1, SS_2, and SS_3. Therefore, herein alarms are included in the safety systems—noting that often the full alarm system is not considered part of a plant's SIS. In this systematic way, the success and failure of each safety system, along with its associated process consequence, is tracked. See Section 2.5 for a more complete discussion of event trees and their paths.

The probability of each consequence can be calculated when the safety system failure probabilities are known; e.g., the probability of a plant shutdown is:

$$\text{Prob}\left(\text{P}^3\right) = x_1 x_2 (1 - x_3) \tag{1}$$

These consequence probabilities are important to process engineers and plant managers, who seek to maintain very high probabilities of continued safe operation with quantifiably very low probabilities of potential plant shutdowns and accidents. The failure probabilities associated with operator action, x_1, and in many cases x_2, are often difficult to estimate. A manufacturing process typically has control systems that can mitigate most disturbances, with few propagated to SCEs. And, when few historical data are available, especially for SS_2, and the failure probabilities of higher-level alarm and safety interlock systems, x_1, x_2, and x_3 estimated from historical data alone have very high uncertainties, not providing useful predictive capabilities (Gelman, Carlin, Stern, & Rubin, 2014). Therefore, there is a need for a method of estimating the failure probabilities from process and alarm data along with our knowledge of the process dynamics and operators' behavior (Jones, Kirchsteiger, & Bjerke, 1999; Mannan, O'Connor, & West, 1999).

In industrial practice, methods such as HAZOP and HAZAN (Crowl & Louvar, 2011; Kletz, 1992) are commonly utilized to make safety and reliability estimates of processes on a unit-operation basis. Equipment–failure probabilities estimated from statistical and equipment–manufacturer data are used to estimate process failure probabilities. But, more recently, dynamic risk analyses have been employed to update these equipment–failure probabilities as real-time data are measured.

This chapter traces the development of dynamic risk analysis over the past 2 decades. First, Yi and Bier (1998) introduced the techniques to use consequence data to estimate dynamically the failure probabilities of safety systems. They used copulas to improve the estimates of failure probabilities of rare events (shutdowns and accidents). Meel and Seider (2006) then extended these methods for a hypothetical plant involving an exothermic continuous stirred tank reactor (CSTR). Pariyani et al. (2012a) and Pariyani, Seider, Oktem, and Soroush (2012b) extended Meel and Seider (2006) using alarm consequence data, and showed the performance of their methods by calculating the failure probabilities of a fluidized catalytic cracking unit. Kalantarnia, Khan, and Hawboldt (2009) and

Kalantarnia, Khan, and Hawboldt (2010) conducted further work on dynamic risk analysis and proposed improved consequence assessment approaches using Bayesian-based failure mechanisms. Khan et al. (2015) and Villa et al. (2016) provided a literature review of risk assessment approaches over the last 2 decades, including dynamic risk analysis. More recently, Moskowitz, Seider, Soroush, Oktem, and Arbogast (2015) and Moskowitz et al. (2016) proposed a method of improving process models and introduced new probabilistic models that describe SCE occurrences and operator response times, allowing for estimating alarm and safety system failure probabilities more accurately.

1.4 Bayesian Analysis

Bayesian analysis is often used to determine the failure probabilities of alarm and safety interlock systems. The central dogma of Bayesian analysis is that parameters of probability distributions (e.g., mean and variance) are themselves distributions. Unlike classical statistics that seeks to capture the true moments of a distribution, Bayesian statistics acknowledges that the moments of a distribution may not be fixed, and seeks to estimate the probability distributions of the moments. This analysis often requires significantly fewer data to make meaningful predictions (Berger, 2013; Gelman et al., 2014). Additionally, as the process dynamics and operators' behavior change with time (because of factors such as process unit degradation and operators' improved skills), real-time data can be collected and used to estimate more accurate failure probabilities in real time.

Bayesian analysis is a statistical approach to reasoning under uncertainty, which is briefly reviewed in this section, as applied to the estimation of failure probabilities. The principal steps in the application of Bayesian analysis include: (i) specifying a probability model for unknown parameter values that includes prior knowledge about the parameters, if available; (ii) updating knowledge about the unknown parameters by conditioning this probability model using observed data; and (iii) evaluating the goodness of the conditioned model with respect to the data and the sensitivity of the conclusions to the assumptions in the probability model.

The uncertainty of the failure probability of a safety system, x, is modeled using a probability distribution function, $f(x)$, called the *prior* distribution. Historical data and/or insights (e.g., based on experience) can then be used to obtain an improved failure probability distribution, $f(x|\text{Data})$, called the posterior distribution using *the Bayes theorem*:

$$f(x|\text{Data}) = \frac{g(\text{Data}|x)f(x)}{\int g(\text{Data}|x)f(x)dx} \tag{2}$$

where $g(\text{Data}|x)$ is the distribution of identically and independently distributed (i.i.d.) data conditional upon x, which is called the likelihood function.

1.4.1 Prior Distributions

In many cases, a prior distribution is approximated by a distribution from a convenient family of distributions, which when combined with the likelihood function gives a posterior distribution in the same family. In this case, the family of the prior distribution is a family of *conjugate priors* to its likelihood family distribution. For example, the Beta distribution is a family of conjugate priors to the Bernoulli distribution. Consider the following Beta distribution as the prior distribution for a random variable, x:

$$f(x) = \frac{\Gamma(a+b)}{\Gamma(a)\Gamma(b)}x^{a-1}(1-x)^{b-1}, \quad a>0, b>0 \tag{3a}$$

where the Gamma function is:

$$\Gamma(a) = \int_0^\infty t^{a-1}e^{-t}dt$$

The expected value and variance of x, denoted by $E(x)$ and $Var(x)$, are:

$$E(x) = \frac{a}{a+b}, \quad Var(x) = \frac{ab}{(a+b)^2(a+b+1)} \tag{3b}$$

For this Beta distribution, Meel and Seider (2006) assumed the data belong to the Bernoulli family, and derived the posterior distribution, utilizing Bernoulli's likelihood function and the Beta prior distribution.

Values of the parameters of the prior distribution function, a and b, can be obtained using historical information, or expert knowledge. In the absence of such information, a *noninformative* (often flat) prior distribution is utilized that weighs all of the parameters equally; e.g., a uniform distribution.

2. DYNAMIC RISK ANALYSIS USING ALARM DATA

Dynamic risk analysis using alarm databases was first introduced by Pariyani et al. (2012a, 2012b) to identify problems and correct them before they result in sizable product and economic losses, injuries, or fatalities. In most industrial processes, vast amounts of alarm data are recorded in their DCSs and ESD systems. The dynamic risk analysis method involves the following steps:

1. Track abnormal events using raw alarm data.
2. Create event trees that show all of the possible paths an abnormal event can take when propagating through the safety systems.
3. Use a set-theoretic framework, such as that developed by Pariyani et al. (2012a) to compact the data into a concise representation of multisets.
4. Perform a Bayesian analysis to estimate the failure probabilities of each safety system, the probability of trips, and the probability of accidents (Pariyani et al., 2012b).

Steps 1–3 are presented in the subsection that follows in which the event trees and set-theoretic formulations allow compaction of massive numbers (millions) of abnormal events. For each abnormal event, associated with a process variable, its path through the safety systems designed to return its variable to the normal operation range is recorded. Event trees, as shown in Fig. 3, are prepared to record the successes and failures of each operation, with on the order of 10^6 paths through event trees stored. As introduced herein, and shown in (Pariyani et al., 2012a), the set-theoretic structure condenses the paths to a single compact data record, leading to significant improvement in the efficiency of the probabilistic calculations and permitting Bayesian analysis of large alarm databases in real time. The latter is in Step 4, which was introduced in Section 1.4.

2.1 Typical Alarm Data

Alarm data are normally comprised of alarm identity tags for the variables, alarm types (low, high, high–high, etc.), times at which the variables cross their alarm thresholds (in both directions), and variable priorities. The associated interlock data, which is of greater consequence, contains trip event data, timer-alert data, and so on. A screenshot of a typical alarm log file for a brief period is shown in Fig. 4. Every row represents a new

A	B	C	D	E	F	G
2008-12-23 6:01 a.m.	Alarm	FI50	HI	ALM	LOW	Flowrate through pump1
2008-12-23 6:01 a.m.	Alarm	PDI10	LO	ALM	HIGH	Press. diff. stand pipe#1
2008-12-23 6:01 a.m.	Alarm	TI25	LO	ALM	MEDIUM	#1 flue gas temperature
2008-12-23 6:02 a.m.	Alarm	PDI10	LL	ALM	HIGH	Press. diff. stand pipe#1
2008-12-23 6:04 a.m.	Alarm	FI50	HI	RTN	LOW	Flowrate through pump1
2008-12-23 6:04 a.m.	Alarm	TI25	LO	RTN	MEDIUM	#1 flue gas temperature

Fig. 4 Screenshot from a typical alarm log file showing alarm entries for a few minutes. From Pariyani, A., Seider, W. D., Oktem, U. G., & Soroush, M. (2012a). Dynamic risk analysis using alarm databases to improve safety and quality: Part I—Data compaction. *AIChE Journal, 58*(3), 812–825. Used with permission.

entry, associated with a process (or quality) variable. Column A displays the times in chronological order, with each entry displaying the "Year–Month–Day Hour: Minute a.m./p.m." Column B indicates the entry type: alarm, change in controller settings, and so on. Column C shows the alarm tag of the variable (defined during the commissioning of the unit). Column D shows the alarm type [LO (low), LL (low–low), HI (high), and so on.]. Column E shows the drift status of the alarms, either ALM (alarm) or RTN (return); that is, whether the variable drifts beyond or returns within the alarm thresholds. Column F shows the alarm priority, and column G presents a brief description of the alarm. Similar data entries exist for the interlock data.

In some industrial-scale processing plants, 5000–10,000 alarm activations may realistically occur daily, with on the order of 10^6 alarm activations over a few months. To carry out Bayesian analysis, it is required to create a compact representation of this big data. Fig. 5 shows a schematic of the steps to create the compact representation, beginning with the raw data at the top and, after the steps described herein, resulting in likelihood data required for Bayesian analysis to estimate the failure probabilities of the safety systems.

2.2 Safety Systems

Next, returning to the discussion of safety systems in Section 1.3, three of the most commonly used systems are further described, noting that the number of systems and their functionalities are dependent on the specific chemical plants.

Operator (machine + human) corrective actions, Level I—SS_1—which refers to human operator-assisted control to keep the variables within their high/low alarm thresholds and to return them to normal operating conditions. When unsuccessful, variables enter into their red-belt zones.

Operator (machine + human) corrective actions, Level II—SS_2—which refers to human operator-assisted control to keep the variables within their high–high/low–low alarm thresholds and to return them to normal operating conditions. These corrective actions are more rigorous than those for level I, because when unsuccessful, the interlocks are activated, often after a short time delay.

Interlock shutdown—SS_3—which refers to an automatic, independent, ESD system that shuts down the unit (or part of the unit).

To assess the reliability of these systems for a process, a framework involving *event-trees* and *multisets* is outlined in this section to provide a compact representation of vast alarm data, which facilitates the statistical analysis using

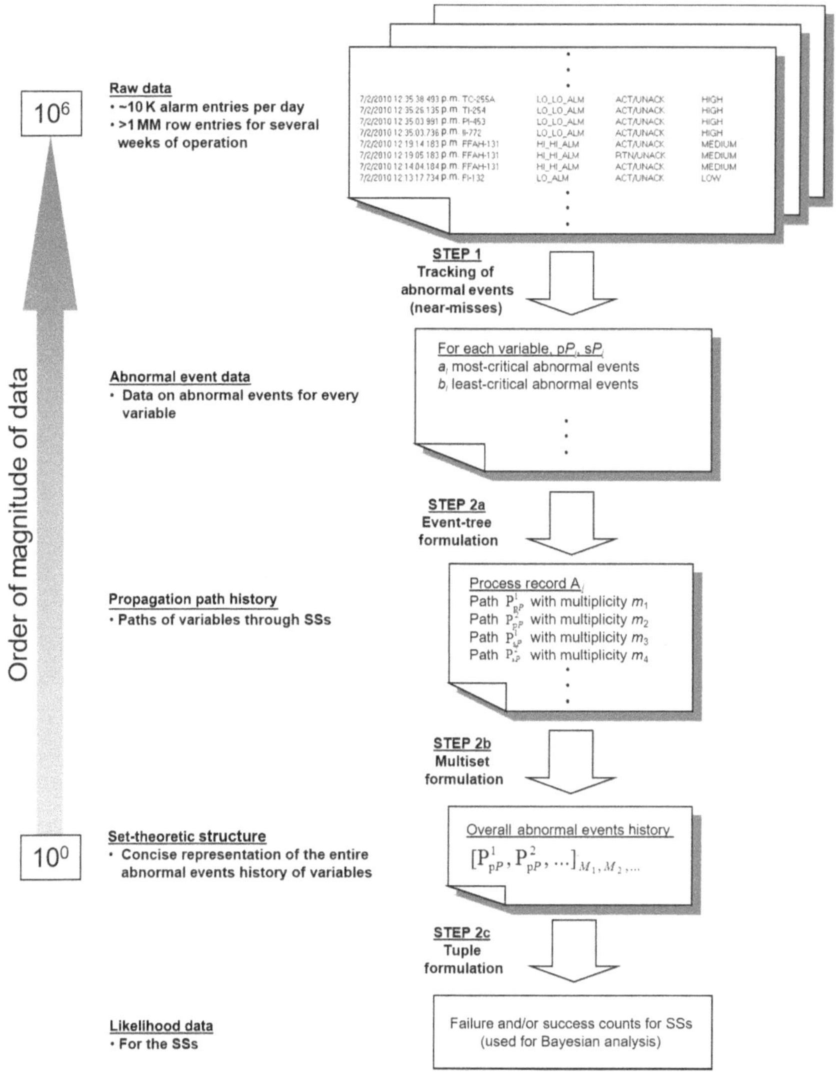

Fig. 5 Steps to prepare compact likelihood data for Bayesian analysis. From Pariyani, A., Seider, W. D., Oktem, U. G., & Soroush, M. (2012a). Dynamic risk analysis using alarm databases to improve safety and quality: Part I—Data compaction. *AIChE Journal, 58*(3), 812–825. Used with permission.

Bayesian theory. The combined framework accounts for the complex interactions that occur between the DCS, human operators, and the interlock system—yielding enhanced estimates and predictions of the failure probabilities of the safety systems and, more importantly, the probabilities of the occurrence of shutdowns and accidents. This causative relationship between the *SSs* is modeled using *copulas* (multivariate functions that represent the dependencies among the systems using correlation coefficients).

2.3 Upset States

A process is said to be in an upset state when process variables move out of their green-belt zones, indicating "out-of-control" or "perturbed" operation. Upset states lead to deterioration in operability, and safety performances of the process. Equations to estimate the operability and safety performances have been proposed in the literature (Pariyani et al., 2010; Ye et al., 2009). The following two upset states are presented here, with additional details available (Pariyani et al., 2010).

Operability upset state (OUS), where at least one of the secondary process variables lies outside its green-belt zone, but all the primary process variables lie within their green-belt zones. In this case, the operability performance deteriorates significantly, whereas safety performance appears to be maintained. This occurs, for example, when the flow rate of a stream (a secondary process variable) moves just above its green-belt zone, but not sufficiently far to move the primary process variables out of their green-belt zones.

Safety upset state (SUS), where at least one of the primary process variables lies outside its green-belt zone. In this case, both safety and operability performances are affected and automatic action is required unless the plant operator takes effective corrective action.

Clearly, plants can move from one upset state to another as disturbances (or special causes, which cause abnormal events) progress.

2.4 Principal Steps in Dynamic Risk Assessment—Summary

The dynamic risk assessment method herein consists of three steps, shown as three regions of the pyramid in Fig. 6. The steps are: (1) near miss tracking, (2) event-tree and set-theoretic formulation, and (3) Bayesian analysis.

Near Miss Tracking refers to identification and tracking of near misses over an extended period of time (weeks, months, and so on). As mentioned earlier, the abnormal events experienced by the process variables are recognized as near misses. Pariyani et al. (2010) presented techniques for tracking abnormal events and analyzing recovery times to quantify, characterize, and track near misses experienced by individual and groups of variables over different periods of time. Using Pareto charts and alarm frequency diagrams, these techniques permit identification of variables that experience excessive numbers of abnormal events, drawing the attention of plant management to

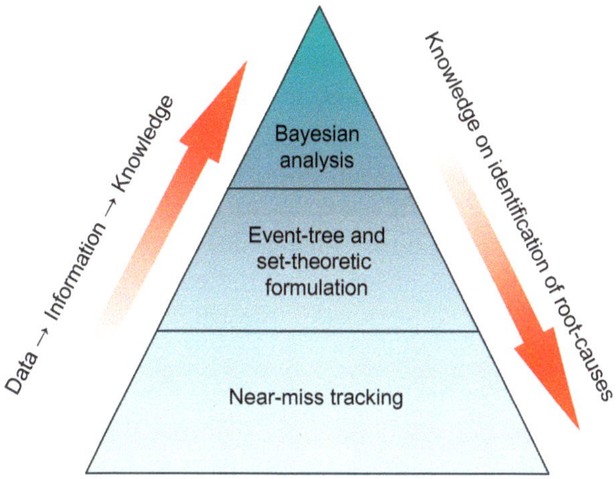

Fig. 6 Dynamic risk assessment pyramid showing different stages. From Pariyani, A., Seider, W. D., Oktem, U. G., & Soroush, M. (2012a). Dynamic risk analysis using alarm databases to improve safety and quality: Part I—Data compaction. *AIChE Journal, 58*(3), 812–825. Used with permission.

potential improvements in control strategies, alarm thresholds, and process designs. They also suggest opportunities to explore the root cause(s) of each abnormal event and reduce the frequency of each abnormal event. These approaches improve upon alarm management techniques by drawing attention to the severity of abnormal events experienced by variables and their associated recovery times (rather than alarm counts only). The results suggest the need to carry out these statistical analyses in real- or near-real time—to summarize for plant operators those alarms associated with the most abnormal events and requiring the most attention; that is, allowing prioritization of the numerous flags raised by the alarms.

Event-tree and set-theoretic formulation permits the transformation of near miss data to *information* on the performances of the safety systems. As presented in the discussion of Fig. 5, in this step, the near miss data that tracks: (a) abnormal events, (b) the propagation of abnormal events, and (c) the attainment of end states, is stored in set-theoretic formulations to represent the branches of event-trees.

As abnormal events arise in processes, they are handled by the *SS*s, whose actions guide the process variables through their green-, yellow-, and red-belt zones, resulting in either continued normal operation (variables in their green-belt zones) or upset states (OUS, SUS). The sequences of responses (that is, successes or failures) of the *SS*s are the *paths* (denoted by P^i) followed by the process variables and are described by the branches of the *event-trees*, as shown in Fig. 3. They track abnormal events to their end states (i.e., normal operation, plant shutdown, accident, and so on). Using a generalized set-theoretic formulation (discussed in Section 2.5.3), these paths are represented in a condensed format—to facilitate Bayesian analysis. Stated differently, near miss data extracted from the alarm data show how variables move among their yellow- and red-belt zones to their end states. From these, event-trees are created and represented with new set-theoretic notations. This permits the systematic utilization of the historical alarm databases in Bayesian calculations to estimate failure probabilities, the probabilities of accidents, and the like.

Bayesian analysis refers to the utilization of the transformed data to obtain *knowledge* (that is, statistical estimates) of the performances (in terms of failure probabilities) and pair-wise interaction coefficients of the safety systems. Also, the probabilities of incidents are estimated. These estimates help to identify the root-causes in the process, for example, the *SS*s with high failure probabilities or variables experiencing high abnormal event rates. In particular, consider a case when the failure probabilities of the operator corrective actions (SS_1 and SS_2) are high—giving operators and managers incentives to identify their root-causes—possibly due to insufficient operator training, stress factors, and so on.

2.5 Data Compaction

2.5.1 Objectives

When compacting millions of alarm data entrees, Pariyani et al. (2012a) introduced path formulations to trace the process variables together with a set-theoretic formulation. This section presents these two formulations.

2.5.2 Event-Tree Formulations for Process (and Quality) Variables

The event-tree formulations depict the actions of the safety systems (*SS*s) as they respond to abnormal events—with their branches representing the paths traced by the process (and quality) variables. Note that each *SS* is represented by a node, with the success or failure of each system denoted by *S* or *F*, respectively, along two branches leaving each node.

Fig. 3 shows an event-tree for an abnormal event (when a *primary* variable leaves its green-belt zone). The tree is illustrated for the three typical *SS*s discussed earlier. Again, primary process variables are denoted by p*P* and secondary process variables by s*P* (and primary and secondary quality variables are denoted by p*Q* and s*Q*).

Depending on the performance (success or failure) of these *SS*s, four paths are possible—with two paths leading to *continued operation*, CO (when variables return to their green-belt zones), one path leading to *emergency shutdown*, ESD (when a primary variable enters its red-belt zone and interlock is activated), and one path leading to *plant accident* (when the *SS*s fail to remove a *primary* variable from its red-belt zone). The latter often leads to loss of life, serious injuries, and major equipment losses, involving major economic losses due to product losses and manpower requirements to return the process to normal operation.

The paths are numbered according to the index of the end state in the event-tree—from top to bottom, using the notation, P^i_{pP} where *i* is the path counter. The primary variables follow the uppermost path, P^1_{pP}, when the basic process control system (BPCS) fails to keep them within their green-belt zones resulting in *least-critical abnormal events*, but the SS_1 successfully returns them to their green-belt zones. Symbolically, the path is represented by S^1_{pP} or simply S^1, where S^k_{pP} or simply S^k denotes the success of safety system *k*. Also, the combined path and its end state are denoted as P^1_{pP}-CO.

The primary variables follow the second path, P^2_{pP}, when they enter their red-belt zones, marking the failure of SS_1, indicating the first level of corrective actions by operators. However, the second-level (more rigorous) corrective actions by the operators successfully return them to their green-belt zones. This path, is represented by $F^1 \rightarrow S^2$, where F^k denotes the failure of SS_k. Together with its end state, this *most-critical abnormal event* is denoted as P^2_{pP}-CO. The primary process variables follow the third path, P^3_{pP}, when they enter their red-belt zones, marking the failures of operator corrective actions (both levels) indicated by SS_1 and SS_2 and successful actions by the interlock, indicated by SS_3, resulting in a shutdown. This path is represented by $F^1 \rightarrow F^2 \rightarrow S^3$. Together with its end state, this *most-critical abnormal event* is denoted as P^3_{pP}-ESD.

The *primary* variables follow the fourth path, P^4_{pP}, when the interlock fails to remove the variables from their red-belt zones or bring plant operations to a shutdown, resulting in a plant accident. This path is represented by $F^1 \rightarrow F^2 \rightarrow F^3$. Together with its end state, this *most-critical abnormal event* is denoted as P^4_{pP}-ACCIDENT.

At times, variables oscillate between their yellow- and red-belt zones, before returning to their green-belt zones. In such cases, the above notation applies—with the criticality of the abnormal events determined by the highest belt zone entered. This concise notation, with the set-theoretic representation in the next subsection, has been shown to represent effectively complex alarm sequences for the Bayesian analysis. As an example, consider an abnormal event with a process variable (assuming no interlock activation): (a) entering its red-belt zone, (b) returning briefly to its yellow-belt zone, (c) reentering its red-belt zone, and (d) returning to its green-belt zone. For this abnormal event, the SS_1 failed to keep the variable within its yellow-belt zone, and the SS_2 succeeded, in its second attempt, in returning the variable to its green-belt zone. Note that, in practice, when variables experience oscillations about their thresholds, *deadbands* (of 2%–5%) are often applied to prevent *nuisance* alarms (Kondaveeti et al., 2010).

For variables (and processes) involving different schemes of *SSs*, the notation remains applicable. For example, if the interlock system (SS_3) is designed to be activated without any time delay (indicating absence of SS_2) for a primary variable, only three paths are possible—with their end states, denoted as: $\mathrm{P}^1_{\mathrm{p}P}$-CO, $\mathrm{P}^3_{\mathrm{p}P}(\text{-II})$-ESD, and $\mathrm{P}^4_{\mathrm{p}P}(\text{-II})$-ACCIDENT. In this notation, the path numbers are unchanged and the event-trees remain applicable when certain *SSs* are not included.

In practice, the interlock system is associated only with critical variables of the process (*primary* variables) to reduce costly shutdowns. The remaining secondary variables involve only corrective actions by the operators to return them to their green-belt zones. Using the earlier steps, event-trees can be constructed for *secondary* variables (s*Ps*) that enter their yellow- or red-belt zones (without any interlock actions).

Note that if one or more primary variables enter their red-belt zones, corrective actions taken by *SSs* are likely to return them and, in turn, the secondary variables to their green-belt zones. Because of interactions between the primary and secondary variables, the effects of their corrective actions are channeled to the latter, causing them to return to their green-belt zones as well. However, the recovery times of p*Ps* and s*Ps* often vary significantly. These interdependent effects due to nonlinear interactions can be handled effectively by implementing multivariable, nonlinear model-predictive controllers (MPCs). Herein, the event-trees do not explicitly account for the *auxiliary* effects of *SSs* on different categories of variables. Also, these event-trees are applicable to only continuous processes, wherein process variables return to their green-belt zones eventually (except when shutdowns or accidents occur). Event trees for batch processes can be developed similarly, with time-varying thresholds. Finally, returning to Fig. 5, in Step 1, abnormal events in the raw alarm data are tracked to extract abnormal event histories for each variable, pP_i, sP_i—involving most- and least-critical abnormal events.

2.5.3 Set-Theoretic Formulation—Case Study 1

To introduce the set-theoretic formulation, consider Case Study 1, which is presented in Table 1 as a *process report* for a typical continuous process over a brief period (consisting of minute-by-minute status updates in which a disturbance drives a few process variables out of their green-belt zones). Between 1:00 and 1:01 p.m., the process enters an OUS. In the next 2 min, it moves from an OUS to a SUS as four of its primary process variables move out of their green-belt zones. However, within the next minute, all the variables are returned to their normal operating ranges, with CO occurring—thanks to levels I and II corrective actions by the operators.

In this case study, four abnormal events (three least-critical and one most-critical) occurred, after the BPCS failed to keep its process variables within their green-belt zones. As the *SS*s responded, pP_2, pP_3, and pP_4, which entered their yellow-belt zones only, followed the path, S^1, whereas, the first primary process variable, pP_1, which entered its yellow- and red-belt zones, followed the path $F^1 \rightarrow S^2$. Thus, based on the event-tree in Fig. 3, two distinct paths were followed: (a) P^1_{pP}—followed by pP_2, pP_3, and pP_4 and (b) P^2_{pP}—followed by pP_1—all leading to the end state, CO. Thus, this *abnormal events history*, which shows several variables experiencing abnormal events over a period of time, is represented by a collection of paths, traced by the variables, leading to the same end state, and therefore, is referred to as a *process record*. Note that an abnormal events history may include more than one process record, corresponding to different end states attained by the variables; e.g., CO, plant shutdown, etc.

Table 1 Process Report—Case Study

1:00 p.m.	Normal operation
1:01 p.m.	Four primary variables pP_1, pP_2, pP_3 and pP_4, enter their yellow-belt zones (4 high alarms go off)
1:02 p.m.	The first primary variable, pP_1, enters its red-belt zone (1 high–high alarm goes off)
1:03 p.m.	Operators successfully diagnose and correct the problem, with four process variables returned to their green-belt zones

Again, returning to Fig. 5, using the abnormal event data created in Step 1, propagation paths through the *SSs* are extracted in Step 2a using the event-tree formulations discussed earlier. Note that for the process report in Table 1, a *process record* is created comprised of paths P^1_{pP} and P^2_{pP}, followed m_1 (=3) and m_2 (=1) times by the four variables.

Next, the key premises of the set-theoretic model are presented:

(1) The paths followed by the variables are modeled as *n-tuples*, where *n-tuples* are *ordered lists* of finite length, *n*. Like *sets* and *multisets* (discussed later), *tuples* contain objects, as discussed by Moschovakis (2006). However, the latter appear in a certain order (which differentiate them from *multisets*) and an object can appear more than once (which differentiates them from *sets*). Herein, for paths of event-trees, modeled as *n-tuples*, *n* denotes the number of *SSs*, and the objects are Boolean variables with permissible values, 0 (FALSE) and 1 (TRUE), for the failure and success of the *SSs*, respectively. When any system is not activated, a null value, φ, is used. For the event-trees discussed earlier, the paths are modeled as three-tuples (for the three *SSs*), given by:

$$P^1_{pP} = (1, \varphi, \varphi);\ \ P^2_{pP} = (0, 1, \varphi);\ \ P^3_{pP} = (0, 0, 1);\ \ P^4_{pP} = (0, 0, 0)$$

This notation is also applicable to event-trees with fewer *SSs*. For example, the three-tuple notation for $P^3_{pP}(\text{-II})$ and $P^4_{pP}(\text{-II})$ are (0, φ, 1) and (0, φ, 0).

(2) The *set* of distinct paths, that is, $\{P^1_{pP}, P^2_{pP}\}$ for Case Study 1, denoted as a_m, followed by the process variables is referred as an *underlying set of paths*. Note that because the elements in a set cannot be repeated (Moschovakis, 2006), the four paths (for four abnormal events) for Case Study 1 are not explicitly shown. To include the number of abnormal events associated with each path, *multisets* (Blizard, 1991) are used herein. In a *multiset*, the elements are repeated with a *multiplicity* equal to the number of repetitions; and the *cardinality* of the multiset is the sum of the multiplicities of its elements. Note that a *set* is a multiset with unique elements. Hence, the abnormal events history in the process report in Table 1 is represented as a process record, A_m, represented using a multiset of cardinality 4, $[P^1_{pP}, P^1_{pP}, P^1_{pP}, P^2_{pP}]$, or in the standard format for multisets, $[P^1_{pP}, P^2_{pP}]_{3,1}$, where the multiplicities of P^1_{pP} and P^2_{pP} are 3, and 1, respectively, with an associated CO end state.

It follows that, using the event-tree and set-theoretic formulation herein, any abnormal events history, comprised of abnormal events involving different process variables, can be represented as process records; that is, *multisets* of paths (modeled as three-tuples herein) traced by the process variables as the *SSs* take actions. Also, for any process record, a unique and nonempty *underlying set of paths* is defined; whose elements are the various paths of the event-trees, as discussed earlier.

Returning to Fig. 5, in Step 2b, the overall abnormal events history is summarized as a *multiset* of paths. The second block from the bottom shows a multiset, for the entire alarm database, showing typical paths, $\mathrm{P}^1_{\mathrm{p}P}$ and $\mathrm{P}^2_{\mathrm{p}P}$, and their multiplicities, M_1 and M_2. Finally, in Step 2c, the likelihood data for the safety systems are obtained from the overall abnormal events history using a tuple formulation, as illustrated for a fluidized catalytic cracking unit next. These contain the failure and/or success counts to be used in Bayesian analysis.

2.5.4 Set-Theoretic Formulation—FCCU Case Study

In this case study, alarm data over an extended time period, associated with a fluid-catalytic-cracking unit (FCCU) at a major petroleum refinery that processes over 250,000 barrels of oil per day, are used. The unit has on the order of 150–200 alarmed variables and as many as 5000–10,000 alarm occurrences per day—indicating 500–1000 abnormal events daily. A total of 2545 abnormal events occurred for the primary variables during the study period. The abnormal events history of the primary variables for the study period is summarized in Table 3 of Pariyani et al. (2012a)—demonstrating that the new set-theoretic framework provides a compact representation in handling thousands of abnormal events depicting success/failure paths followed by the process variables through the *SSs*. This framework facilitates the Bayesian analysis to compute failure probabilities of the *SSs* and incident probabilities.

2.6 Bayesian Analysis Using Copulas

As discussed by Pariyani et al. (2012b), significant interactions between the performances of the *SSs*, measured as failure probabilities, are due to nonlinear relationships between the variables and behavior-based factors. In their Bayesian analysis, they use copulas to account for these interactions.

To introduce copulas, the Sklar (1973) shows that a copula function can model the *cumulative joint distribution*, $F(x_1, \ldots, x_n)$, between random variables when the correlation between them is unknown:

$$F(x_1, \ldots, x_n) = C[F_1(x_1), \ldots, F_n(x_n)],$$
$$u_j = F_j(x_j) = \int_0^{x_j} f_j(x_j)\,dx_j, \quad j = 1, \ldots, n, \tag{4}$$

where C is the copula function, and $F_1, \ldots, F_n$ denote the marginal cumulative distributions for random variables, $x_1, \ldots, x_n$, respectively. The joint distribution is independent of the type of random variable and only depends upon the degree of correlation. Similarly, the joint distribution, $f(x_1, \ldots, x_n)$, can be obtained as a function of the *copula density*, c, and the marginal distributions ($f_1, \ldots, f_n$):

$$f(x_1, \ldots, x_n) = f_1(x_1) \times \cdots \times f_n(x_n) c[F_1(x_1), \ldots, F_n(x_n)], \tag{5}$$

where c is defined as:

$$c(u_1, \ldots, u_n) = \frac{\partial^n C}{\partial u_1, \ldots, \partial u_n}, \quad u_j = F_j(x_j), \quad j = 1, \ldots, n. \tag{6}$$

Several multidimensional copula families have been reported (Nelsen, 1999; Yi & Bier, 1998), one of which is the Cuadras and Auge copula:

$$F(x_1, \ldots, x_n) = \prod_{h=1}^{n} F_h(x_h) \prod_{k=1}^{n-h} \left(1 - \theta_{h-(h+k)}\right), \tag{7}$$

where θ_{j-q} expresses the degree of correlation between the random variables x_j and x_q, and $\prod_{k=1}^{n-h}\left(1 - \theta_{h-(h+k)}\right) = 1$ when $n - h = 0$. In three dimensions, Eq. (6) becomes:

$$C(u_1, u_2, u_3) = u_1^{(1-\theta_{1-2})(1-\theta_{1-3})} u_2^{(1-\theta_{2-3})} u_3 \tag{8a}$$

and the copula density (Eq. 6) is:

$$\begin{aligned} C(u_1, u_2, u_3) &= \frac{\partial^3 C}{\partial u_1 \partial u_2 \partial u_3} \\ &= (1 - \theta_{1-2})(1 - \theta_{1-3}) \\ &\quad \times (1 - \theta_{2-3}) u_1^{(1-\theta_{1-2})(1-\theta_{1-3})-1} u_2^{-\theta_{2-3}}. \end{aligned} \tag{8b}$$

Spearman's ρ-rank correlation coefficients are often used to represent the degree of correlation in joint distributions using copulas. These are a function of the copula alone and are independent of the marginal distributions of the correlated random variables. Many copulas have closed-form expressions for the ρ-rank correlation coefficients as a function of their correlating parameters, as discussed by Nelsen (1999); for example, the ρ-rank correlation coefficient, $\rho(x_j, x_q)$, and the correlation parameters θ_{j-q}, in the Cuadras and Auge copula, are related by:

$$\rho(x_j, x_q) = \frac{3\theta_{j-q}}{4 - \theta_{j-q}} \quad \text{for } j \neq q. \tag{9}$$

To understand the usage of the Cuadras and Auges copula to estimate the failure probabilities, the reader is referred to Meel and Seider (2006). In that paper, four Bayesian models are implemented for the same CSTR consequence data, each model differing in the use of the Cuadras and Auges copula. Therein, prior joint distributions for safety systems are expressed using the copula density in Eq. (5). The failure probabilities of various safety systems (associated with the CSTR) are updated in real time, and the reader gains an appreciation for the dependency of these probabilities on the copula structure. Also, matrices are provided to specify the ρ-rank correlation coefficients that represent the interactions between the failure probabilities associated with safety systems. Finally, the parameters of the Beta prior distributions and the resulting posterior distributions are provided. These latter distributions depend on the actions of the prior safety systems in the event-tree.

2.6.1 FCCU Results

For the FCCU, Pariyani et al. (2012b) use the Normal and Cuadras and Auges copulas. Unfortunately, due to space limitations, the detailed correlation matrix and discussion are not included herein, as well as the details of the prior distributions. Instead, results are shown in Fig. 7 for the failure probabilities of the *SS*s.

The mean posterior failure probability (and associated variance) of the operator's level I corrective actions is fairly low (=0.074)—indicating their robust performance. However, for operator's level II corrective actions, the mean failure probability is high (=0.851), indicating their difficulties in keeping the variables within their orange-belt zones—note they use an additional safety system (referred as *override controller*) with an additional *orange-belt zone* between the yellow- and red-belt zones. Stated differently, the probability that the primary variable moves from its yellow-belt zone to its orange-belt zone is just 7.4%, while the probability of moving from its orange-belt zone to its red-belt zone is 85%—indicating ineffective actions of level II operator corrections. Hence, for the FCCU, to prevent variables from moving to their red-belt zones, it is important to prevent their crossing into orange-belt zones.

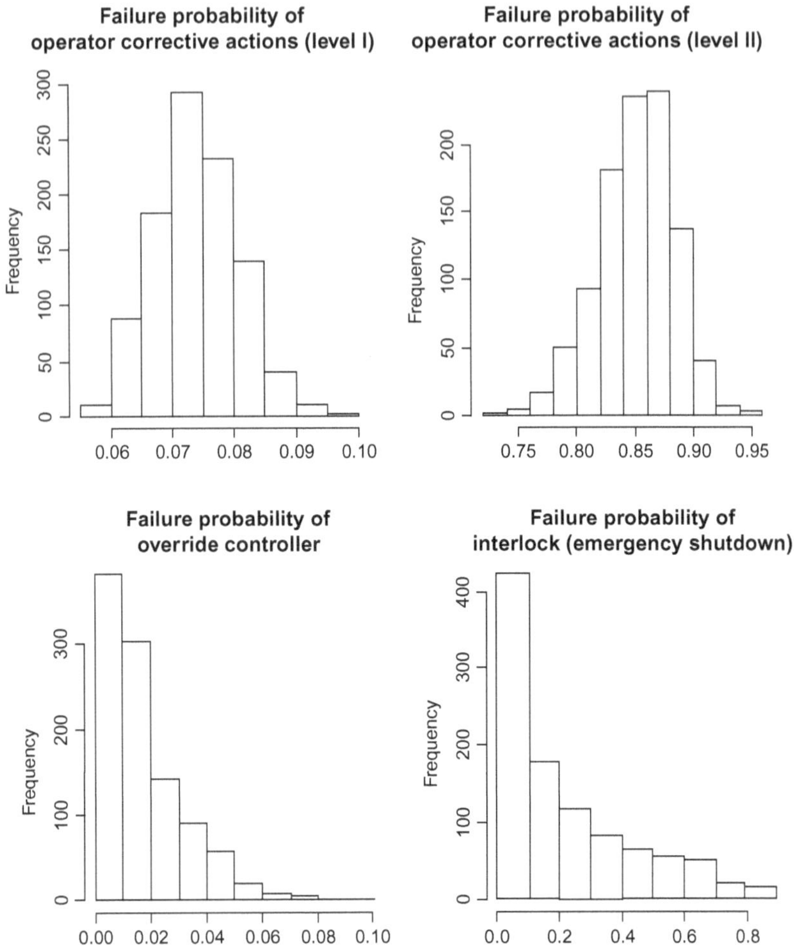

Fig. 7 Marginal posterior distributions using multivariate normal copula. From Pariyani, A., Seider, W. D., Oktem, U. G., & Soroush, M. (2012a). Dynamic risk analysis using alarm databases to improve safety and quality: Part II—Bayesian analysis. *AIChE Journal, 58*(3), 826–841. Used with permission.

The mean failure probability of the override controller is also quite low (=0.017). However, its variance is high, as compared to that of previous safety systems. Also, the failure probability of the ESD system is relatively uncertain, due to the availability of few data points. In these cases, past performances (over several months and years) and/or expert knowledge regarding their failures (often supported by system test results), are desirable. The latter can be derived from the dynamic risk analysis of near miss data from related plants. Note that similar results can be obtained for individual time periods.

3. INFORMED PRIOR DISTRIBUTIONS

Next, techniques for constructing *informed prior distributions* when carrying out Bayesian analyses to determine the failure probabilities of safety systems are presented (Moskowitz et al., 2016, 2015). This is to circumvent the inaccuracies introduced because rare-event data are sparse having high-variance likelihood distributions. When these are combined with typical high-variance prior distributions, the resulting posterior distributions naturally have high variances preventing reliable failure probability predictions. As discussed in this section, the techniques use a repeated-simulation method to construct informed prior distributions having smaller variances, which in turn lead to lower variances and more reliable predictions of the failure probabilities of alarm and safety interlock systems.

3.1 Constructing Informed Prior Distributions

The method involves the eight steps listed in Table 2. In Steps 1–3, a robust, dynamic, first-principles model of the process incorporating the

Table 2 Steps to Construct an Informed Prior Distribution

1. Develop a dynamic first-principles process model
2. Incorporate control system into the dynamic process model
3. Incorporate the alarm and safety interlock system into the dynamic process model
4. Postulate potential special-cause events to be studied
5. For each special-cause event, construct a distribution for the event magnitudes, A_{SC} (i.e., for a postulated pressure decrease, construct a probability distribution for a decreasing magnitude).
6. For each special-cause event, construct a distribution for operator response time, τ.
7. For each special-cause event, conduct the simulation study according to the algorithm described in Fig. 8 to simulate the range of possible event magnitudes, A_{SC}, and operator response time, τ.
8. Using the method of Gelman et al. (2014), estimate parameters of a distribution model (e.g., Beta distribution) representing the data generated in Step 7—this is used as an informed prior distribution.

control, alarm and safety interlock systems, is built. The model can then be simulated using a simulator such as gPROMS v.3.6.1 (2014), which is used herein. The control system in the model mimics the actual plant control system, with consistent control logic and tuning. Likewise, the alarm and safety interlock systems in the model mimic those in the plant. For operator actions, this can be difficult, as operators often react differently to alarms. In particular, expert operators may take into account the state of the entire process when responding to alarms. When creating a model, the likelihood of operator actions must be considered. Either the modeler can use the action most commonly taken by operators or a stochastic simulation can be set-up in which the different actions are assigned probabilities.

With these models, SCEs are postulated in Step 4. The list of SCEs can be developed from various sources: HAZOP or LOPA analysis, observed accidents in the plant (or a similar plant), near miss events at the plant (or in a similar plant), or from risks suggested in first-principles models of the plant. For each SCE, an event magnitude distribution is created in Step 5. A distribution for operator response time, τ, is created in Step 6. These three distributions are used along with the dynamic simulation in Step 7 to obtain simulation data. Lastly, in Step 8, the simulated data are used to regress parameters for the informed prior distribution. The algorithm used to generate simulation data (Step 7) and regresses informed prior distribution parameters (Step 8) is described in the paragraph below, and represented pictorially in Fig. 8.

The script that manages the dynamic simulations starts by sampling A_1 from the event magnitude distribution created in Step 5. Note that Fig. 8 shows a normal distribution centered at μ_{SC} with variance σ^2_{SC}, however any distribution can be used. Assign the number of safety system failures, i, to $i=0$. With this A_1, the user script samples τ_1 from the distributions created in Step 6. Although Fig. 8 shows uniform distributions (with the maximum operator response time at τ_{max}), any distribution can be used. With A_1 and τ_1, a dynamic simulation is run. If the safety system fails to avoid a plant-wide shutdown, then $i=i+1$; if the safety system is successful, i is not incremented. When $n<N$, $n=n+1$; i.e., for sampled A_i and τ_i, a dynamic simulation is run, and i is adjusted when necessary. After N iterations, $j_1=i/N$ is calculated, in the range [0, 1]. Then m is incremented and A_m sampled, the inner loop is reexecuted, and j_m is calculated. When the outer loop has been completed ($m=M$), a vector of M elements (j_1, ..., j_M) has been accumulated. The average and variance of this vector are used to calculate α and β of the Beta distribution. Note that

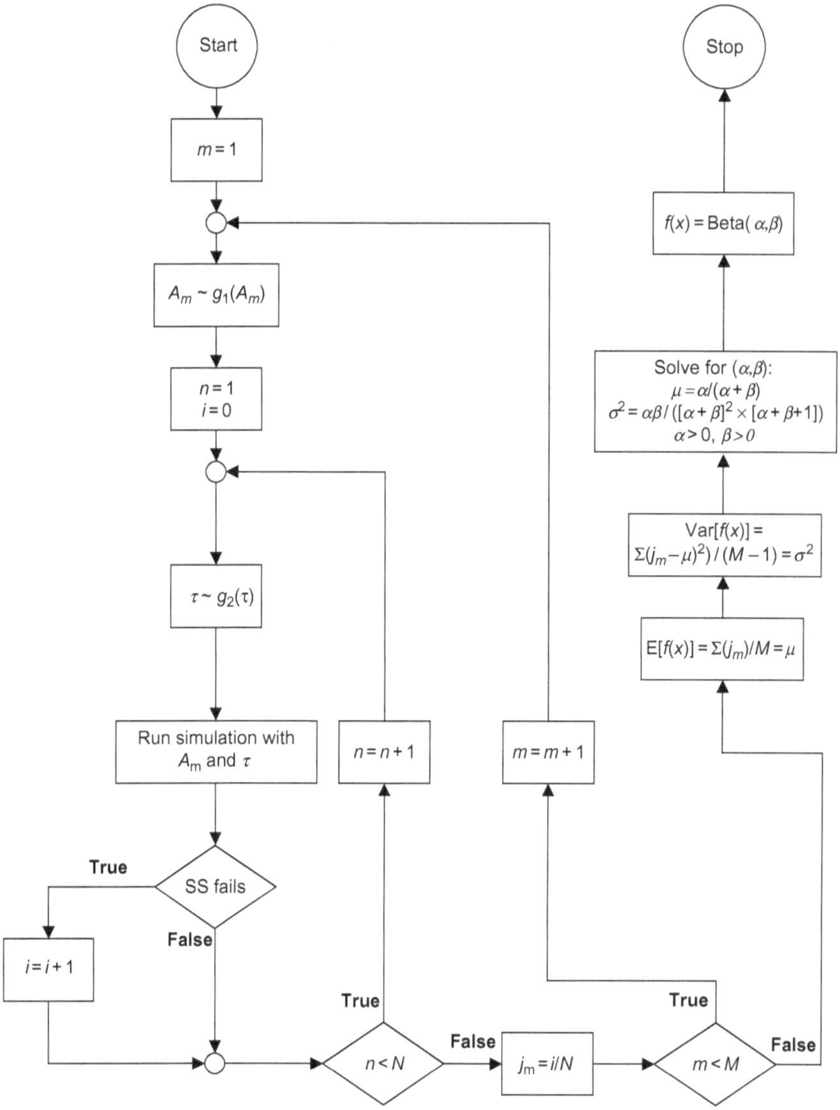

Fig. 8 Sampling algorithm used in Steps 7 and 8 in Table 2. From Moskowitz, I. H., Seider, W. D., Arbogast, J. E., Oktem, U. G., Pariyani, A., & Soroush, M. (2016). Improved predictions of alarm and safety system performance through process and operator response-time modeling. AIChE Journal, 62*(9), 3461–3472. Used with permission.*

because the Beta distribution is the conjugate prior of the binomial likelihood distribution, it is the recommended choice. The number of simulations, $M \times N$, is chosen, recognizing that more simulations yield a smaller prior distribution variance.

3.2 Testbed—Steam-Methane Reforming (SMR) Process

A typical SMR process is shown in Fig. 9. After pretreatment, natural gas feed (70) and steam (560) are mixed before entering the process tubes of an SMR unit (90), where hydrogen, carbon monoxide, and carbon dioxide are produced. This hot process gas (100) is then cooled and sent to a water–gas shift converter (110), where carbon monoxide and water are converted to hydrogen and carbon dioxide. The process gas effluent (120) is cooled in another heat exchanger, producing stream 170, which is sent to two water extractors. Note that the last section of this heat exchanger is used to transfer heat to a boiler feed water makeup stream in an adjacent process. The gaseous hydrogen, methane, carbon dioxide, and carbon monoxide, in stream 210, are sent to PSA beds. Here, high-purity hydrogen is produced (220), and the PSA-offgas is sent to a surge drum. Stream 800 from the surge drum is mixed with hot air (830) and a small amount of natural gas makeup (815), and sent to the furnace side, where it is combusted to provide heat to the highly endothermic process-side reactions. Its' hot stack gas (840) is sent through an economizer, where it is used to heat steam (520), some of which are used on the process side (560), with the rest available for use or sale as a steam product (570).

In modeling for process safety, emphasis should be placed on units that present the greatest risk; i.e., have the largest probabilities of incidents

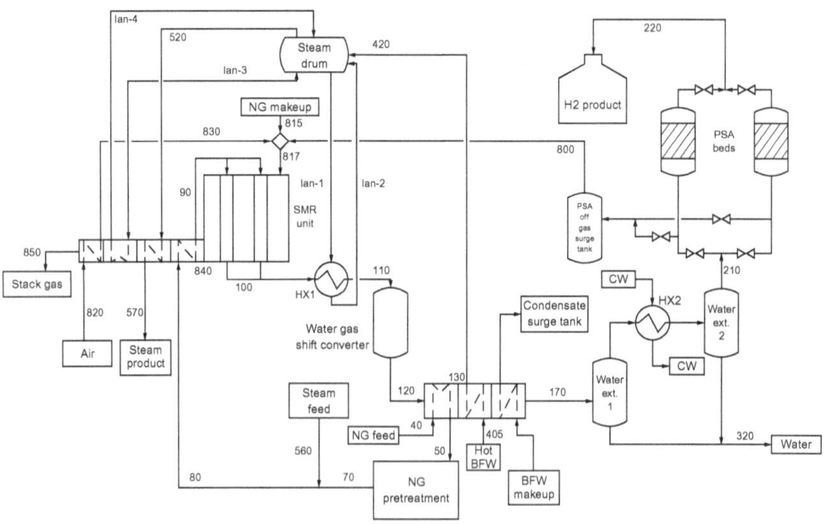

Fig. 9 SMR process flow diagram. From Moskowitz, I. H., Seider, W. D., Soroush, M., Oktem, U. G., & Arbogast, J. E. (2015). Chemical process simulation for dynamic risk analysis: A steam-methane reformer case study. Industrial and Engineering Chemistry Research, 54(16), 4347–4359. Used with permission.

multiplied by incident impact (Kalantarnia et al., 2009). In an SMR process, temperatures rise above 1300 K with pressures over 20 atm. Because overheating can lead to process-tube damage and failure, potentially leading to safety concerns, its model should receive special attention. In Moskowitz et al. (2015), partial differential and algebraic equations (PDAEs), that is, momentum, energy, and species balances, accounted for variations of pressure, temperature, and composition in the axial direction for both the process- and furnace-side gases. For the reforming tubes, the rigorous kinetic model of Xu and Froment (1989) was used, while the furnace–gas combustion reactions were modeled using a parabolic heat-release profile. Convection and radiation were modeled on the furnace side, where view factors were estimated using Monte Carlo simulations and gray–gas assumptions. The heat transfer on the process side was modeled by convection only, assuming a pseudo steady state between the process gas and catalyst. Details of the models are presented by Moskowitz et al. (2015).

The PSA beds represent a cyclic process, with beds switched from adsorption to regeneration on the order of every minute. This type of separation scheme induces oscillatory behavior throughout the SMR process. As the flow rates, compositions, and pressures fluctuate in effluent streams from the PSA beds, variables throughout the entire plant fluctuate as well. In processes with such cyclical units, buffer tanks are often used to dampen fluctuations. However, typical buffer-tank sizes (comparable to SMR-unit sizes) reduce the amplitude of these fluctuations by on the order of 50%. In Moskowitz et al. (2015), the SMR process test-bed involves four PSA beds, which operate in a four-mode scheme, with each bed undergoing adsorption, depressurization, desorption, and repressurization steps.

In the full safety process model used by Moskowitz et al. (2015), the SMR-unit and PSA-bed models are used in conjunction with dynamic models for the water–gas shift reactor, water extractor, surge tank, heat exchanger, and steam drum. Furthermore, the controls used with the dynamic process model are consistent with those used in the Air Liquide process. The full process is modeled using the software package, gPROMS. A challenging aspect of the full process model involves convergence of the PSA-offgas recycle loop. Note that this process model combines SMR and PSA-bed units within a plant-wide scheme with PSA-offgas recycle. The results computed by gPROMS are consistent with the process data from the industrial plant. This plant-wide model is extremely useful for building leading indicators and prior distributions of alarm and safety interlock system failure probabilities.

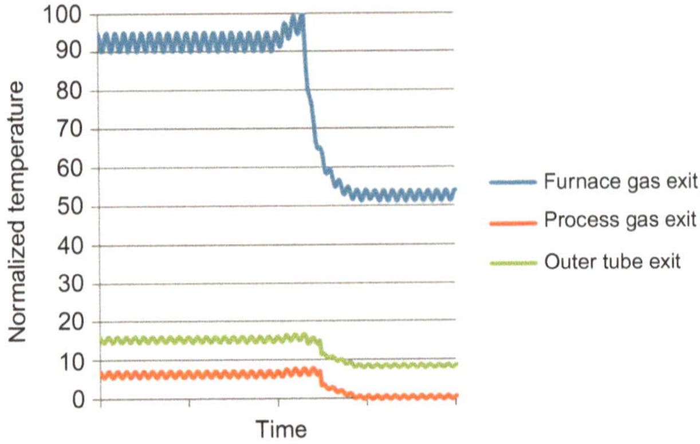

Fig. 10 SMR effluent temperatures for a 10% decrease in the Btu content of the natural gas feed. From Moskowitz, I. H., Seider, W. D., Soroush, M., Oktem, U. G., & Arbogast, J. E. (2015). Chemical process simulation for dynamic risk analysis: A steam-methane reformer case study. Industrial and Engineering Chemistry Research, 54(16), 4347–4359. Used with permission.

With this dynamic process model, process engineers can simulate SCEs and track variable trajectories. Consider an unmeasured 10% decrease in the Btu rating, due to a composition change of the natural gas feed (40), in Fig. 9. Note that the makeup stream (815) on the furnace side is relatively small and is not changed in the simulation. Initially, because the process stream contains less carbon, less H_2 is produced. Because these reactions are endothermic, less heat from the furnace is consumed by the reactions and the furnace temperature rises, as shown in Fig. 10. Also, the process-side temperature increases. Eventually, the low-carbon PSA-offgas reaches the SMR furnace. With less methane for combustion, the furnace temperature decreases, as does the temperature of the process gas. This effect is shown in Fig. 6. Note that the temperatures oscillate due to the natural gas oscillation in stream 800 from the PSA-offgas surge tank—due to the cyclic nature of the PSA process.

3.3 SMR Informed Prior Distributions

For the SMR process in Fig. 9, a loss in steam pressure to the reformer side (stream 560), was simulated with the responses of the safety systems monitored (Moskowitz et al., 2015). For small disturbances, the process control system handled the effect of steam pressure decreases. There is a flow controller on the steam line, whose set point is generated using a linear equation involving the flow of natural gas into the SMR process side, seeking to achieve a constant steam-to-carbon ratio in the process tubes. This control system normally arrests typical fluctuations in steam pressure and flow rate, but for large steam pressure decreases, feedback control alone is insufficient. In this case, the control valve is wide open, with a flow rate insufficient to accompany natural gas fed to the SMR unit. For this reason, an investigation was undertaken to assess the effectiveness of the alarm systems associated with the SMR steam line. When the steam flow rate is below its L-alarm threshold, the steam-to-carbon ratio drops, accompanied by an increase in the process-side temperature and potential tube failure. Because of these operating limits, an interlock was placed at the HH-alarm threshold with a time delay. This time delay, of several seconds, reflects that the temperature threshold may be exceeded in this case for a short period of time and permits the operator to respond rapidly in an attempt to bring the furnace temperature below its HH-alarm threshold.

Three operator responses to the alarm are simulated: (1) the valve on the steam line is opened, (2) the valve on the makeup fuel line is pinched, and (3) the dampers associated with air flow in the furnace are opened (effectively increasing air flow rate). When the operator is able to bring the furnace temperature below the HH-threshold before the interlock delay times out, an automatic shutdown is avoided. If, however, the operator is unable to do so, the interlock is activated and a plant shutdown occurs. The simulated abnormal event leads to either a success of SS_2 (interlock is avoided), or a failure of SS_2 (interlock is activated). It is desirable to have a reliable estimate of x_2, the probability that the operator is not successful, despite only a few activations of this HH-alarm during the recorded history over several years.

Herein, a pressure decrease in the steam line to the process side of the SMR unit was simulated. The magnitude of the pressure decrease was a random variable, sampled from a normal distribution centered at 50% of stream pressure. The response time of the operator was taken as a random variable, sampled from a uniform distribution ranging from 0 to 15 s. The operators three responses were all incorporated into the simulation as step changes in valve settings. One thousand simulations were run, and the effectiveness of the operator's response in each simulation was tracked. In some simulations, the operator successfully reduced the furnace temperature below the interlock threshold in the allotted time before the automatic shutdown. In others, the operator failed and the plant was shutdown. In Fig. 11, a temperature trajectories for events resulting in an interlock activation and in a near miss are shown. In the scenario where SS_2 succeeds, the temperature is brought below the interlock threshold within the interlock delay time. Note that the action of the control system was observed early in the trajectory, but it was insufficient to avoid the abnormal event and eventual plant shutdown.

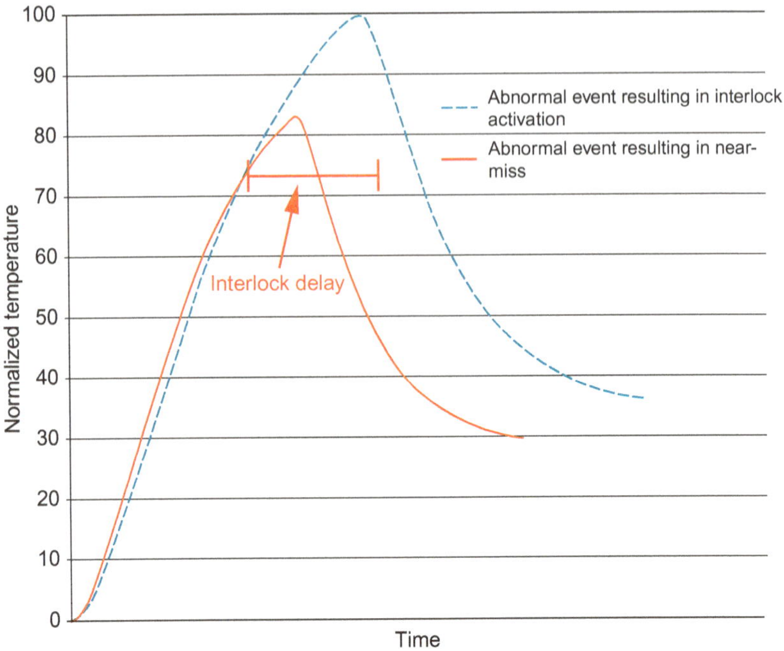

Fig. 11 Furnace outlet temperature for a decrease in steam pressure. From Moskowitz, I. H., Seider, W. D., Soroush, M., Oktem, U. G., & Arbogast, J. E. (2015). Chemical process simulation for dynamic risk analysis: A steam-methane reformer case study. Industrial and Engineering Chemistry Research, 54(16), 4347–4359. Used with permission.

The average number of safety system failures was recorded for the simulations, as well as the failure variance, which were used to generate a Beta distribution to describe the failure probabilities. The Beta distribution, which has just two parameters, was created easily and is supported only in the range [0, 1], which bounds the failure probability.

This informed prior distribution was built using dynamic simulations with first-principle models. Even with no data available to update the distribution, process engineers and plant operators can make improved risk predictions (Leveson & Stephanopolous, 2014). The alarm data are used to build a likelihood distribution, in this case a binomial likelihood distribution of a few trials, all of which are successes. Fig. 12 shows the prior and posterior distributions. The informed posterior is shifted to the left of the prior distribution by the 0% failure rate observed in the data. Unlike the commonly used uninformed prior distributions, its posterior distribution has a similar shape to its informed prior distribution. The posterior distribution generated using the informed prior distribution can alert process engineers that a significant decrease in steam pressure has the high probability (>20%) of

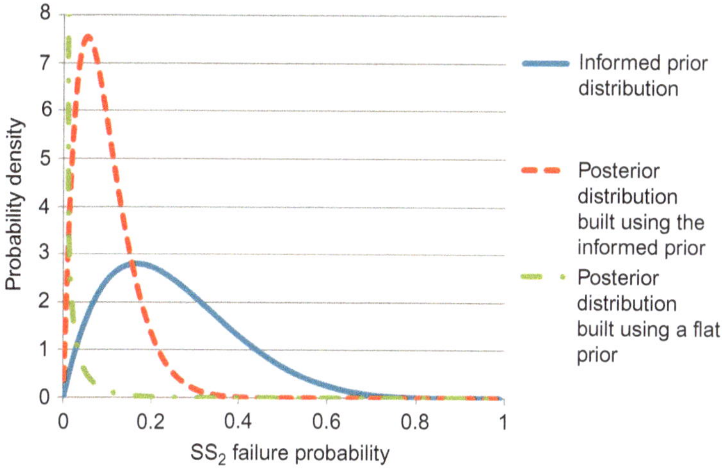

Fig. 12 Prior and posterior distributions generated by dynamic simulations. From Moskowitz, I. H., Seider, W. D., Soroush, M., Oktem, U. G., & Arbogast, J. E. (2015). Chemical process simulation for dynamic risk analysis: A steam-methane reformer case study. Industrial and Engineering Chemistry Research, 54(16), 4347–4359. Used with permission.

causing a plant shutdown. This may lead operators to pay special attention to the steam pressure measurements, and may lead process engineers to install a more robust controller on the steam line.

3.4 Improvements Using Process and Response-Time Monitoring

To improve the predictions of the informed prior distribution, Moskowitz et al. (2016) proposed refinements in the models used to construct them. They began by using SS_1 activations, which are infrequent, but still provide sufficient data for studying the propagation of SCEs. The activation of SS_2 is rarer, occurring $1/x_1$ times less frequently—often resulting in very sparse data. While the data associated with SS_2 are insufficient alone to analyze the performance of the safety system, the similarities between the safety systems can be utilized. The activations of both safety systems originate from a control system failure, which, for example, can be due to the large magnitude of the disturbance, the inability of the control system to handle the disturbance, and/or the occurrence of an electrical or mechanical failure. It is assumed that the same group of operators is involved. If highly skilled, they should arrest the special causes at a high rate (Chang & Mosleh, 2007; Meel et al., 2007; Meel, Seider, & Oktem, 2008).

Clearly, the need for urgent responses of SS_2 is greater. Also, when operators take action (e.g., as furnace temperatures become elevated), the need to respond within the interlock delay times dominate their concerns and actions. This would normally stimulate a strong reaction to avoid automatic shutdown.

A sufficiently accurate first-principles process model is needed to implement the method of Moskowitz et al. (2016). Also, the automated safety system models should be sufficiently accurate. The second model, represented by $g_1(A_m)$ in Fig. 8, is the distribution of SCE magnitudes to be simulated. The operator behavior model, $g_2(\tau)$, unlike automatic safety systems, must reflect human behaviors. Here, the speed and effectiveness of operator responses often depend on the state of the process, the number of competitive active alarms, distractions, personal health and conflicts, and the like.

In the method proposed by Moskowitz et al. (2016), SS_1 models are first constructed and validated with plentiful data. After constructing these models, and validating them with measured SS_1 data, they are modified to handle SS_2 activations (recognizing that their rare occurrences do not allow for reliable model validation). In Sections 3.4.1–3.4.3, all three models are described with respect to SS_1. Their modifications to handle SS_2 activations are then described. Lastly, the failure probability estimates generated by using the SS_2 informed prior distributions are presented.

3.4.1 Dynamic Process Models

Because dynamic first-principles process models are widely used, approaches to model development are not considered here. Instead, this section focuses on model evaluation and improvement for constructing informed prior distributions.

Often, process engineers have developed dynamic process models for control scheme testing during the design phase. These are commonly used initially for carrying out dynamic risk analysis. However, process models used for process design and control are normally developed to track responses in their typical operating regimes (green-belt zones)—but may *not* respond to SCEs with sufficient accuracy; i.e., their predictions far from set points may be poor for risk analysis. Consequently, dynamic process models should often be improved to construct informed prior distributions.

For the SMR process shown in Fig. 9, four dynamic process models are constructed, as summarized in Fig. 13. The first, Process Model A, is the same as the one summarized in Section 3.2. This model includes constitutive equations to model the endothermic reformer reactions, the furnace that provides their heat, the exothermic water–gas shift reaction, the separation of hydrogen product from offgas in adsorption beds, and the production of

		Heat transfer	
		Radiation + convection	Convection only
Kinetics	Froment and Xu Kinetics	Process Model A	Process Model B
	Elementary kinetics	Process Model C	Process Model D

Fig. 13 Steam-methane reforming process models. From Moskowitz, I. H., Seider, W. D., Arbogast, J. E., Oktem, U. G., Pariyani, A., & Soroush, M. (2016). Improved predictions of alarm and safety system performance through process and operator response-time modeling. *AIChE Journal, 62*(9), 3461–3472. Used with permission.

steam (for process heating or sale), as well as models of associated PID controllers.

In Process Model A, to model the radiative heat transfer (~90% of the total heat transfer to the tubes), view factors are estimated from each surface or volume zone to each other zone, with the dynamics of radiative heat-transfer modeled between all discretized zones (Hottel & Sarofim, 1967). The remaining convective heat transfer is simpler, because heat transfer only occurs between physically adjacent zones (Lathem, McAuley, Peppley, & Raybold, 2011).

In Process Models B and D, only convection heat transfer is modeled, with radiative heat transfer accounted for by overstating the heat-transfer coefficients between the furnace gases and tube surfaces. To estimate the overstated heat-transfer coefficients, 50 steady-state windows were identified in the historical process data (Moskowitz et al., 2016). Each window corresponds to a duration of operation, on the order of a day, where process variables are at steady state. Many different steady-state windows exist in the process data due to different demand rates of hydrogen and steam, different feed ratios of steam to natural gas, and natural aging of the catalyst (in the reformer as well as the water–gas shift reactor). The heat-transfer coefficient was estimated from the temperature and flow rate of the process and furnace gas inlet and outlet measurements.

In Process Models A and B, the reforming reaction kinetics proposed by Xu and Froment (1989), which had been shown to be quite accurate over a broad range of temperatures and reactant concentrations, are used. Note that, due to the presence of a complex denominator in the kinetic equations, the spatially distributed SMR model can be difficult to converge. Accurate guess values for the concentration of the reactants and products along the axial direction of the reformer tubes must be available, or generated using homotopy-continuation techniques, to converge the steady-state model.

However, in Process Models C and D, elementary reaction kinetic equations are simpler to converge. The rate constants of the elementary reactions are estimated, similar to the convection heat-transfer coefficients. Using the data in the 50 steady-state windows, along with measured hydrogen product flow rates and offgas concentrations, the rate constants were estimated (Moskowitz et al., 2016).

Initially, the four process models were compared in the 50 steady-state windows. Beginning with the measured inlet temperatures and flow rates for each mode, predicted and measured outlet temperatures were compared for each model. The root-mean square outlet temperature differences are shown in Fig. 14. For this steady-state evaluation, Process Model A provided the best agreement with the data, whereas Process Model D was least accurate.

However, because the models were used to estimate the responses to SCEs, agreement with dynamics data is more important. Fifty dynamic

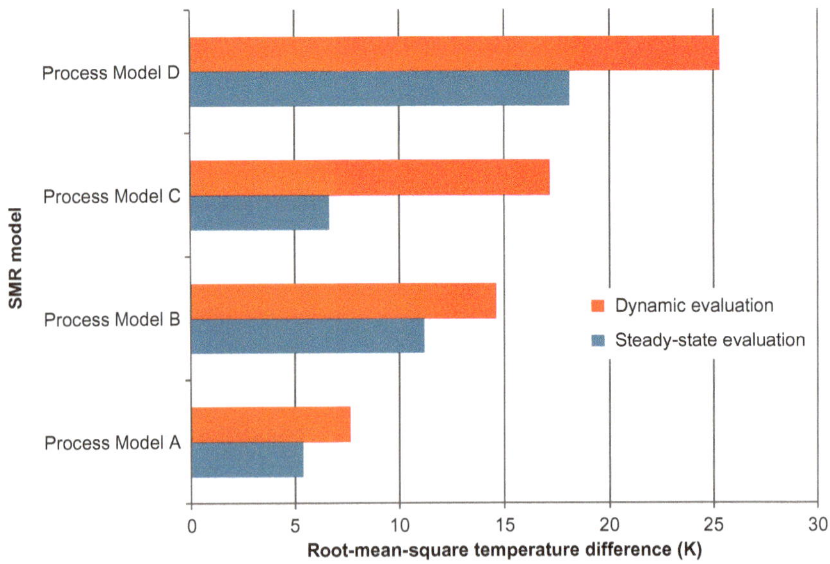

Fig. 14 Process model goodness-of-fit using steady-state and dynamic evaluations. From Moskowitz, I. H., Seider, W. D., Arbogast, J. E., Oktem, U. G., Pariyani, A., & Soroush, M. (2016). Improved predictions of alarm and safety system performance through process and operator response-time modeling. *AIChE Journal, 62*(9), 3461–3472. Used with permission.

windows were identified in the historical process data—periods of time where the process variables describing the operation of the steam-methane reforming reactor are transient. These windows are on the order of minutes to hours, and typically arise when hydrogen or steam demand rates change, feed ratios of steam to natural gas change, or operational changes occur in another process unit (such as a pair of pressure-swing adsorbent beds are taken offline). For each of the 50 dynamic windows, inlet temperature and flow rate trajectories were input to each model, with model-predicted outlet temperature trajectories compared to measured outlet temperature trajectories. Dynamic predictions are typically less accurate than the steady-state ones. Here, Process Model B outperformed Process Model C, but when comparing just steady-state outlet temperature differences, Process Model C provided a closer fit to the data. Clearly, Process Model B should be selected, rather than Process Model C, when constructing informed prior distributions.

Next, the four process models were used to construct informed prior distributions for the failure of SS_1—using the uniform distributions for special-cause magnitude and operator behavior used in Section 3.3. The 300 measured SS_1 failures/successes were then used to construct a low-variance binomial likelihood distribution (see Eq. 4 in Moskowitz et al., 2016). The four informed prior distributions for the failure of SS_1 and the binomial distribution are shown in Fig. 15. To compare the four informed prior distributions with this likelihood distribution, the index

$$\xi_{i,m} = 1 - \frac{1}{2}\int_0^1 |f_m(x) - f_i(x)|dx \qquad (10)$$

was used, where i represents the model (i=A, B, C, D) and m represents the data-based likelihood distribution. This index ranges from [0, 1], where unity corresponds to perfect matching between the informed prior distribution of model i and the measured likelihood distribution m. As shown in Table 3, this index is consistent with the dynamic model accuracies in Fig. 14, but low levels of agreement are obtained. Note that using more detailed operator response-time models, in the next subsection, the performance indices are improved significantly.

3.4.2 SCE Occurrence Model

When constructing informed prior distributions to estimate the failure probabilities of safety systems that act infrequently, it is important to assess the SCEs that can activate specific safety systems. Given that process units fail

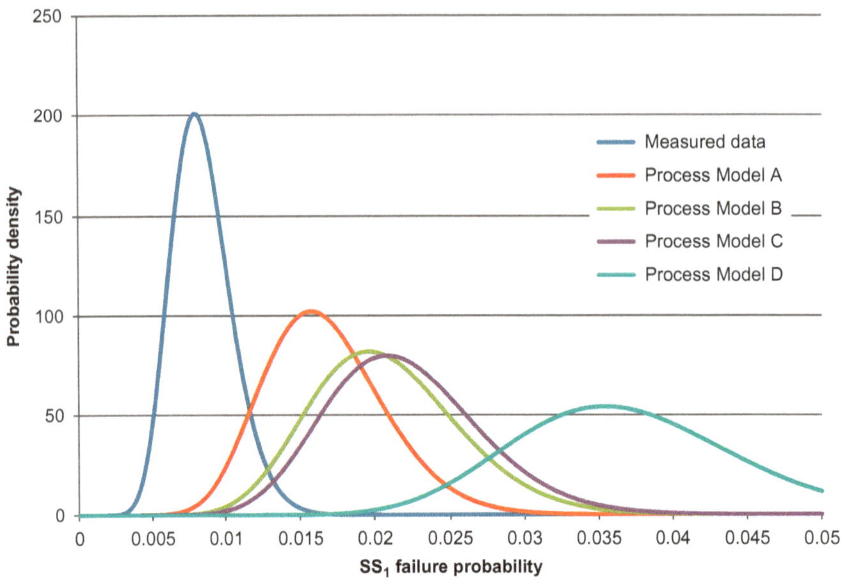

Fig. 15 Informed prior distributions created using the four process models, as well as the binomial likelihood distribution created using the measured alarm data. From Moskowitz, I. H., Seider, W. D., Soroush, M., Oktem, U. G., & Arbogast, J. E. (2015). Chemical process simulation for dynamic risk analysis: A steam-methane reformer case study. Industrial and Engineering Chemistry Research, 54(16), 4347–4359. Used with permission.

Table 3 Performance Index for Process Models A–D

	Process Model A	Process Model B	Process Model C	Process Model D
$\xi_{i,m}$	0.159	0.063	0.045	0.026

in many ways (e.g., as inlet stream compositions, temperatures, flow rates, and pressures vary; controllers experience measurement bias; valves malfunction; controller electronic mechanisms fail), special-cause modeling deserves attention. Clearly, for specific SCEs, known to trigger safety systems, it is crucial to account for them when creating informed prior distributions. Some are known to have a high likelihood of occurrence over the lifetime of a plant, while others may be *unobserved* locally, having occurred at other plant sites or even related plants. For all potential SCEs, a probability distribution should be constructed, even when likelihood data are unavailable.

When developing occurrence models to estimate safety system failure probabilities, SCEs must be selected and their magnitudes must be investigated, as SCEs are likely to have devastating consequences (e.g., propagation of runaway reactions, leading to explosions). To identify these events,

HAZOP and LOPA analyses, especially, are particularly helpful. HAZOP is the industry standard for postulating all possible SCEs.

The effect of an SCE depends on its magnitude. SCE models (e.g., $g_1(A_m)$ in Fig. 8) are needed to estimate failure probability distributions. SCE models having low expected values are most representative and rarely activate second-level alarms, while SCE models having high expected values represent extreme cases, allowing for the study of second-level alarms. Various steam pressure decrease magnitudes are shown in Fig. 16—each being a delta function centered at its corresponding point on the abscissa. The expected value of the SS_1 failure probability, j_m, is graphed accordingly.

3.4.3 Operator Response-Time Models

In the simulation of a SCE, Moskowitz et al. (2016) show that the behavior of the operator must be well understood. In Fig. 8, $g_2(\tau)$ represents operator response times in taking action following activated alarms associated with SS_1. An initial construction of $g_2(\tau)$ can be made using the histogram of operator response times to high-frequency alarms. This provides valuable information about how the operators tend to act when a variable is under alarm (Bendoly, Donohue, & Schultz, 2006; Macwan & Mosleh, 1994; Stylios & Groumpos, 1999). In some cases, when operators anticipate that the alarm thresholds have been set conservatively, they are slower to respond

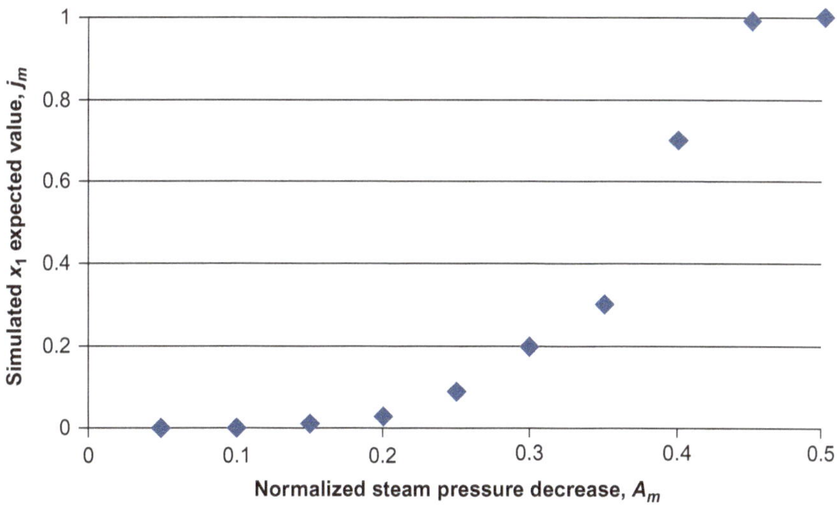

Fig. 16 SS_1 failure probability as a function of a steam pressure decrease. From Moskowitz, I. H., Seider, W. D., Arbogast, J. E., Oktem, U. G., Pariyani, A., & Soroush, M. (2016). Improved predictions of alarm and safety system performance through process and operator response-time modeling. *AIChE Journal, 62*(9), 3461–3472. Used with permission.

to expected nuisance alarms. On the other hand, when operators recognize that alarms tend to trigger a flood of alarms elsewhere in the process, they view these alarms as critical—even though they just signal entry into the yellow-belt zone. To the extent possible, it is important to make quantifiable justifications when modeling operator effectiveness (Hollnagel, 1998; Reason, 2000).

A histogram of approximately 300 observed operator response times to the H-alarm associated with the SMR furnace effluent temperature (Moskowitz et al., 2016) is shown in Fig. 17. The operator response times were collected using the alarm data log, which recorded the time of each alarm activation and the time of each operator manipulation. The time between an alarm activation and the initial operator manipulation is the operator response time. This calculation method provides the most accurate data on operator response time. Alarm data are convenient to work with, but without alarm data, process data can be sampled to obtain operator response times. A script can be written to record the times process variables cross their thresholds, as well as the times controlled-variable set points or actuators undergo step changes (considered to be operator actions) (Pariyani et al., 2012a). In either case, the data sampling interval is important to consider.

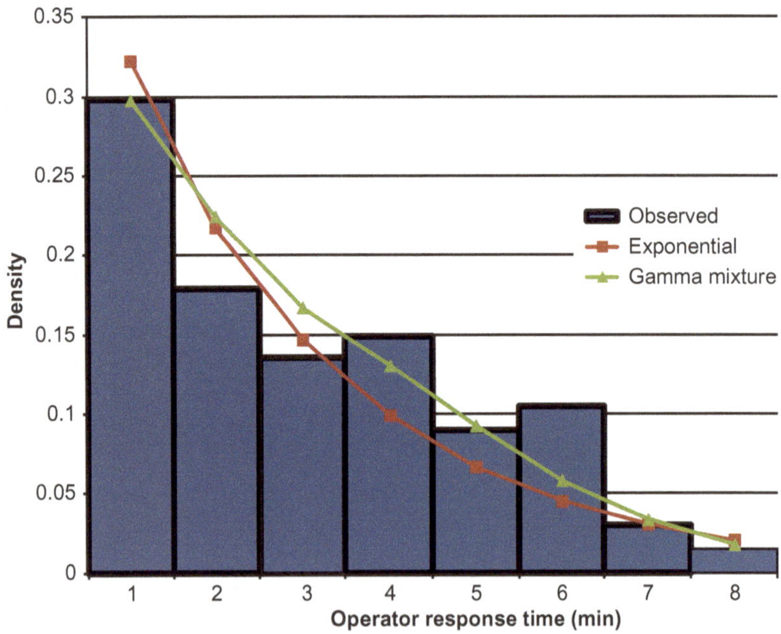

Fig. 17 Operator response-time histogram. From Moskowitz, I. H., Seider, W. D., Arbogast, J. E., Oktem, U. G., Pariyani, A., & Soroush, M. (2016). Improved predictions of alarm and safety system performance through process and operator response-time modeling. *AIChE Journal, 62*(9), 3461–3472. Used with permission.

If data are sampled or recorded infrequently (such as by a composition analyzer), operator response times may be inaccurate. Depending upon the frequency of process data points, reasonable estimates of operator response times can be obtained. The wide range of operator response times, nearly all well represented, suggest many kinds of operator actions. The highest number of responses is associated with the shortest response times—with operators taking action in less than 1 min. Nearly all of the operator response times lie between 0 and 6 min, with far fewer of longer duration. Past a 6 min response time, the number of responses decreases rapidly, with just three response times beyond 8 min.

Two parametric distributions were used to model the operator response-time distribution. The first distribution is an exponential distribution, which is called Operator Response-Time Model A:

$$g_{2_A}(\tau) = \lambda e^{-\lambda\tau} \tag{11}$$

where λ is a parameter to be estimated by maximizing the likelihood function:

$$L(\lambda|\tau) = \prod_{i=1}^{n} \lambda e^{-\lambda\tau_i} \tag{12}$$

where i is the response counter, n is the total number of response times measured (on the order of 300), and τ_i is response-time i. It is often convenient to maximize the log-likelihood function instead of the likelihood function:

$$LL(\lambda|\tau) = \sum_{i=1}^{n} (\ln[\lambda] - \lambda\tau_i) \tag{13}$$

The maximum of the log-likelihood function has a simple analytical form:

$$\lambda = \frac{1}{\overline{\tau}} \tag{14}$$

where $\overline{\tau} = \frac{\sum_{i=1}^{n} \tau_i}{n}$ is the sample mean of the measured response times.

The second is a weighted sum of three gamma distributions (Operator Response-Time Model B):

$$g_{2_B}(\tau) = \sum_{j=1}^{3} \theta_j \frac{b_j^{a_j}}{\Gamma(a_j)} \tau^{a_j - 1} e^{-b_j\tau}, \quad \sum_{j=1}^{3} \theta_j = 1, \quad 0 \le \theta_j \le 1 \tag{15}$$

where each θ_j is a weighting coefficient for gamma distribution j, and where a_j and b_j are the parameters of gamma distribution j. While any number of gamma distributions can be used, here the fourth distribution gives a negligible increase in the likelihood function (compared to the impact of the third distribution). The eight parameters in Eq. (16) are estimated by maximizing the likelihood function:

$$L(a_1, a_2, a_3, b_1, b_2, b_3, \theta_1, \theta_2 | \tau) = \prod_{i=1}^{n} \left(\sum_{j=1}^{3} \theta_j \frac{b_j^{a_j}}{\Gamma(a_j)} \tau_i^{a_j - 1} e^{-b_j \tau_i} \right) \tag{16}$$

Using Newton's optimization method with an analytical Hessian matrix, the parameter values in Table 4 were estimated.

While Operator Response-Time Models A and B represent operator response times well, they do not account for the rate of change of each variable crossing its alarm threshold, as well as the number of activated alarms being monitored by an operator(s). Clearly, operator responses gain urgency, and often speed, when a variable crosses one of its thresholds rapidly. Also, as the number of active alarms decreases, operators are less distracted and respond more rapidly.

To account for the rate of change of each variable when the variable crosses an alarm threshold, using the SMR plant data, operator response times are displayed in Fig. 18 as a function of the furnace effluent temperature derivative, dx/dt, as the temperature crosses its high-alarm threshold (Moskowitz et al., 2016). The dependence of the operator response time, τ, on the rate of change is well described by:

$$\hat{\tau}_C = \kappa_1 e^{-\nu_1 (dx/dt)} \tag{17}$$

where κ_1 and ν_1 are the model parameters. These parameters are estimated by minimizing the sum of the squared errors:

$$SSE_1 = \sum_{i=1}^{n} \left(\kappa_1 e^{-\nu_1 |dx/dt|_i} - \tau_i \right)^2 \tag{18}$$

The estimated values of κ_1 and ν_1 as well as the corresponding SSE_1 value are given in Table 5.

Table 4 Parameters for Operator Response-Time Models A and B

λ	a_1	a_2	a_3	b_1	b_2	b_3	θ_1	θ_2	θ_3
0.39	1.83	3.56	4.51	0.68	1.04	0.88	0.48	0.07	0.45

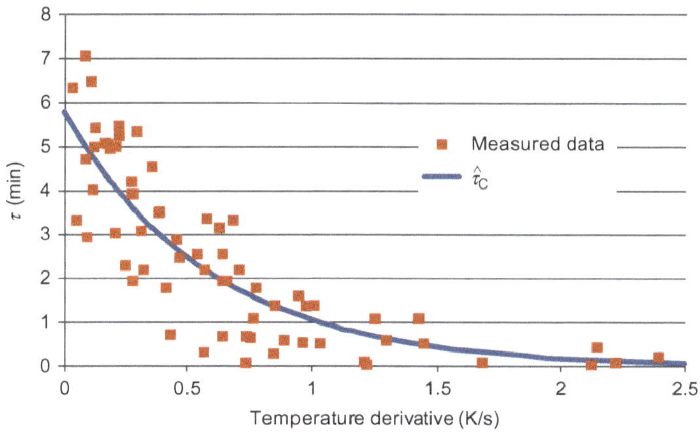

Fig. 18 Operator response time as a function of temperature rate of change (plant data and model prediction). From Moskowitz, I. H., Seider, W. D., Arbogast, J. E., Oktem, U. G., Pariyani, A., & Soroush, M. (2016). Improved predictions of alarm and safety system performance through process and operator response-time modeling. *AIChE Journal, 62*(9), 3461–3472. Used with permission.

Table 5 Parameters Used for Operator Response-Time Models C–E

κ_1 (min)	ν_1 (s/K)	SSE_1	κ_2 (min)	ν_2 (1/no. of alarms)	SSE_2
4.33	1.54	302	3.8	4.2	413

To construct a $g_2(\tau)$ that accounts for the alarmed variable time derivative, a stochastic component must be maintained—because if $g_2(\tau)$ were purely deterministic (like $\hat{\tau}_C$), the variance of the safety system failure probability with respect to A_m cannot be calculated. While $\hat{\tau}_C$ is an estimate for the operator response time, it must be incorporated into a random variable distribution for τ. One choice for $g_2(\tau)$ is the exponential distribution having an expected value equal to $\hat{\tau}_C$. The exponential distribution in Eq. (17) is known to have an expected value of $1/\lambda$. Therefore, Operator Response-Time Model C is proposed:

$$g_{2_C}(\tau) = \frac{1}{\kappa_1 e^{-\nu_1 (dx/dt)}} \exp\left[\frac{-\tau}{\kappa_1 e^{-\nu_1 (dx/dt)}}\right] \tag{19}$$

A similar method can be used to account for the effect of multiple alarm activations in the process. When many alarms are active, competing for operator(s) attention, response times are expected to increase. Here, also, the exponential distribution is appropriate:

$$\hat{\tau}_D = \kappa_2 e^{-\nu_2 \gamma} \tag{20}$$

where γ is the reciprocal of active alarms. The parameters κ_2 and ν_2 are estimated by minimizing the sum of the squared errors:

$$SSE_2 = \sum_{i=1}^{n} \left(\kappa_2 e^{-\nu_2 \gamma_i} - \tau_i\right)^2 \tag{21}$$

The estimated values for κ_2 and ν_2 as well as the corresponding SSE_2 value are given in Table 6.

Using this logic, the Operator Response-Time Model D is formulated by setting the expected value of an exponential distribution equal to $\hat{\tau}_{\mathrm{D}}$:

$$g_{2_{\mathrm{D}}}(\tau) = \frac{1}{\kappa_2 e^{-\nu_2 \gamma}} \exp\left[\frac{-\tau}{\kappa_2 e^{-\nu_2 \gamma}}\right] \tag{22}$$

Finally, the Operator Response-Time Model E is formulated that incorporates both $\hat{\tau}_C$ and $\hat{\tau}_D$. The effectiveness of $\hat{\tau}_{\mathrm{C}}$ and $\hat{\tau}_{\mathrm{D}}$ can be compared by their associated *SSE*s. In Moskowitz et al. (2016), the expected value of Operator Response Model E is set equal to:

$$\mu_{3\mathrm{E}} = \frac{SSE_1 \hat{\tau} + SSE_2 \phi}{SSE_1 + SSE_2} \tag{23}$$

where the weighting coefficients for each distribution are proportional to their *SSE*s. This yields Operator Response-Time E model:

$$g_{2_{\mathrm{E}}}(\tau) = \frac{SSE_1 + SSE_2}{SSE_1\left(\kappa_1 e^{-\nu_1 (dx/dt)}\right) + SSE_2\left(\kappa_2 e^{-\nu_2 \gamma}\right)} \exp\left[-\frac{SSE_1 + SSE_2}{SSE_1\left(\kappa_1 e^{-\nu_1 (dx/dt)}\right) + SSE_2\left(\kappa_2 e^{-\nu_2 \gamma}\right)} \tau\right] \tag{24}$$

Next, Operator Response-Time Models A–E are used to construct informed prior distributions for the failure probability of SS_1, along with dynamic Process Model A. The results are shown in Fig. 19, along with the binomial likelihood distribution of the measured SS_1 data, Eq. (4). The model performance index, $\xi_{i,m}$, in Eqs. (3a) and (3b) is used to quantify the model performance of Operator Response-Time Models A–E, with results shown in Table 6. Operator Response-Time Models A and B, which are independent of the process state, describe the measured alarm data

Table 6 Performance Index for Operator Response-Time Models A–E With Process Model A

	ORTM A	ORTM B	ORTM C	ORTM D	ORTM E
$\xi_{i,m}$	0.029	0.030	0.633	0.701	0.812

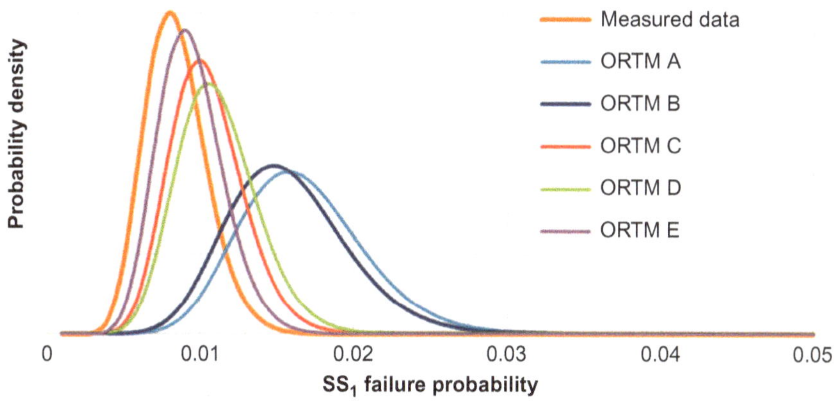

Fig. 19 SS_1 informed prior distributions constructed using the five operator response-time models (ORTMs) with dynamic Process Model A . From Moskowitz, I. H., Seider, W. D., Arbogast, J. E., Oktem, U. G., Pariyani, A., & Soroush, M. (2016). Improved predictions of alarm and safety system performance through process and operator response-time modeling. *AIChE Journal, 62*(9), 3461–3472. Used with permission.

poorly. As expected, Operator Response-Time Model B, with eight parameters, performs better than Operator Response-Time Model A, with just a single parameter. The incorporation of $\hat{\tau}_C$ and $\hat{\tau}_D$ in Operator Response-Time Models C and D, clearly improves the informed prior distributions, with Model C performing better than Model D—expected because $SSE_1 < SSE_2$. Of the five models, Model E is in the closest agreement with the likelihood data. Given preferred Model E, the choice of process model can be revisited. In Fig. 20, the four dynamic process models are used to build informed prior distributions with Operator Response-Time Model E. The model performance indices are shown for Process Models A–D in Table 7. Once again, Process Model A yields the best agreement with the observed likelihood data and can be considered the most appropriate process model.

3.5 Modeling SS_2 Failures Using Models With Parameters Estimated From SS_1 Failures

Once the three types of models (process, SCE, and operator response-time models) are chosen and their parameters are estimated, Moskowitz et al. (2016) use them to estimate the failure probabilities of SS_1. These models must then be adjusted to handle simulations that involve the activation of SS_2 (i.e., after the failure of SS_1). While it is desirable to keep the models intact, a few adjustments are recommended (Moskowitz et al., 2016).

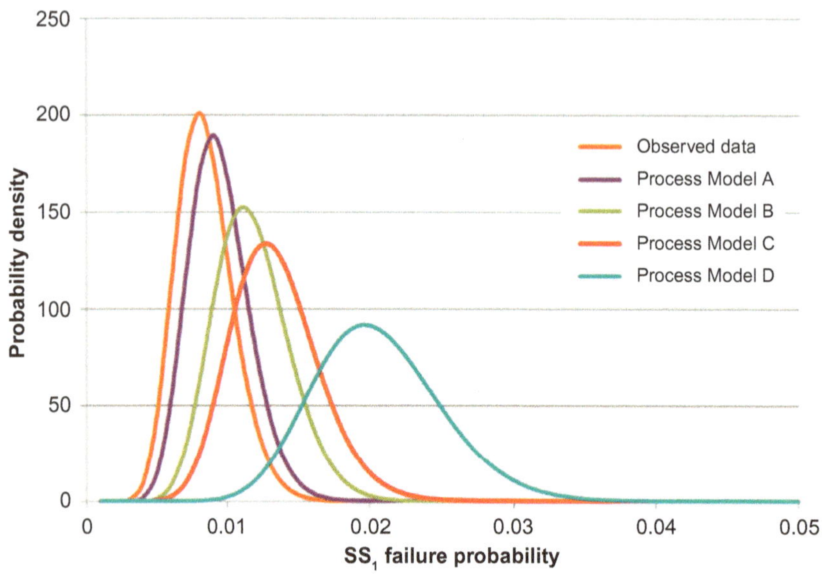

Fig. 20 SS_1 informed prior distributions constructed using the four process models with Operator Response-Time Model E . From Moskowitz, I. H., Seider, W. D., Arbogast, J. E., Oktem, U. G., Pariyani, A., & Soroush, M. (2016). Improved predictions of alarm and safety system performance through process and operator response-time modeling. *AIChE Journal, 62*(9), 3461–3472. Used with permission.

Table 7 Performance Index Revisited for Process Models A–D Using Operator Response-Time Model E

	Process Model A	Process Model B	Process Model C	Process Model D
$\xi_{i,m}$	0.812	0.481	0.325	0.051

The dynamic process model, Process Model A, should not be altered much to simulate SS_2 activation events for the SMR—because the simulations from high and high–high alarms, and beyond, are similar, with small changes in physical properties as temperatures rise. In general, Process Model A adjustments would be required when physical and chemical phenomena change abruptly—for example, with shifts from laminar to turbulent flows, or the introduction of two-phase flows.

The SCE Occurrence Model needs significant adjustment because the $g_1(A_m)$ distribution used for simulating L/H alarm activations infrequently activates LL/HH alarms. To achieve this, a normal distribution is chosen for $g_1(A_m)$, having mean $\overline{A_m}$ and standard deviation σ. A lower-tail, bounded by θ_L and an upper-tail bounded by θ_U, each two standard deviations from the mean, are defined:

$$\theta_L = \overline{A_m} - 2\sigma;\quad \theta_U = \overline{A_m} + 2\sigma \tag{25}$$

noting that θ_L is a special-cause magnitude closer to zero; that is, closer to typical operation. The normal distribution is described by θ_L and θ_U, rather than the typical mean and standard deviation:

$$g_1(A_m) = \frac{1}{\sqrt{2\left(\frac{\theta_U - \theta_L}{4}\right)^2 \pi}} \exp\left(-\frac{\left(A_m - \frac{\theta_U + \theta_L}{2}\right)^2}{2\left(\frac{\theta_U - \theta_L}{4}\right)^2}\right) \tag{26}$$

The lower bound of θ_L is set at A_m values for which the L/H alarms failed in simulation—with smaller choices of θ_L yielding many simulations where the LL/HH alarms are not activated. Referring to Fig. 16, with SS_1 failures frequently observed when $A_m \geq 0.4$, 0.4 is a good lower bound for θ_L. θ_U is set such that the SCEs are of interest and relevance. Three SCE occurrence models are shown in Fig. 21, each sharing $\theta_L = 0.4$. SCEM A has θ_U closest to θ_L, and samples SCEs that are most likely (closest to typical operation), yet have the least potential for SS_2 failures. The other extreme is SCEM C, which has θ_U furthest from θ_L. SCEM B, having $\overline{A_m} = 0.56$, is chosen as an interior

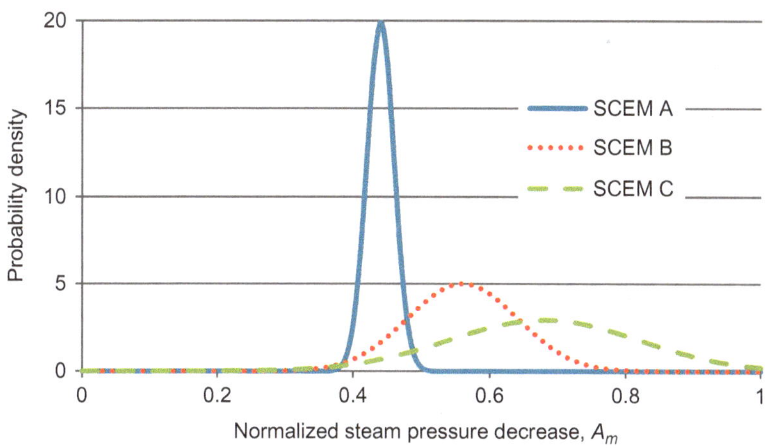

Fig. 21 Special-cause event models for SS_2 simulation. From Moskowitz, I. H., Seider, W. D., Arbogast, J. E., Oktem, U. G., Pariyani, A., & Soroush, M. (2016). Improved predictions of alarm and safety system performance through process and operator response-time modeling. *AIChE Journal, 62*(9), 3461–3472. Used with permission.

candidate to analyze the risk of SS_2 failures. The choice of $g_1(A_m)$ has a significant impact on the estimated failure probabilities—the user must keep this in mind when analyzing the simulation results and making statements about the risk of the process. The failure probability estimate attempts to describe the probability of failure while undergoing a SCE sampled from $g_1(A_m)$, which is very different than typical day-to-day process fluctuations.

The Operator Response-Time Models also need adjustment, it being expected that operators react quicker to alarm activations that are associated with more urgent consequences. Given an immediate threat of an automatic plant shutdown, operator actions should be accelerated. To account for this, when SS_1 is activated to SS_2, the time response to the SS_2 activation is divided by the ratio of the 90% percentile of operator response to a SS_1 activation, $\tau_{90\%}$, over the interlock shutdown time, Δt_{int}. For this case, the Operator Response-Time Model E takes the form:

$$g_{2_E}(\tau) = \frac{SSE_1 + SSE_2}{SSE_1\left(\kappa_1 e^{-\nu_1 (dx/dt)}\right) + SSE_2\left(\kappa_2 e^{-\nu_2 \gamma}\right)} \frac{\Delta t_{\text{int}}}{\tau_{90\%}} \exp\left[-\frac{SSE_1 + SSE_2}{SSE_1\left(\kappa_1 e^{-\nu_1 (dx/dt)}\right) + SSE_2\left(\kappa_2 e^{-\nu_2 \gamma}\right)}\left(\frac{\Delta t_{\text{int}}}{\tau_{90\%}}\tau\right)\right] \tag{27}$$

Having constructed, chosen, and regressed the three types of models using the H-alarm data, Moskowitz et al. (2016) make adjustments to model SS_2 activation events. An informed prior distribution is then constructed for the failure probabilities of SS_2. Fig. 22 shows the resulting informed prior and associated posterior distributions describing the failure probability of SS_2. The sparse alarm data are used to build binomial likelihood distributions that modify the prior distributions to form posterior distributions. It can be seen that the most accurate posterior distribution (formed using Process Models A and Operator Response-Time Model E with an *urgency* adjustment for time responses in SS_2) is not shifted as dramatically as the posterior distribution formed using the simple informed prior distribution. This indicates that the most accurate posterior distribution is more effective at handling the special-cause simulations—leading to more accurate failure probability predictions.

The interpretation of the posterior distribution is that during a severe loss in steam pressure, the probability that the process will undergo an automatic shutdown is on the order of 5%. This allows engineers responsible for setting reliability estimates to have quantifiable justification when they do so. An estimate for the reliability for various SCEs can provide engineers with a broader understanding of the events that pose the greatest odds of an interlock activation, thus motivating different designs to handle these events.

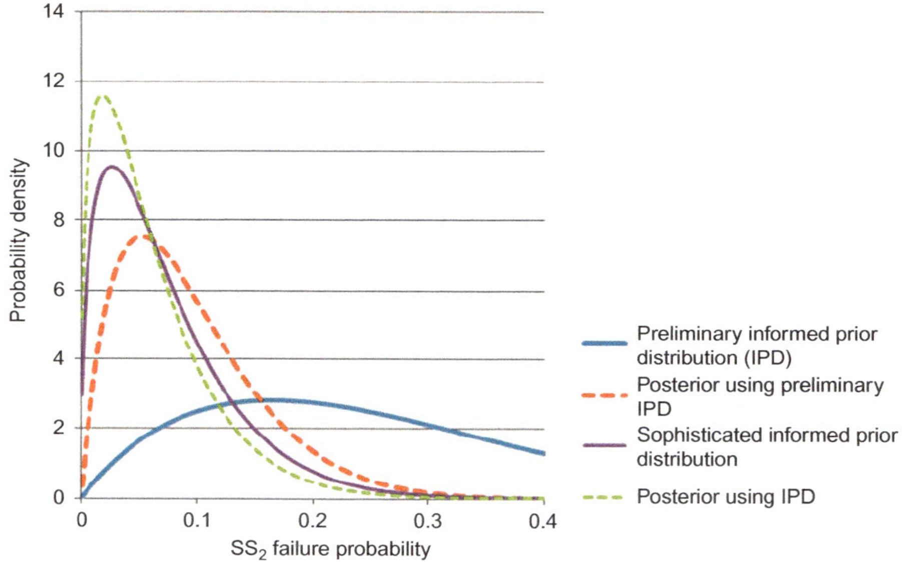

Fig. 22 Informed prior distributions and associated posterior distributions describing the failure probability of SS_2. From Moskowitz, I. H., Seider, W. D., Arbogast, J. E., Oktem, U. G., Pariyani, A., & Soroush, M. (2016). Improved predictions of alarm and safety system performance through process and operator response-time modeling. *AIChE Journal, 62*(9), 3461–3472. Used with permission.

The operator also benefits from these distributions, as he/she becomes aware that, during such severe pressure drops, his/her reactions have been projected to result in interlock activations after on the order of 5% of occurrences. If this value is perceived to be too high, the operator may be motivated to act more urgently during this type of event.

4. SUMMARY

The rapid development of dynamic risk analyses, over the past 2 decades, has been reviewed, principally to introduce many techniques to chemical process safety practitioners/researchers. Initial strategies involved Bayesian analysis with copulas to model the failure probabilities of safety systems when responding to rare alarms. For safety interlock systems and their associated alarms, statistical techniques on the sparse records of activations are alone insufficient to make meaningful evaluations of their failure probabilities. With process models, the use of operator response data associated with the relatively frequent alarm activations (e.g., H alarms), that is, plentiful near miss operator data, are examined. Better estimates were obtained using improved process/operator models to inform *prior* failure probability distributions. The latter are very promising for improving the failure probability estimates of safety systems associated with less frequently activated alarms (e.g., HH alarms) and safety interlock systems; that is, to provide more reliable safety risk analyses.

REFERENCES

Abimbola, M., Khan, F., & Khakzad, N. (2014). Dynamic safety risk analysis of offshore drilling. *Journal of Loss Prevention in the Process Industries*, *30*, 74–85.

Anand, S., Keren, N., Treter, M. J., Yang, Y., O'Connor, T. M., & Mannan, M. S. (2006). Harnessing data mining to explore incident databases. *Journal of Hazardous Materials*, *130*(1–2), 33–41.

ANSI/ISA 18.2. (2009). *Management of alarm systems for the process industries.*

Bendoly, E., Donohue, K., & Schultz, K. L. (2006). Behavior in operations management: Assessing recent findings and revisiting old assumptions. *Journal of Operations Management*, *24*, 737–752.

Berger, J. O. (2013). *Statistical decision theory and Bayesian analysis*. Berlin, Germany: Springer Science & Business Media.

Blizard, W. D. (1991). The development of multiset theory. *Modern Logic*, *1*(4), 353.

CCPS (Center for Chemical Process Safety). (1999). *Guidelines for chemical process quantitative risk analysis*. New York: AIChE.

CCPS (Center for Chemical Process Safety). (2007). *Guidelines for risk based process safety*. Hoboken, NJ: Center for Chemical Plant Safety, AIChE, Wiley.

Chang, Y. H. J., & Mosleh, A. (2007). Cognitive modeling and dynamic probabilistic simulation of operating crew response to complex system accidents: Part 3. IDAC operator response model. *Reliability Engineering and System Safety*, *92*, 1041–1060.

Crowl, D. A., & Louvar, J. F. (2011). *Chemical process safety: fundamentals with applications* (3rd ed.). Upper Saddle River, NJ: Prentice-Hall.

EEMUA, Engineering Equipment & Materials Users' Association Publication 191. (2007). *Alarm systems—A guide to design, management, and procurement* (2nd ed.). EEMUA.

Elliott, M. R., Wang, Y., Lowe, R. A., & Kleindorfer, P. R. (2004). Environmental justice: Frequency and severity of U.S. chemical industry accidents and the socioeconomic status of surrounding communities. *Journal of Epidemiology and Community Health, 58*(1), 24–30.

Ferdous, R., Khan, F., Veitch, B., & Amyotte, P. R. (2009). Methodology for computer-aided fuzzy fault-tree analysis. *Process Safety and Environmental Protection, 87*, 217–226.

Gelman, A., Carlin, J. B., Stern, H. S., & Rubin, D. B. (2014). *Bayesian data analysis* (2nd ed.). Boca Raton, FL: Chapman & Hall/CRC.

gPROMS. (2014). *gPROMS, version 3.6.1 [computer software]*. London: Process Systems Enterprise (PSI).

Hollnagel, E. (1998). *Cognitive reliability and error analysis (CREAM)*. Oxford, UK: Elsevier.

Hottel, H. C., & Sarofim, A. F. (1967). *Radiative transfer*. New York: McGraw-Hill.

Jones, S., Kirchsteiger, C., & Bjerke, W. (1999). The importance of near miss reporting to further improve safety performance. *Journal of Loss Prevention in the Process Industries, 12*(1), 59–67.

Kalantarnia, M., Khan, F., & Hawboldt, K. (2009). Dynamic risk assessment using failure assessment and Bayesian theory. *Journal of Loss Prevention in the Process Industries, 22*, 600–606.

Kalantarnia, M., Khan, F., & Hawboldt, K. (2010). Modeling of BP Texas City refinery accident using dynamic risk assessment approach. *Process Safety and Environment Protection, 88*, 191–199.

Khakzad, N., Khan, F., & Amyotte, P. (2013a). Quantitative risk analysis of offshore drilling operations: A Bayesian approach. *Safety Science, 57*, 108–117.

Khakzad, N., Khan, F., & Amyotte, P. (2013b). Dynamic safety analysis of process systems by mapping bow-tie into Bayesian network. *Process Safety and Environment Protection, 91*, 46–53.

Khan, F., Rathnayaka, S., & Ahmed, S. (2015). Methods and models in process safety and risk management: Past, present, and future. *Process Safety and Environment Protection, 98*, 116–147.

Kleindorfer, P. R., Belke, J. C., Elliott, M. R., Lee, K., Lowe, R. A., & Feldman, H. I. (2003). Accident epidemiology and the U.S. chemical industry: Accident history and worst-case data from RMP*Info. *Risk Analysis, 23*(5), 865–881.

Kletz, T. (1992). *Hazop and Hazan* (3rd ed.). Rugby, Warwickshire: IChemE.

Kondaveeti, S. R., Izadi, I., Shah, S. L., Shook, D. S., & Kadali, R. (2010). Quantification of alarm chatter based on run length distributions. In *49th IEEE conf. on dec. and control, Atlanta, GA*.

Lathem, D. A., McAuley, K. B., Peppley, B. A., & Raybold, T. M. (2011). Mathematical modeling of an industrial steam-methane reformer for online deployment. *Fuel Processing Technology, 92*, 1574–1586.

Lauridsen, K., Kozine, I., Markert, F., & Amendola, A. (2002). *Assessment of uncertainties in risk analysis of chemical establishments*. Summary report on ASSURANCE project Roskilde, Denmark: Risk National Laboratory.

Leveson, N., & Stephanopolous, G. (2014). A system-theoretic control-inspired view and approach to process safety. *AIChE Journal, 60*(1), 2–14.

Macwan, A., & Mosleh, A. (1994). A methodology of modeling operator errors of commission in probabilistic risk assessment. *Reliability Engineering and System Safety, 45*, 139–157.

Mannan, M. S., O'Connor, T. M., & West, H. H. (1999). Accident history database: An opportunity. *Environmental Progress, 18*(1), 1–6.

Markowski, A. S., Mannan, M. S., & Bigowszewska, A. (2009). Fuzzy logic for process safety analysis. *Journal of Loss Prevention in the Process Industries, 22*, 695–702.

Meel, A., O'Neill, L. M., Levin, J. H., Seider, W. D., Oktem, U. G., & Keren, N. (2007). Operational risk assessment of chemical industries by exploiting accident databases. *Journal of Loss Prevention in the Process Industries*, *20*(2), 113–127.
Meel, A., & Seider, W. D. (2006). Plant-specific dynamic failure assessment using Bayesian theory. *Chemical Engineering Science*, *61*(21), 7036–7056.
Meel, A., Seider, W. D., & Oktem, U. G. (2008). Analysis of management actions, human behavior, and process reliability in chemical plants. I. Impact of management actions. *Process Safety Progress*, *27*(1), 7–14.
Moschovakis, Y. N. (2006). *Notes on set theory* (2nd ed.). New York: Springer.
Moskowitz, I. H., Seider, W. D., Arbogast, J. E., Oktem, U. G., Pariyani, A., & Soroush, M. (2016). Improved predictions of alarm and safety system performance through process and operator response-time modeling. *AIChE Journal*, *62*(9), 3461–3472.
Moskowitz, I. H., Seider, W. D., Soroush, M., Oktem, U. G., & Arbogast, J. E. (2015). Chemical process simulation for dynamic risk analysis: A steam-methane reformer case study. *Industrial and Engineering Chemistry Research*, *54*(16), 4347–4359.
Nelsen, R. B. (1999). An introduction to copulas. *Lecture notes in statistics*. New York: Springer-Verlag.
Pariyani, A., Seider, W. D., Oktem, U. G., & Soroush, M. (2010). Incidents investigation and dynamic analysis of large alarm databases in chemical plants: A fluidized-catalytic-cracking unit case study. *Industrial and Engineering Chemistry Research*, *49*, 8062–8079.
Pariyani, A., Seider, W. D., Oktem, U. G., & Soroush, M. (2012a). Dynamic risk analysis using alarm databases to improve safety and quality: Part I—Data compaction. *AIChE Journal*, *58*(3), 812–825.
Pariyani, A., Seider, W. D., Oktem, U. G., & Soroush, M. (2012b). Dynamic risk analysis using alarm databases to improve safety and quality: Part II—Bayesian analysis. *AIChE Journal*, *58*(3), 826–841.
Reason, J. (1990). *Human error*. Cambridge: Cambridge Univ. Press.
Reason, J. (2000). Human error models and management. *BMJ*, *320*, 768–770.
Risk World. (2013). *Risk-related software*. www.riskworld.com/software/sw5sw001.htm.
Sklar, A. (1973). Random variables, joint distribution, and copulas. *Kybernetika*, *9*, 449–460.
Stylios, C. D., & Groumpos, P. P. (1999). A soft computing approach for modeling the supervisor of manufacturing systems. *Journal of Intelligent and Robotic Systems*, *26*, 389–403.
Valle, D. L. (2009). Bayesian copula distributions, with application to operational risk management. *Methodology and Computing in Applied Probability*, *11*, 95–115.
Villa, V., Paltrinieri, N., Khan, F., & Cozzani, V. (2016). Towards dynamic risk analysis: A review of the risk assessment approach and its limitations in the chemical process industry. *Safety Science*, *89*, 77–93.
Vinnem, J. E. (2010). Offshore risk assessment: Principles, modelling and applications of QRA studies. *Springer series in reliability engineering* (2nd ed.). Springer.
Xu, J., & Froment, G. F. (1989). Methane steam reforming, methanation and water-gas shift: 1. Intrinsic kinetics. *AIChE Journal*, *35*(1), 88–96.
Yang, X., Rogers, W. J., & Mannan, M. S. (2010). Uncertainty reduction for improved mishap probability prediction: Application to level control of distillation unit. *Journal of Loss Prevention in the Process Industries*, *23*(1), 149–156.
Ye, L., Liu, Y., Fei, Z., & Liang, J. (2009). Online probabilistic assessment of operating performance based on safety and operability indices for multimode industrial processes. *Industrial and Engineering Chemistry Research*, *48*, 10912–10923.
Yi, W., & Bier, V. M. (1998). An application of copulas to accident precursor analysis. *Management Science*, *44*(12), S257.

CHAPTER SIX

Regulation of Chemical Process Safety: Current Approaches and Their Effectiveness

Daryl Attwood[1]
Lloyd's Register EMEA, London, UK
[1]Corresponding author: e-mail address: daryl.attwood@lr.org

Contents

Methods in Chemical Process Safety, Volume 1
ISSN 2468-6514
http://dx.doi.org/10.1016/bs.mcps.2017.01.003

1. GLOSSARY

ABS	American Bureau of Shipping
AGA	American Gas Association
ALARP	As low as reasonably practicable
AMSA	Australian Maritime Safety Authority
AoC	Acknowledgment of Compliance (Norway)
API	American Petroleum Institute
ASME	American Society of Mechanical Engineers
BOEM	Bureau of Ocean Energy Management
BPCS	Basic process control system
BSEE	Bureau of Safety and Environmental Enforcement
BV	Bureau Veritas
CA	Certifying Authority
CE	European Certificate
CFR	Code of Federal Regulations (USA)
CGA	Canadian Gas Association
CGSB	Canadian General Standards Board
CNLOPB	Canada—Newfoundland and Labrador Offshore Petroleum Board

CNPC	China National Petroleum Corporation
CNOOC	China National Offshore Oil Corporation
CNSOPB	Canada—Nova Scotia Offshore Petroleum Board
CoF	Certificate of Fitness
COOOSO	China Offshore Oil Operation Safety Office
CSA	Canadian Standards Association
DEA	Danish Energy Agency
DEPA	Danish Environmental Protection Agency
DMP	Western Australian Department of Mines and Petroleum
DNV-GL	Det Norske Veritas—Germanischer Lloyd
DWEA	Danish Working Environment Authority
DCR	Design and Construction Regulations
DMA	Danish Maritime Agency
DPR	Department of Petroleum Resources
EPA	Environmental Protection Agency
EPIC	Engineering, procurement, installation, commissioning
ESD	Emergency shutdown
EU	European Union
FARSI	Functionality, availability, reliability, survivability, interdependencies
FEED	Front end engineering and design
FLNG	Floating liquefied natural gas
FMEA	Failure mode and effect analysis
FPSO	Floating production, storage, and offloading facility
GTI	Gas Technology Institute
HAZID	Hazard identification study
HAZOP	Hazard and operability study/analysis
HSE	Health and Safety Executive
HVAC	Heating, ventilation, and air conditioning
ICP	Independent and competent person (United Kingdom)
IGS	Inert gas system

ISA	Instrument Society of America
LNG	Liquefied Natural Gas
LR	Lloyd's Register
LTIF	Lost time incident frequency
MAE	Major accident event
MAR	Management and Administration Regulations
MDR	Master Document Register
MOC	Management of change
NEB	National Energy Board
NFPA	National Fire Protection Association
NHSER	Norwegian health, safety, and environmental regulations
NIDR	Norwegian Information Duty Regulations
NOPSEMA	National Offshore Petroleum Safety and Environmental Management Authority
NPD	Norwegian Petroleum Directorate
OPA	Offshore Petroleum Act
OPGGS(S)R	Offshore Petroleum and Greenhouse Gas Storage (Safety) Regulations 2009
OSCR	Offshore Safety Case Regulations
OSDR	Offshore Safety Case Directive Regulator
OSHA	Occupational Safety and Health Administration
OSH	Occupational Safety and Health
PFEER	Prevention of fire and explosions, and emergency response
PFP	Passive Fire Protection
PPM	Process, Plant, and Machinery
PS	Performance Standard
PSA	Petroleum Safety Authority
PTW	Permit to work
QA	Quality assurance
QC	Quality control
QRA	Quantitative (or qualitative) risk assessment

SAWS	State administration of work safety
SCC	State Council of China
SCE	Safety critical element
SCR	Safety case regulations
SI	Statutory instrument
SIL	Safety integrity level
SIMOPS	Simultaneous operations
SIS	Safety instrumented system
SoV	Scope of Validation
TEMA	Tubular Exchanger Manufacturers Association
TEMPSC	Totally enclosed motor propelled survival craft
TRIR	Total recordable incident rate
UARB	Utilities and Review Board
UK	United Kingdom
USCG	United States Coast Guard

2. SUMMARY

Approaches to chemical process safety regulation differ from one country to the next. The two basic approaches are (i) those based on specific prescriptive requirements and (ii) the currently more prevalent safety case/goal setting regimes, comprising owner-defined major accidents, safety critical elements (SCEs), and performance standards (PSs). The key principle of the latter is that risks to individuals are expected to be reduced to a level known as ALARP, as low as reasonably practicable.

Even in countries where the approach is primarily prescriptive, elements of safety case methodology are gradually being introduced. Slightly different terminology is sometimes used, for example, safety program instead of safety case, and items critical to safety instead of SCEs.

This chapter describes the origins and history of the safety case approach and the basics of safety case generation. The 1988 explosion of the North Sea oil and gas platform Piper Alpha and the subsequent investigation led by

Lord Cullen were the key drivers that changed global preferences from prescription to safety cases and goal setting.

Some of the typical equipment, structures, and systems used to ensure process safety are described. There are two main types. First, that for which its failure could lead to a major accident, for example, pressure vessels and primary structure, and second, that which is intended to prevent or mitigate the effects of an accident, for example, gas detection, deluge systems, and lifeboats.

The approaches used to regulate chemical process safety in eight countries—Canada, the United States, the United Kingdom, Australia, Denmark, Norway, Nigeria, and China are described. The regimes in the countries chosen show the full range in approach from primarily prescriptive (China, Nigeria) to full safety case regimes (the United Kingdom, Australia).

Several infamous chemical process industry disasters (Piper Alpha, Bhopal, Texas City, Flixborough, England, Pasadena, Texas, and ARCO, Channelview, Texas) are described, particularly from the perspective of how an effective and systematic regulatory process might have averted them. The root causes are different for the events, but in general they all included failure to comply with the principles of currently mandatory safety cases and quality management systems. Common themes included inadequate communications, inadequate attention to detail during modifications and repairs, ineffective adherence to maintenance programs, and breakdowns in permit to work (PTW) implementation.

The role of the classification societies in chemical process safety is discussed. The societies provide a completely independent overview of facilities, without the internal cost and schedule pressures experienced by owners. Some details are provided of how one of the societies (Lloyd's Register) organizes itself to manage megaprojects.

The chapter is organized as shown in Table 1.

3. INTRODUCTION

Oil and gas have provided energy for human activity for centuries. Large-scale extraction from beneath the earth's surface and oceans has been undertaken for more than 80 years. Protecting property and keeping people safe while conducting this potentially dangerous work have challenged engineers for just as long.

Participation in the early days of oil and gas extraction was an extremely dangerous vocation, considered a badge of courage by some. Hollywood's

Table 1 Summary

Section	Content
One	Glossary
Two	Summary
Three	Introduction
Four	Overall approaches, history, and the basic process of generating a safety case. The definitions of and processes used to establish major accident hazards, safety critical elements, and performance standards are described
Five	Brief descriptions of some equipment, structures, and systems utilized to make personnel safe in the chemical process industry, primarily related to oil and gas production. The types of equipment can be divided into two categories—first, equipment which, if it failed or was compromised, might lead to a major accident event (MAE), and second, equipment or structures which are meant to prevent or mitigate the effects of an MAE
Six	Descriptions of the regulatory regimes, from top level legislation to practical requirements, utilized in Canada, the United States, the United Kingdom, Australia, Denmark, Norway, Nigeria, and China. The descriptions included do not represent an exhaustive list, and they are primarily related to the oil and gas industry. Nevertheless, the Americas, Europe, and Asia are all represented in the sample chosen. Variations of the approaches described are used in many other countries
Seven	Review of six historical chemical process accidents (Piper Alpha, Bhopal, Texas City, Flixborough, Pasadena, Texas, and ARCO, Texas), and descriptions of how effective, or more effective, regulation could have averted them
Eight	Role of classification societies in chemical process safety regulation. Brief descriptions of four prominent societies are provided: Lloyd's Register, Det Norske Veritas/Germanischer Lloyd, The American Bureau of Shipping, and Bureau Veritas. Lloyd's Register's approach to organization, project management, and quality assurance on megaprojects is described
Nine	Concluding remarks
	References

portrayals included films such as "There Will Be Blood," starring Daniel Day-Lewis, and "Hellfighters," which starred John Wayne and was loosely based on the life of legendary oilfield fire-fighter Red Adair. Other documentaries and TV programs (e.g., "Dallas") continued to romanticize the macho world of the oil and gas business.

But thankfully, those "bad old days" have come to a close in most locations. Fig. 1 shows the continuing improvement in the oil and gas industry's total recordable incident rate (TRIR*) between 2005 and 2014.

Furthermore, statistics show that by some measures, the oil and gas business is now safer than many other industries such as mining, agriculture, and even education and retail, and certainly safer than participation in many forms of leisure activity such as American football, surfing, and mountain climbing. Fig. 2 shows a comparison of lost time incident frequency (LTIF**) in several industries.

*TRIR—the number of recordable injuries (fatalities + lost work day cases + restricted work day cases + medical treatment cases) per million hours worked.

**LTIF—the number of lost time injury (fatalities + lost work cases) incidents per million hours worked.

Some would even say that the pendulum has swung too far, that health and safety measures have in some cases strangled creativity and threatened to stagnate progress. But the motives are beyond criticism. There is never a good reason to be injured or killed, but having a serious accident while trying to do one's job is no longer tolerated by workers or management in the process industry.

Some statistics show that about 25% of oil and gas industry fatalities are related to the production process. Fig. 3 shows that 11 of the 45 fatalities reported in 2014 by the International Association of Oil and Gas Producers were related to explosions, burns, pressure releases, or confined space entries. The remainder were related to nonprocess events such as being struck by or caught between objects, falls from height, electrocution, and drowning.

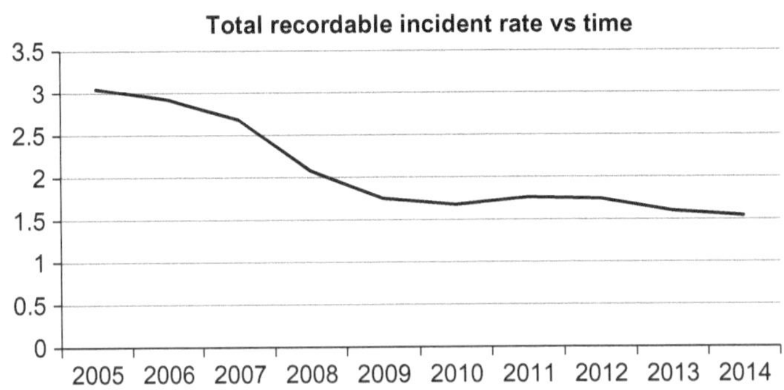

Fig. 1 Overall TRIR in the oil and gas industry from 2005 to 2014. *International Association of Oil & Gas Producers. (2015). Safety performance indicators, 2014 data.*

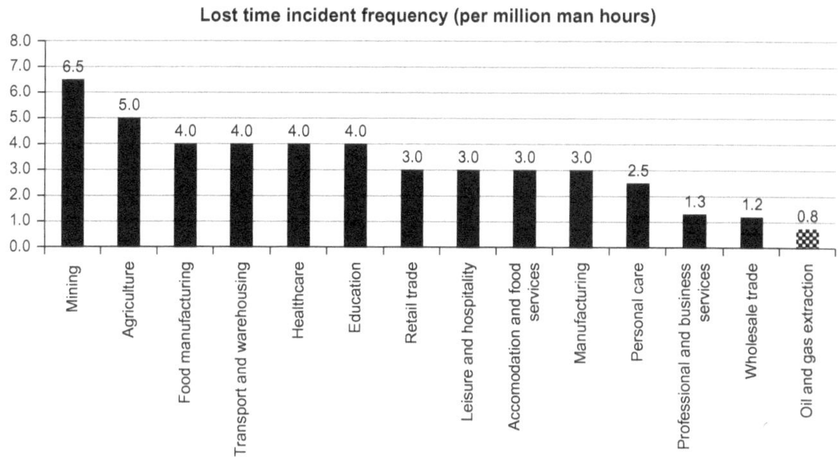

Fig. 2 Lost time incident frequency in several industries. *Energyindepth.org. (2016).*

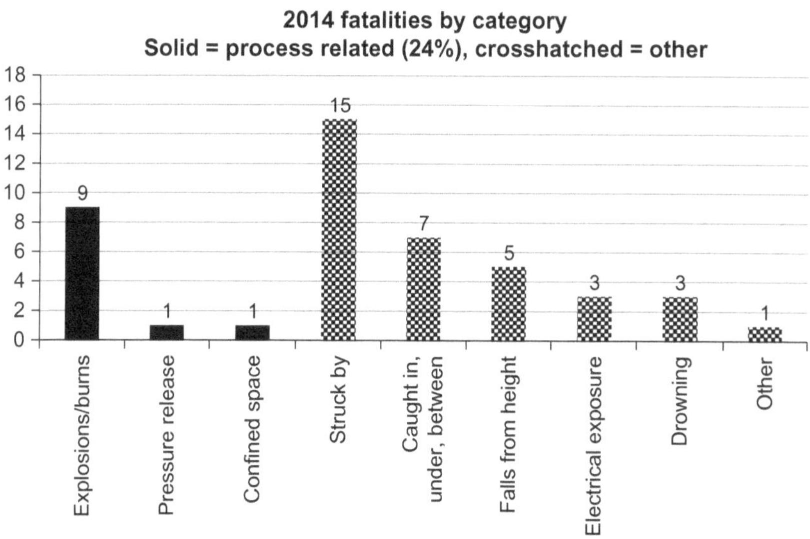

Fig. 3 2014 fatalities in the oil and gas industry by category. *International Association of Oil & Gas Producers. (2015). Safety performance indicators, 2014 data.*

One of the positive elements used to drive this continuous improvement in chemical process industry safety has been the regulatory process, which itself has evolved significantly from the early days of strict prescriptive regulation to today's more popular approach of operator conceived, generated, and enacted safety case regimes. Lord Cullen's investigation of the 1988

Piper Alpha disaster and subsequent recommendations were the main drivers in the gradual migration of offshore oil and gas safety approaches from prescriptive regimes to the more modern safety case approaches.

The key concept in the safety case approach is that risk to individuals should be reduced to "ALARP," as low as reasonable practicable. All stakeholders recognize that there will inevitably be things that could be done to make any facility safer—extreme examples being building causeways from distant offshore platforms to shore or replacing floating installations with artificial islands irrespective of water depth. However, budgets are not unlimited and therefore reasonable and practical expectations of personal safety and individual risk have been adopted. The ALARP principle, when applied to proposed safety improvement measures, has come to be interpreted in the industry as: a proposed change should not be implemented if its time or monetary cost is grossly disproportionate to the risk reduction actually achieved.

The types of equipment governed by the regulations can be divided into two categories. First, equipment which, if it failed or was compromised, might lead to an MAE, for example, primary structure, pressure vessels, process piping, and electrical wiring; and second, equipment or structures which are meant to prevent or mitigate the results of an existing event, for example, gas detection, temporary refuges, lifeboats, fixed and portable fire-fighting equipment, and blast walls.

Some countries (Australia and the United Kingdom) operate primarily goal setting/safety case regimes, while others (Nigeria and China) continue to use mostly prescriptive approaches. A hybrid approach is taken in countries (Canada, the United States, Norway, and Denmark) where elements of safety case approaches are introduced into existing prescriptive regulations. In general, the movement globally is toward more goal setting methods which place more responsibility on the duty holder, who is considered to be better placed to understand the specific requirements of the project and its location-specific hazards. The gradual reduction of accident frequency and general improvement in most safety statistics over time validates the direction of this philosophical movement.

A review of historical accidents is useful in helping to decide future approaches to chemical process safety, either from the regulator's or the duty holder's perspective. A rigorous and systematically implemented regulatory program could have provided the tools necessary to prevent some infamous process related disasters, such as the Piper Alpha production facility explosion and fire in the UK North Sea and well known process related disasters in Texas City, United States, and Bhopal, India. In every case more than one root cause can be identified and good arguments can be made that an effectively implemented regulatory program may have prevented the accidents. Some of the common root causes were lack of PTW systems, noncode-compliant design, improper modifications, failure to perform or improperly performed maintenance, general noncompliance with safety procedures, lack of appropriate management/supervisory sign-off of work, and inadequacies in hazard identification and mitigative measures. These elements are key components in both the goal setting/safety case and prescriptive regimes utilized in today's process industry.

Classification societies play an important role in the regulation of process safety. In some cases, the societies offer consultancy services and technical expertise directly to clients, but more often they provide an independent assessment of compliance with codes, standards, and regulations. In most cases this is mandated by law, but occasionally the verification is done as part of the duty holder's due diligences process. The societies can offer a view based solely on the technical details of the facility, unfettered by schedule and cost constraints. The process is based on (i) a review of the facility's design plans and (ii) surveys during construction, commissioning, and installation. Four prominent classification societies are: The UK-based Lloyd's Register (LR), the Norway/Germany-based DNV-GL, formed by the merger of previously Norwegian Det Norske Veritas and German Germanischer Lloyd, The French-based Bureau Veritas, and the Houston-based American Bureau of Shipping. Each society has specific strengths, and it is quite common to see more than one organization employed on megaprojects, as operators capitalize on their respective abilities.

4. OVERALL APPROACHES: PRESCRIPTIVE AND SAFETY CASE REGIMES

4.1 Origins of Safety Case Regimes

On July 6, 1988, the Piper Alpha platform, located in the United Kingdom sector of the North Sea, exploded and burned, resulting in the deaths of 165 of the 226 personnel on board, and £1.7 billion worth of property damage. At the time, the platform's production accounted for approximately 10% of North Sea output. The structure was almost completely destroyed within a few hours and the direct cause was a lack of containment of flammable hydrocarbon vapors.

Prior to the Piper Alpha tragedy, offshore process safety was regulated by prescriptive regimes, wherein a series of regulations specified exactly what was required to meet a country's safety expectations. This approach, still used in some countries (e.g., Nigeria and China), is unpopular with many operators due to its inflexible nature. Operators, however, do see at least one overall benefit in the approach—the perceived retention by governments of accident liability when operators comply with all the specific requirements of the prescribed regulations.

A downside of prescriptive regulation occurs when operators install equipment they consider unnecessary or inappropriate for the sole purpose of complying with regulatory requirements and obtaining a classification certificate, permit to operate, or similar. Doing so can mean missing opportunities to apply best industry practice and/or installing the latest and best safety equipment and systems.

Following the Piper Alpha disaster, a public inquiry, chaired by Lord Cullen, was initiated by the United Kingdom government. In addition to his original mandate to investigate the event's causes, Lord Cullen offered 106 recommendations to prevent reoccurrences. In view of its failure to prevent the disaster, the previous prescriptive regulatory approach was rejected in favor of a safety case regime for offshore oil and gas installations.

Lord Cullen's recommendation to adopt a safety case/goal setting regime was accepted in the United Kingdom and has gradually been utilized to different degrees and with different formats in other locations. Under these regimes, the operator is completely responsible for any and all accidents and is required to establish a safety case to ensure that the risk to personnel due to MAEs is ALARP.

The ALARP principle is a key element of the safety case approach and was based on a judgment by Lord Justice Asquith in the United Kingdom in 1949 and subsequently confirmed by the Australian high court. It is interpreted in the industry as meaning that the cost, in terms of either time or money, to further reduce an identified level of risk is disproportionate to the risk reduction potentially achieved.

The goal setting/safety case approach is popular in that it provides flexibility in how the zero accident goal is achieved, but the duty holders have realized there is nowhere to hide—by definition any accident is a failure in the safety case's suitability.

In some countries retaining primarily prescriptive regulatory approaches, for example, the United States and Canada, some form of safety case, or at least the basic elements of one, is often one of the prescriptive requirements.

4.2 Safety Case Development

The format for safety cases is similar globally, but differences in emphasis exist between countries and operators. In the United Kingdom, for example, the template requires the inclusion of the following main sections:

- Introduction.
- Description of safety systems.
- Safety management system.
- Management of major hazards.
- Justification for continued operation (including following repairs, modifications, or unexpected events).

Examples of the steps, elements, and terminology used to develop and execute a safety case are as follows:

- *Safety reviews*: These consider the overall safety of an installation and usually comprise multiday workshop style meetings attended by qualified and knowledgeable safety representatives of the owner and any of its contractors familiar with the facility's safety arrangements.
- *Hazard and operability studies (HAZOPs)*: These are formal, facilitated, documented systematic workshops intended to identify and classify the facility's hazards and their causes. They are most effective when representatives from many departments and disciplines are included.
- *Hazard identification studies (HAZIDs)*: Similar to HAZOP studies, these are formal documented workshops intended to systematically identify all credible hazards in a facility or plant. They benefit from multidiscipline participation, including, for example, the responsible design engineer, project manager, installation manager, maintenance engineer, and several project engineers. The HAZID output is usually a list of hazards which can be used, for example, in quantitative or qualitative risk assessments (QRAs).
- *Development and population of a hazard/safety risk register*: This is usually the output from safety reviews, HAZOPs and HAZIDs. Experienced facilitators are often employed to ensure that sufficient rigor is applied to the process and all hazards have been evaluated.
- *Quantitative/qualitative risk assessment studies (QRAs)*: The purpose of these activities is to assess the risk, comprised of both probability of occurrence and consequence, of each identified event.

  Quantitative approaches involve calculating risk by multiplying together the assigned numerical values of probability of occurrence and consequence. The process has drawn criticism from some industry participants who questioned the accuracy and suitability of applying numbers to some risk elements, for example, the value of a human life. Conclusions based on the result of multiplying what are sometimes very small numbers together to produce an even smaller number were also considered by some to be somewhat unconvincing and difficult to accept.

Qualitative approaches have become more popular, wherein probability of occurrence is subjectively evaluated along a range, for example, from "remote" to "certain," and consequences are categorized from "negligible" to "catastrophic." Plotting both results along orthogonal axes then leads to an overall risk assessment, with conclusions potentially including actions such as "tolerate," "monitor controls," and "stop all production until corrective action is effectively implemented." An example of this system is shown in Fig. 4.

- *Generation of a list of potential MAEs*: These are project specific and can include natural events. Most often they are defined as an event having the potential to cause multiple fatalities. In some cases, however, the definition is expanded to include, for example, environmental damage, asset damage, and even corporate reputational damage. It is noteworthy that in many cases events having the potential to result in only a single fatality are excluded from safety case analyses. These would then include the events which, statistically, produce the most offshore injuries and fatalities—slips, trips, and falls.
- *"Bow-tie" analysis and workshop*: The title originates from the shape of a pictorial description of the work (see Fig. 5): The MAE is at the center of

Probability of occurrence	Certain	Moderate	High	Intolerable	Intolerable	Intolerable
	Probable	Low	Moderate	High	High	Intolerable
	Possible	Low	Low	Moderate	High	Intolerable
	Unlikely	Tolerable	Low	Low	Moderate	High
	Remote	Tolerable	Tolerable	Low	Low	Moderate
		Negligible	Low	Significant	Major	Catastrophic
		Consequence				

Fig. 4 Qualitative risk approach. Actions: Tolerable—proceed with caution; Low—monitor controls; Moderate—monitor existing controls and consider additional controls; High—implement additional controls including a permit to work system; Intolerable—stop work. Do not proceed until the risk is reduced.

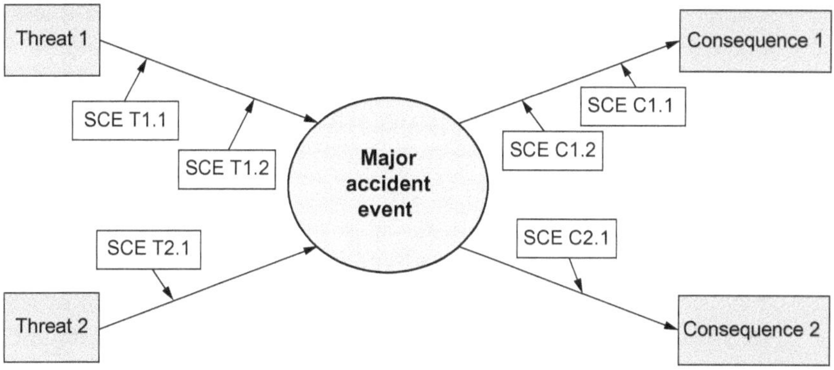

Fig. 5 Bow-tie analysis.

the "tie," with threats and consequences forming the opposite sides. The SCE form barriers on either side of the MAE, with those on the threat side representing steps taken to ensure that the threats do not cause a MAE, and those on the consequence side representing the effective performance of SCEs meant to mitigate the effects of MAEs.

One of the outputs of the bow-tie analysis is a list of SCEs. There are two types:

- Structure, plant, equipment, system, or other elements, the failure of which could lead to a MAE. Examples would be pressure vessels, main structure, process piping, and electrical cabling.
- Equipment or systems whose purpose is to prevent or limit the consequences of a MAE. Examples of this type of SCE would be gas detection equipment, deluge systems, fire pumps, life boats, and temporary refuges.

Care needs to be taken to include both types of SCEs in safety case discussions and considerations, and not only the second type, which are traditionally considered to comprise a project's safety equipment.

- *Establishment of PSs, which, if complied with, ensure SCE effectiveness*: Each SCE has an associated specific PS or series of PSs. In theory, if the PS is complied with, the SCE is effective in doing its part to prevent the occurrence or mitigate the effects of a MAE. Examples of PSs might be:
  - Firewater pumps to provide a given flow rate.
  - Main structure to withstand the effects of a hurricane of a given strength for a given time period.
  - Temporary refuge to provide safe cover for personnel for a given time period following the most intense credible identified fire or explosion.

  - Cables supplying power to emergency equipment to remain operational for a given time period following initiation of the most intense identified fire.
  - Hull structure to meet all requirements of a recognized classification society's applicable rules.
- *PS requirements are of five types*: functionality, availability, reliability, survivability, and interdependencies, often referred to by the acronym "FARSI." These terms are defined as follows:
  - *Functionality*: The specific functions required to be performed by the SCE—for example, hydrocarbon carrying piping should withstand, without loss of containment, all static and dynamic loads applied by internal fluid motion and external supports.
  - *Availability*: The ability of the SCE to perform its function under the specific expected conditions—for example, systems should not be adversely affected by electromagnetic interference generated by other proximate equipment.
  - *Reliability*: The probability that the equipment will operate without failure under the expected conditions. In some cases, the probability is defined in project-specific programs such as QRA or a defined safety integrity level (SIL). In others, the probability of failure is defined in external codes and standards.
  - *Survivability*: The ability of the SCE to operate for as long as is required following an event—for example, a temporary refuge is required to provide protection for occupants for as long as is required to safely evacuate all persons on board under the most extreme fire and/or explosion event. Similarly, passive fire protection (PFP) would be required to prevent any collapse of the structure which would jeopardize evacuation for as long as it would take to safely evacuate all personnel under the most extreme fire and/or explosion event.
  - *Interdependencies*: Any other systems upon which an SCE depends to perform properly—for example, process emergency shutdown (ESD) valves require that the cabling carrying signals to them operate properly under the expected conditions, and that the control logic governing their use has been correctly developed and coded within the ESD system. Additionally, an ESD system typically requires interfaces with other SCE systems such as ignition prevention, fire and gas detection and signaling, and emergency power.
- It is usually a regulatory requirement that compliance with the PSs is verified. This needs to be done by the builders and operators as a normal

part of project due diligence. There is also usually a regulatory requirement to have an independent third party assess the arrangements and confirm PS compliance. This role is usually performed by one of the classification societies.

5. SOME SCEs RELATED TO PROCESS SAFETY

There are two types of SCE:

- Those, which are not traditionally considered to be "safety equipment," for which their failure could lead to a MAE.
- Those, which are more frequently considered to comprise a facility's safety equipment, whose function is to either prevent or mitigate the effects of an MAE.

Following are some examples of SCEs related to process equipment. The below is not an exhaustive list.

5.1 SCEs, the Failure of Which Could Lead to a MAE

- *Main structure supporting process equipment*: The structures (e.g., beams supporting decks, decks themselves, and pipe supports) supporting process equipment need to be sufficiently robust to allow the equipment to operate under normal operating conditions. Failure to do so could lead to, for example, piping or pressure vessel failure, subsequent release of hydrocarbon, and, upon ignition, an explosion or fire resulting in multiple fatalities and/or other serious events.
- *Pressure vessels, heat exchangers, fired heaters*: This equipment needs to be designed and constructed properly, thereby preventing either sudden catastrophic explosive failures or process leaks. The former would likely lead to serious fires and/or explosions, the latter, in combination with a source of ignition, could likewise lead to serious explosions and/or fires.
- *Storage tanks*: Leaks in storage tanks, in combination with a source of ignition, can quickly lead to serious fires and explosions. If under pressure, undetected fabrication or design flaws in tanks can lead to more sudden explosions.
- *Piping systems*: Failures in piping systems, whether induced by design or fabrication flaws, can lead to MAEs either through sudden explosions or the production of standing hydrocarbons, which provide a fuel source for fires/explosions. Pipe supporting arrangements are critical in the

prevention of piping systems induced MAEs. For floating offshore installations, forces induced by the movement of the carried fluids combined with those related to vessel movement all need to be considered in pipe support design and installation.

5.2 SCEs Expected to Prevent or Mitigate the Effects of a MAE

5.2.1 SCEs Expected to Prevent the Occurrence of a MAE

- *Hazardous area ventilation*: The purpose is to prevent the formation of hazardous flammable gas mixtures in hazardous areas such as paint stores, battery storage rooms, and areas around process equipment with some likelihood of the presence of leaked gaseous hydrocarbons. Equipment used to ensure effective hazardous area ventilation management include, for example, heating, ventilation, and air conditioning (HVAC) dampers, fire dampers, and forced draft ventilation.
- *Inert gas system*: The system provides an inert gas (e.g., nitrogen) blanket in tanks in order to displace oxygen and prevent the development of a flammable atmosphere, which could enable a fire or explosion.
- *Relief system*: Comprised of, for example, pressure relief valves and/or rupture disks on individual process lines and/or vessels. The purpose is to ensure that line pressures do not build-up to the point where an explosion could occur.
- *ESD system*: The system is designed to shut down appropriate process equipment upon detection of abnormal situations within the system—for example, high pressure, temperature, or level within individual vessels, or detection of smoke, fire, or gas anywhere on the installation. The shutdown philosophy forms the basis for verifiable cause and effects charts. Systems usually include manual hand switches allowing operator intervention in the event of emergencies.
- *Fire and gas detection*: These systems use various methods to detect the presence of, for example, flames, heat, smoke, or flammable gas mixtures. Manual call points are also frequently part of the system. Outputs from the system, depending on the situation, can be alarms or shutdown signals to the appropriate process equipment.
- *Flame retardant and fire resistant cabling*: Specific types of cables are specified dependent on their use. For example, different requirements would apply for cables supplying safety-related equipment such as fire and gas detectors than for cabling supplying nonsafety critical equipment such as entertainment appliances.

5.2.2 SCEs Designed to Mitigate the Effects of a MAE

- *Main structure*: As stated earlier, this is also an SCE of the first type. The main structure also needs to maintain sufficient structural integrity during and after an MAE for long enough to allow all personnel to evacuate safely. In this way, it mitigates the effects of an MAE that has already happened.
- *Flare and vent system*: Flare towers on offshore facilities can be greater than 150 m in height. Their purpose is to allow the safe removal of hydrocarbon inventory, thereby eliminating a fire's fuel source and minimizing damage to equipment and danger to personnel.
- *Deluge and water monitor systems*: These systems include pipework, control instrumentation, and fixed and oscillating monitors. Their purpose is to prevent escalation of fires by supplying cooling water to ignited hydrocarbons when called upon either automatically or manually.
- *Structural fire and explosion protection*: This usually refers to bulkheads, structural decks, doors, etc. The structural items are designed to contain explosions and/or limit the spread of fires for defined periods. Their ability to maintain integrity is usually proved and documented by type approval tests. These tests are performed remotely and subject test structures of the same design and construction as provided on the facility to fires having defined intensities and periods.
- *Temporary refuge*: This is meant to provide an initial safe haven for personnel for sufficient time to allow their safe evacuation from the facility to be planned and executed.
- *Passive fire protection (PFP)*: Although it can take several forms, PFP is usually a sprayed on coating which limits or delays the effects of fire on the structure, piping, or vessels on which it is applied. The extent and thickness of PFP application is determined by the overall safety philosophies of the installation/plant. The overriding philosophy is that PFP should prevent failure of, for example, structures, piping, or pressure vessels for sufficient time to allow personnel to be safely evacuated.
- *Water mist systems*: These are usually used to extinguish fires in equipment rooms such as diesel generator rooms, etc. The equipment usually includes pipework, pumps, water supply tanks, and control equipment. The purpose of the system is to prevent the escalation of fires in the protected area.
- *Fixed foam systems*: These are usually utilized at helicopter landing facilities and can be initiated either manually or automatically. The systems extinguish or limit the effects of hydrocarbon fuel fires, essentially by

cutting off the supply of oxygen to the fire. Components include pipework, pumps, instrumentation, storage tanks, and fixed and oscillating monitors.

- *Process and ESD valves*: The purpose of these valves is to isolate the process or riser inventory feed and thereby limit the time that fuel is provided to an existing fire. The equipment is usually operated pneumatically or hydraulically, upon an either automatically or manually initiated instrument signal.
- *Lifeboats and life rafts*: These are the ultimate means of personnel protection when the decision has been made to evacuate an offshore platform. Modern lifeboats are usually of the totally enclosed motor propelled survival craft (TEMPSC) type, need to be available for all persons on board, and usually rely only on gravity for deployment. Life rafts are deployed at strategic locations on offshore facilities, both to comply with statutory requirements and provide alternatives to personnel not able to get to the primary evacuation location for any reason.
- *Emergency power*: A dedicated generator is designed and installed to start automatically and provide power to essential and emergency services in the event that main power is lost. Special arrangements are made to protect the space enclosing the emergency generator from the most intense identified fire and explosion events.

6. SOME EXAMPLES OF REGULATORY REGIMES

6.1 Overview

The process whereby chemical process safety is regulated varies from country to country. Some, for example, Nigeria and China, operate primarily prescriptive regimes, wherein detailed safety requirements are specified in regulations. Others, for example, the United Kingdom and Australia, regulate process safety within safety case regimes. For the latter, the operator generates and is responsible for all aspects of a safety case, wherein a series of SCEs and PSs are defined. Theoretically, if the PSs are complied with, a level of safety ALARP is achievable. Other countries, for example, Canada, the United States, Norway, and Denmark, operate hybrid regimes wherein many of the elements present in safety cases are embedded within prescriptive regulations.

Key elements and philosophies of the regimes in Canada, the United States, the United Kingdom, Australia, Denmark, Norway, Nigeria, and

China are described in this section. Because of differences in the various regimes' structures and the style of information available in the literature describing each, the descriptions are not completely consistent in their presentation. Emphasis has been placed on any distinctive aspects of the countries' contributions to process safety regulation, for example, the UK's influence on all the other regimes resulting from its formal adoption of a safety case approach, and Australia's definition of Validation as an activity to be kept separate and distinct from all other aspects of facility regulation.

Table 2 summarizes each country's primary methods of regulation and their overarching acts and main regulations.

6.2 Canada: Offshore—Certificate of Fitness; Onshore—Permits to Construct and Operate

6.2.1 Offshore Certificate of Fitness Regime

Canada has operated joint federal—provincial Certificate of Fitness (CoF)-based safety regimes for its offshore oil and gas industry since its beginnings with the Cohasset Panuke project, near Sable Island, southeast of Nova Scotia, in 1991. Since then, other projects, including Newfoundland's Hibernia, Terra Nova, White Rose, and Hebron, and Nova Scotia's Sable and Deep Panuke, have been regulated under the same regime, using a CoF approach. The regulatory agencies and governing regulations for offshore facilities are different than those for pipelines, as described later. Platforms and floating installations are overseen by the Canada—Nova Scotia (and Newfoundland) Offshore Petroleum Boards (CNSOPB, CNLOPB), pipelines by a combination of the foregoing and the National Energy Board (NEB).

The overarching legal instrument governing offshore activity in Nova Scotia is the Canada—Nova Scotia Offshore Petroleum Resources Accord Implementation Act, Part III, Petroleum Operations (Croston, 2016). Similar legislation governs offshore activity in other provinces. Under this act, the Governor in Council is authorized, for the purposes of safety and the protection of the environment, to make various regulations covering, for example:

- The production, processing, and transportation of petroleum.
- The management and control of petroleum production.
- The removal of petroleum from the offshore area.
- The design, construction, operation, and abandonment of pipelines within the offshore area.

Table 2 Country Summary

Country	Primary Method of Regulation	Overarching Acts/Main Regulations
Canada	Offshore: Compliance with certificate of fitness regulations	Offshore: Canada—Nova Scotia Offshore Petroleum Resources Accord Implementation Act, Part III Petroleum Operations (or similar acts/regulations for other provinces (Newfoundland, British Columbia))
	Onshore and pipelines: Compliance with regulations leading to permits to construct and operate	Onshore and pipelines: Nova Scotia Energy Resources Conservation Act National Energy Board Act Nova Scotia Pipelines Act Gas Plant Facilities Regulations Department of Energy Code of Practice (or similar acts/regulations for other provinces)
The United States	Compliance with Code of Federal Regulations	Code of Federal Regulations 29 CFR 1910.119. Process Safety Management Code of Federal Regulations 40 CFR 68.65. Risk Management Program
The United Kingdom	Verification of safety case	UK Health and Safety at Work Act Offshore regulations: Offshore Installations (Offshore Safety Directive) (Safety Case) Regulations Prevention of Fire and Explosions, and Emergency Response on Offshore Installations Regulations Offshore Installations and Wells (Design and Construction) Regulations Offshore Installations and Pipeline Works (Management and Administration) Regulations
Australia	Validation and verification of safety case	Offshore Petroleum Act Offshore Petroleum and Greenhouse Gas Storage (Safety) Regulations Petroleum Pipelines Act (Onshore) Petroleum and Geothermal Energy Resources Act
Denmark offshore	Goal setting and risk based within a prescriptive Act	Danish Subsoil Act The Offshore Safety Act

Table 2 Country Summary—cont'd

Country	Primary Method of Regulation	Overarching Acts/Main Regulations
Norway	Safety case type approach (although a safety case is not specifically mandated)	The Petroleum Activities Act The Framework Health, Safety, and the Environment Regulations The Management Regulations The Facilities Regulations The Activities Regulations The Technical and Operations Regulations
Nigeria	Primarily prescriptive	Petroleum Act Oil Pipelines Act Oil in Navigable Waters Act Mineral Oils (Safety) Regulations Petroleum Regulations Petroleum (Drilling and Production) Regulations
China	Primarily prescriptive	Safety Production Law Offshore Oil Safety Production Regulations Detailed Rules on Offshore Oil Safety Management Provisional Regulations on the Supervision and Administration of Oil and Gas Pipelines Safety Safety Rules for Coalbed Methane Surface Mining Health, Safety and Environmental Management Systems for Oil and Gas Industries

The primary regulation governing nonpipeline offshore activity is the overarching "Certificate of Fitness Regulations," which, in turn, references specific clauses from the next tier of regulations, including:

- Installations regulations.
- Occupational safety and health regulations.
- Production and conservation regulations.

These regulations reference specific codes and standards related to chemical process safety, which are required to be complied with. Such compliance is expected to minimize the likelihood of process related accidents.

While the regime is primarily prescriptive, goal-oriented approaches are being introduced. There are specific requirements for the operator to generate and comply with a concept safety analysis, a safety plan, and to set safety target levels such as the risk to life. Unlike the systems utilized in the United Kingdom and Australia, however, these and other goal setting requirements do not comprise the regime's overarching philosophy. Some specific prescriptive requirements similar to safety case elements are as follows:

- The concept safety analysis is required to encompass all phases of the facility's life, including design, construction, installation, commissioning, operation, and abandonment.
- A determination of probability of occurrence and consequences are required for each potential accident.
- Provision for the safe evacuation of all personnel from the production installation is required.

The future direction of the regime's philosophy can be anticipated by the 2009 publishing of Safety Plan Guidelines, which include concepts similar to those required in the United Kingdom safety cases, stating, for example, that:

- "While the concept 'as low as reasonably practicable' (ALARP) is not discussed in the regulations, this will be a factor when considering a safety plan under the regulations."
- "Industry may demonstrate incorporation of ALARP into their risk reduction and associated mitigating measures through a number of means, including by a combination of qualitative analysis, quantitative analysis, and good industry practice."
- "The safety plan is to include a listing of all structures, facilities, equipment, and systems critical to safety (analogous to SCEs)."
- "The methodology used to develop the list of safety critical items should include links to the overall risk assessment, design processes, and ALARP studies."

The operator is required to obtain a permit to operate prior to commencing production, and a CoF is a prerequisite for obtaining this. The petroleum boards rely on classification societies, acting as Certifying Authorities (CAs), to perform sufficient design appraisal and construction surveillance to confirm to the boards that the requirements of the CoF regulations have been complied with.

After production has commenced, the ongoing validity of the CoF is confirmed by the CA via a series of periodical surveys, the frequency of which is determined by the complexity of the installation. The most common frequency is annual, but for larger and more complicated installations, semiannual or quarterly visits are made.

6.2.2 Pipelines

In addition to CNSOPB/CNLOPB oversight for offshore pipelines, all pipelines are subject to regulation by the NEB. The primary regulatory documents are the NEB Act and the Canadian Onshore Pipeline Regulations. The regulations set down prescriptive requirements related to other Canadian national standards, and in some high risk cases (e.g., as defined by Canadian standard Z662, or within 500 m of a railway or paved road) also require submission of a documented risk assessment.

The regulations also specify the submission of several manuals and programs which are very similar to those required by safety case regimes—for example:

- Construction safety manual.
- Maintenance safety manual.
- Emergency procedures manual.
- A training program including safety regulations, procedures, and working practices.
- Safety program.

6.2.3 Onshore Facilities: Permits to Construct and Licenses to Operate

6.2.3.1 Introduction and Organizations Offering Applicable Codes and Standards

Onshore facilities in Nova Scotia, including liquefied natural gas (LNG) plants, are subject to the requirements of the Nova Scotia Energy Resources Conservation Act, which in turn references the Pipeline Act (Act) (Province of Nova Scotia, 2000), the Gas Plant Facilities Regulations (Regulations) (Province of Nova Scotia, 2013), and the Nova Scotia Department of Energy Code of Practice for Liquefied Natural Gas (Code of Practice) (Nova Scotia Department of Energy, 2005). These regulations, enforced in Nova Scotia by the Nova Scotia Utilities and Review Board (UARB), require the operator to obtain a permit to construct prior to construction commencement and a license to operate prior to startup. The Code of Practice includes references to various organizations' codes and standards applicable to chemical process safety, as follows.

- American Gas Association for purging procedures.
- American Petroleum Institute for construction of low pressure storage tanks and associated relief devices.
- American Society of Mechanical Engineers for design of piping and pressure vessels and associated relief devices.
- Canadian Gas Association for cryogenic liquid standards.

- Canadian General Standards Board for piping systems and qualification of personnel conducting nondestructive examination.
- Canadian Standards Association for LNG facilities and equipment.
- Gas Technology Institute for information on modeling of LNG releases.
- Instrument Society of America for design of safety instrumented systems (SISs).
- National Fire Protection Association for fire protection systems.
- Tubular Exchanger Manufacturers Association for design of heat exchangers.

Requirements also include the submission and approval of the proposed facility's quality assurance (QA) plans and policies, and procedures for design, material control, construction, and operation.

6.2.3.2 Prescribed Safety Case Requirements

Many of the main elements of a safety case are required prescriptive elements for issuance of a license to operate, including the following:

- *Hazard Operability (HAZOP) Report*: Describing the hazard identification activity performed during detailed engineering design, including a list of any changes made after the HAZOP and a copy of the action plan showing the resolution of all action items.
- *Emergency Response Plan*: Developed in conjunction with the facility security assessment.
- *Prestartup Safety Review Report*: Summarizing the review and documentation of the identification and subsequent resolution of any action items.
- *Operational Phase Process Safety Management (PSM) System procedures*: A complete set of PSM procedures to be followed during operation; specifically the remaining procedures not already developed during design and construction, including:
  - Prestartup safety review following any changes.
  - Process hazard analysis—revalidation of the HAZOP.
  - Incident investigation.
  - Emergency planning and response.

6.2.3.3 Some Example Process Safety Equipment Requirements

The Code of Practice includes specific requirements for many process safety equipment items, for example:

- *Building ventilation*: Air intakes to be provided with gas detectors which shutdown air handling units and inhibit startup in the event of gas detection, in order to prevent the introduction of toxic or flammable gas.

Control room HVAC system to be designed to cater for the maximum possible thermal radiation flux.

- *Relief devices*: Sizing to be based on fire exposure, process upsets, thermal expansion, sudden change in barometric pressure, thermal expansion, control malfunction, improper valve position, mechanical failure, and loss of utilities.
- *Boil off recovery system*: To be installed to collect LNG boil off. Vapors to be safely disposed of through reliquefaction, used as fuel, returned to storage tanks or marine tankers, recompression to a gas pipeline network, or as a last resort, flared or released to atmosphere.
- *Flare and venting system*: Sized for the maximum expected accidental gas flow.
- *Check valves*: To be installed downstream of block valves in the direction of flow. In this manner, any LNG trapped between the check valve and block valve will be relieved on expansion into the downstream piping.
- *ESD valves*: Closure time to be adjusted to ensure closing does not result in pressure surges that could cause piping or pipe support failure. ESD valves to meet the following criteria:
  - Valves that may be operated for process control by the basic process control system (BPCS) shall only be operated for emergency reasons by the SIS.
  - Valves shall fail in a safe position as identified in the HAZOP.
  - Valves which are used also to isolate inventories of flammable materials shall be designed as fire safe.
- *Emergency power*: Designed to ensure, in case of main power supply failure, continued availability of all vital safety functions.
- *Safety instrumented system (SIS)*: Designed to protect equipment and personnel from process upsets and emergency conditions. Required to automatically detect unsafe conditions and activate automatically or manually the appropriate equipment, unit, process, or ESD. Main functions to include:
  - Activate automatically the appropriate shutdown system and protection equipment.
  - Inform the operator of an incident.
  - Inform the BPCS of SIS activation.
  - Monitor and control the protection equipment (e.g., fixed fire protection systems).
  - Monitor and control protection system auxiliaries (e.g., fire pumps, foam agent pumps, and fire water system valves).

6.2.3.4 Approval Process and Classification Societies

The onshore approval process includes the direct engagement by the UARB of a classification society to act as a CA. The essential scope is to certify that the applicable requirements of the regulations have been complied with during the design, construction, operation, and abandonment phases of the facility's life cycle. The CA scope of work can include the following:

- A review of QA plans, policies, and procedures prepared by project proponents, in order to confirm that they reflect the applicable standards referenced in the Regulations.
- A review of, and provision of advice to the UARB regarding, applications made by project proponents for permits to construct. The advice can include recommendations regarding terms and conditions which should be attached to the permit.
- Prior to issuance of the UARB's permit to construct, certification that the LNG plant has been designed in accordance with the applicable provisions of the Regulations, including all applicable codes and standards.
- A review of the QA and quality control (QC) programs and procedures of major equipment fabrication, to ensure they reflect all applicable standards referred to in the Regulations. The CA attends at the manufacturing plants for such time at it considers necessary in order to meet the requirements of this section.
- Provision of a report to the UARB describing the work performed and certifying that the fabricated equipment fully meets the requirements of the Act, Regulations, and all applicable codes and standards.
- Monitoring, on both a scheduled and random audit basis, of all plant construction activities including the implementation of field testing programs, and communication of any concerns to the UARB.
- A review of, and provision of advice to the UARB regarding, applications made by project proponents for licenses to operate. The advice can include recommendations regarding terms and conditions which should be attached to the permit.
- Prior to issuance of the UARB's license to operate, certification that the LNG plant has been constructed in accordance with the applicable provisions of the Regulations, including all applicable codes and standards.
- Provision of requested advice, assistance, reports, certifications, or other related work required in relation to abandonment of facilities, including support for any permits or licenses which the UARB may need to issue.

6.3 The United States: Code of Federal Regulations

Crowl and Louvar (2002) describe the United States government process for generating and enacting laws and regulations, as summarized below.

6.3.1 Establishment of Laws

The United States laws are put in force using the following process:

1. A member of congress proposes a bill, which is a document that, if approved, becomes a law, or act.
2. If both houses of congress approve the bill, it is sent to the president, who has approval or veto rights. If approved, it becomes a law.
3. The complete text of the law is published in the US Code (USC), which is the official record of all federal laws.

6.3.2 Establishment of Regulations

Laws do not specify detailed requirements to be complied with. These are provided in regulations, which are generated by organizations so authorized by congress, for example:

- Environmental Protection Agency (EPA).
- Occupational Safety and Health Administration (OSHA).
- Bureau of Ocean Energy Management (BOEM).
- Bureau of Safety and Environmental Enforcement (BSEE).

The process for generating regulations is as follows:

1. The authorized organization or agency decides when a regulation is needed and then researches, develops, and proposes the regulation document. The proposal is listed in the federal register for public evaluation and comment. The comments are used to finalize the regulation.
2. After a regulation is finalized, it is posted to the federal register as a final rule, and it is simultaneously codified by publishing it in the Code of Federal Regulations (CFR).

Many acts and regulations are in force related to chemical process safety. The number of federal laws related to process safety has increased from less than 10 in 1950 to more than 50 today. Two of the more prominent and widely used regulations, both of which were generated in response to the Bhopal chemical accident in 1985, are:

- 29 CFR 1910.119—Process Safety Management.
- 40 CFR 68.65—Risk Management Program.

Some details of these two important chemical process safety documents are offered below.

6.3.3 Content of the PSM Regulations

The PSM regulation, formally titled "Process Safety Management of Highly Hazardous Chemicals," is comprised of the following 14 sections:

1. Employee participation.
2. Process safety information.
3. Process hazard analysis.
4. Operating procedures.
5. Training.
6. Contractors.
7. Prestartup safety review.
8. Mechanical integrity.
9. Hot work permits.
10. Management of change (MOC).
11. Incident investigation.
12. Emergency planning and response.
13. Audits.
14. Trade secrets.

Details of each section are included in Crowl and Louvar (2002). This regulation is primarily designed to protect personnel inside a process plant. It is noteworthy that within this primarily prescriptive document, many of the elements required by a safety case are specified (process safety information, process hazard analysis, prestartup safety review, incident investigation, emergency planning and response).

There is also significant consistency between the requirements prescribed within many of the regulation's sections and those of most safety cases utilized globally, for example:

- The process hazard analysis is required to be performed by experts, including duly qualified engineers and industrial hygienists. For complex processes, a HAZOP is required, and for less complicated processes, "what if" scenarios, fault trees and failure mode and effects analyses (FMEAs) are utilized.
- The prestartup review is conducted before restarting a facility following modifications or repairs. Elements of the review include design appraisal, confirmation that appropriate safety and emergency procedures are in place, and that suitable training has been conducted.
- A MOC process is required to be developed and effectively implemented. Changes need to be evaluated, particularly with respect to safety, prior to their implementation. Employee training and a prestartup review are both postchange requirements.

6.3.4 Content of the Risk Management Plan

Protection of citizens outside a process plant is provided by the risk management plan. Its primary elements, detailed in Crowl and Louvar (2002), are as follows:

1. Hazard assessment.
2. Prevention program.
3. Emergency response program.
4. Site-based documentation maintained on site, submitted to federal, state, and local authorities, and shared with the local community.

Again, elements of safety case content are present in these prescriptive requirements (hazard assessment, emergency response program). Some noteworthy comments are as follows:

- Both worst case and more likely scenarios are required to be assessed in the hazard assessment. The use of risk matrices is encouraged in the analysis.
- The prevention program includes many of the same elements as the Process Safety Management Regulations, for example, a prestartup review, MOC process, emergency response, and safety audits.
- The emergency response program includes the requirement to notify local authorities and agencies in the event of an accident. This is similar to notifications processes required by both prescriptive and safety case-based regimes in other countries.

6.3.5 US Offshore Regulation

The US offshore oil and gas industry is overseen by the BSEE. The regime can be generally classified as prescriptive, requiring the issuance of United States Coast Guard (USCG) certificates of compliance and inspection. A classification certificate issued by a recognized classification society can be accepted as partial input to the USCG certificate of compliance.

6.4 The United Kingdom: Safety Case/Verification Regime

6.4.1 The United Kingdom: Health and Safety at Work Act

Under the United Kingdom (UK) legislative process, an Act of Parliament is a law, enforced in all applicable areas of the UK. The Act is approved by a majority in the UK Parliamentary House of Commons and House of Lords, the two separate houses of the UK Parliamentary system, and formally agreed by the UK monarch.

The Health and Safety at Work Act (Act) is the top level UK health and safety legislation (Russell, 2016). This Act places a duty on all employers to ensure, as far as is reasonably practicable, the health, safety and welfare at work of their employees.

6.4.2 Piper Alpha and the Cullen Enquiry

Following the 1988 Piper Alpha disaster, Lord Cullen's UK government-initiated enquiry (described in more detail earlier) rejected prescriptive regulation in favor of a safety case regime for offshore oil and gas installations. This was implemented into legislation in 1992 through the Offshore Installations (Safety Case) Regulations, Statutory Instrument (SI 1992/2885).

The Safety Case Regulations require that specific safety cases and associated safety requirements be defined and established by the duty holder for each offshore oil and gas installation. The safety case has to be accepted by the UK Health and Safety Executive (HSE).

6.4.3 Supporting Regulations

Lord Cullen recommended that the offshore safety case regime should not rely solely on the facility-specific safety cases, and that the Safety Case Regulations should be complemented by other regulations dealing with specific features of offshore safety. This led to the following three additional sets of offshore-specific regulations:

- Prevention of Fire and Explosions, and Emergency Response on Offshore Installations Regulations (PFEER).
- Offshore Installations and Wells (Design and Construction) Regulations (DCR).
- Offshore Installations and Pipeline Works (Management and Administration) Regulations (MAR).

The UK HSE then used the combined requirements of the Offshore Installations (Safety Case) Regulations, PFEER, DCR, and MAR, under the overall UK Health and Safety at Work Act requirements for all employers, "to ensure, so far as is reasonably practicable, the health, safety and welfare at work," of all UK employees.

6.4.4 Subsequent Amendments

In 2005 the Offshore Installations (Safety Case) (OSCR) Statutory Instruments were finalized, and came into force on April 6, 2006. They replaced and revoked the previous 1992 Regulations along with certain transitional arrangements.

In 2015 the Offshore Installations (Offshore Safety Directive) (Safety Case) Regulations (SCR 2015), Statutory Instruments were finalized and came into force on July 19, 2015. They apply to oil and gas operations in the territorial sea adjacent to Great Britain and any designated area within the UK continental shelf. They replaced the 2005 Offshore Installations (Safety Case) Regulations (SCR 2005) in these waters, subject to certain transitional arrangements.

A change under the 2015 Safety Case Regulations requires that safety cases be submitted for assessment to the competent authority, the Offshore Safety Case Directive Regulator (OSDR).

The 2015 Safety Case Regulations incorporate additional safety-related requirements of certain European Union (EU) Directives together with a requirement to ensure improved incident response mechanisms.

6.4.5 Verification by an Independent Body

The 2015 Safety Case Regulations require that a verification scheme be established by the duty holder. Furthermore, an "independent and competent person" (ICP) is required to be engaged by the duty holder to confirm that the SCEs are suitable and remain in good order. The ICP role is usually fulfilled by one of the classification societies. The independence of the ICP organization is defined in the regulations, paraphrased as follows:

- The ICP scope is not to include examination of something for which the ICP bears or has borne responsibility, or where the ICP's objectivity may be compromised.
- The ICP is required to be sufficiently independent of a management system which has, or has had, any responsibility for any aspect of something liable to be examined by the ICP, so as to ensure objectivity in carrying out the ICP function under the scheme.

Some activities of a classification society acting as an ICP are also suggested in the regulations, as below:

- Examination, including testing where appropriate, of the SCEs.
- Examination of any design, specification, certificate, marking, or standard relating to the SCEs.
- Examination of work in progress.
- The creation of reports covering the examination and testing carried out, findings, and any remedial action recommended.
- The documentation of appropriate action by the duty holder following a report.
- The reporting to the duty holder of any instances of noncompliance with the scheme's standards.

6.5 Australia: Safety Case Validation/Verification Regime

6.5.1 Offshore/Onshore/Pipelines Regulatory Summary

Onshore and offshore regulatory regimes in Australia are similar, relying on goal setting and safety cases and their associated SCEs and PSs. In addition to the management of major accidents, risk-based regulation in Australia also includes occupational accident management, which is somewhat unusual. Australian guidance on safety case generation recognizes the benefits of both qualitative and quantitative approaches to risk management.

All offshore facilities and pipelines within Australian commonwealth waters fall under The Offshore Petroleum Act (OPA). This act is administered by the National Offshore Petroleum Safety and Environmental Management Authority (NOPSEMA).

Supporting this act are the Offshore Petroleum and Greenhouse Gas Storage (Safety) Regulations which set out the requirements for the contents of safety cases and general requirements for Validation and Verification. Further applicable regulations are as follows:

- Petroleum (Submerged Lands) (Occupational Safety and Health) Regulations.
- Petroleum (Submerged Lands) (Management of Safety on Offshore facilities) Regulations.
- Petroleum (Submerged Lands) (Management of Well Operations) Regulations.
- Petroleum (Submerged Lands) (Pipelines) Regulations.
- Petroleum (Submerged Lands) (Diving Safety) Regulations.

Onshore activities are regulated by the Resources Safety Department of the Western Australian Department of Mines and Petroleum (DMP). The overarching act is the Petroleum and Geothermal Energy Resources Act. The governing regulations, which have very similar requirements to their offshore counterparts, are as follows:

- Petroleum and Geothermal Energy Resources (Occupational Safety and Health) Regulations.
- Petroleum and Geothermal Energy Resources (Management of Safety) Regulations.

Pipelines (onshore and offshore) fall under the Petroleum Pipelines Act. The governing regulations are:

- Petroleum Pipelines (Management of Safety of Pipeline Operations) Regulations.
- Petroleum Pipelines (Occupational Safety and Health) Regulations.

Since the onshore and offshore regulatory regimes are similar, the remaining discussion in this section will specifically describe the offshore requirements. The distinctive element of the Australian regime is the differentiation between Validation and Verification, which, in other regimes, are jointly known as Verification. The differences will be described in detail below.

6.5.2 Steps in the Validation Process

The steps whereby NOPSEMA approval is achieved are as follows:

1. NOPSEMA requests Validation for new facilities and significant modifications.
2. NOPSEMA and the operator reach agreement on the Scope of Validation (SoV), specifically which SCEs are required to be Validated and against which codes and standards.
3. The SoV is agreed by NOPSEMA in writing and Validation commences.
4. The operator submits the facility safety case.
5. Safety case assessment is underway (including formal comment from NOPSEMA).
6. Validation is finalized and a "clean" statement delivered.
7. Final safety case decision making is delivered.

Verification is a separate process which applies throughout, in parallel with and continuing after Validation.

6.5.3 The Validation and Verification Processes

The separation of Validation and Verification is unique to Australian safety regulation. The Australian versions of the two activities are generally captured under the term Verification in other regimes.

Validation, as applied in Australia, is a documentation review exercise that can, in theory, be completed before or by the early stages of construction, provided sufficient documentation on the design, construction, and installation of the SCEs has been made available to the Validator and all comments resolved.

Verification requires a physical check to ensure that safety critical equipment has been installed correctly and is fit for its function and use. Verification is a survey process, including routine surveillance, witness and hold points, and attendance at final factory and site acceptance tests.

Safety case regulations detail Verification requirements that must be addressed in the safety case. NOPSEMA require that Validation can be documented as a separate assurance process, provided by the regulations and tied to safety case decision making.

More details describing the two processes are as follows.

6.5.3.1 Validation

Validation is a process undertaken by an independent competent party, known as the Validator, which is required to be separate from the duty holder. The process is meant to ensure that the design, construction, and installation of the facility's SCEs will incorporate appropriate measures to protect the health and safety of personnel. Validation activities related to construction and installation are limited to a review of related documentation (e.g., inspection and test plans and welding instructions) and do not include physical survey of equipment.

Validation is required for an operator to obtain safety case acceptance. The operator produces a SoV and obtains NOPSEMA approval. A Validator then performs Validation in accordance with the approved SoV and issues a Validation statement. The statement allows NOPSEMA to be assured that the design, construction, and installation of safety critical systems will incorporate measures that protect the health and safety of persons to the extent required by the SoV, provided the agreed codes and standards are implemented as specified in the documentation reviewed by the Validator. Validation of design, construction, and installation can be defined as follows:

- Validation of design is the review of design documentation against nominated codes and standards in order to form a view about the appropriateness of the standards for the SCEs and to confirm that they are consistent with the safety case.
- Validation of construction and installation are reviews of documentation to ensure that the intent of the design is specified to be carried forward into actual facility build. Typically, this would include construction and installation specifications such as piping or structural fabrication specifications, procedures such as anchor pile installation procedures, and inspection and test plans. Again, the codes and standards must be appropriate and consistent with the safety case.

Since it relies upon forward-looking project documentation, it is possible for Validation to be completed before construction is complete and prior to installation.

The scheduling and relationships between the SoV, the Validation and safety case acceptance are illustrated in Fig. 6.

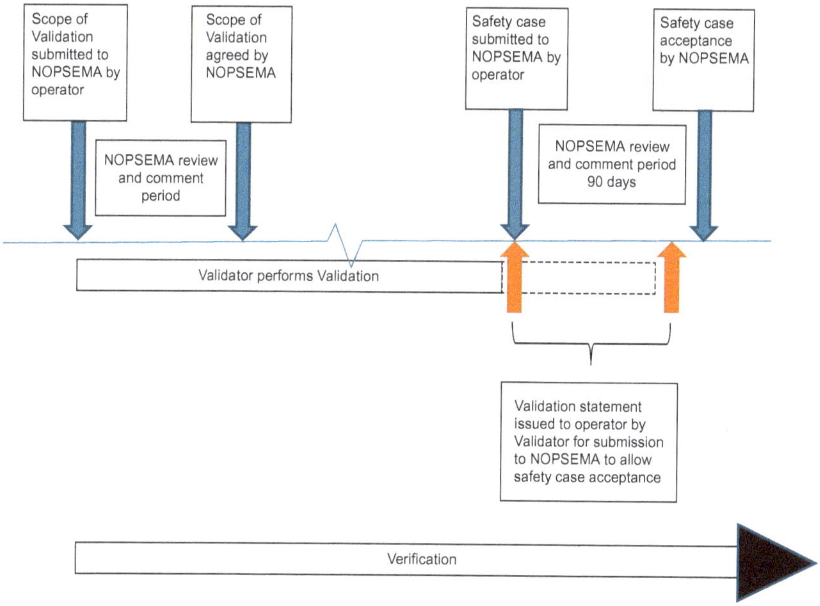

Fig. 6 Australian verification and validation processes. *Lloyd's Register. (2016). Factsheet: Validation versus verification, Australian regulatory summary.*

6.5.3.2 Verification

NOPSEMA require that Verification is documented as a separate and distinct process from Validation and generally should be addressed within the safety case for the facility rather than documented in combination with Validation. Fig. 6 shows that Verification can take place along an independent timeline from Validation, at the discretion of the duty holder.

Verification is not required in relation to the acceptance of a safety case but is a requirement within the safety case. The safety case will contain a commitment from the operator that a suitable verification process has been performed. Verification is the determination that "the SCEs have been designed, constructed, installed, tested, and commissioned in accordance with the nominated codes and standards." It is the physical survey of the facility's onshore procurement, construction, testing and precommissioning and offshore installation, testing, and final commissioning.

6.6 Denmark Offshore

Energy supply administration in Denmark is the responsibility of the Danish Energy Agency (DEA), which is part of the Ministry of Energy,

Utilities, and Climate (Ens.dk (2016)). The Minister grants licenses to produce oil and gas when all legislation has been complied with. The overarching legislation is the Danish Subsoil Act together with its accompanying executive orders, guidelines, etc. Safety of offshore installations is regulated by the Offshore Safety Act (Act). The regime can be considered to be primarily prescriptive, in that the requirements of the Act must be complied with. However, many of the Act's requirements are similar to those defined within operator-generated safety cases utilized in other regimes. This section describes the Danish offshore safety regime only.

6.6.1 Definitions and Requirements Included in the Offshore Safety Act

The purposes of the Offshore Safety Act are "to promote a high level of health and safety offshore which is in accordance with technical and social developments in society" and "to create a framework that allows enterprises themselves to address health and safety issues offshore." The latter is consistent with the "operator is completely responsible" fundamental philosophy utilized in fully goal setting/safety case regimes. Several key definitions and requirements of the Act, which are similar to those seen in operator-defined safety cases, are as follows:

Definitions

- "Emergency response plan" is a document. A plan is required to be prepared and submitted and must have a strategy to prevent escalation and/or limit the consequences of a major accident.
- "Risk" is defined as the product of the probability of occurrence of an event and its consequences.
- "Safety critical elements" are defined as "parts of an installation, equipment, or components, including computer programs, the purpose of which is to prevent or limit the consequences of a major accident, or the failure of which could cause or contribute substantially to a major accident."
- "Major accidents" are defined as either of three types of incidents:
  - Incidents involving explosions, fires, loss of well control, or release of oil, gas, or dangerous substances or materials with potential to cause fatalities or serious personal injury.
  - Incidents leading to serious damage to the facility and having the potential to cause fatalities or serious personal injury.
  - Any other incident having the potential to cause fatalities or serious injury to five or more persons.

It is noteworthy that the definitions include details of the types of incidents, make reference to facility damage, and even specify the number of persons to be seriously injured. In regimes considered to be totally goal setting in nature, the operator would be responsible for defining such things.

Requirements

- A "safety and health document" is required to be submitted by the owner. The document is required to contain or comply with many of the same elements included in operator-generated safety cases in goal setting regimes. Examples are:
  - Identification and assessment of safety risks.
  - Demonstration that the risks have been reduced to a level ALARP.
  - Provision of efficient and controlled methods for safe escape, evacuation, and rescue of all persons on board during emergency situations.
  - Documentation that the management system ensures compliance with all of the Act's and Regulations' requirements.
  - Description of the independent verification scheme.
  - An emergency response plan.
  - Requirement that the document is reviewed and updated at least once every 5 years.
- The operator is required to establish and maintain a safety management system.
- A scheme for independent verification is required to be established, intended to provide assurance that the SCEs are appropriately tested and will meet their objectives (usually referred to as PSs in goal setting regimes).

  Independent verification is defined as an assessment and confirmation of validity of written statements by an enterprise or organizational part of the operator, not under the control of or influenced by the enterprise or organizational part using those statements. The definition is interesting in that it excludes the usually required physical survey from verification, and it also allows a part of the owner's organization to act as the verifier.
- The design plans for both new installations and any major reconstructions are required to be submitted to the supervisory authority.
- The owner is required to obtain approval, via a permit, before any material modifications are made. The safety document must be updated and resubmitted in such cases.

6.6.2 Control Measures and Inspections

Several key elements of safety supervision and control are utilized, including the following:

- The owner's self-monitoring process, documented under the safety management system. Two key objectives of the program are confirmation of regulatory compliance and minimization of both personal and property risk.
- Independent verification by classification societies and/or certification companies. The verification process is described elsewhere but basically includes a combination of design appraisal and site visits, executed throughout a facility's lifecycle.
- Inspection visits by the Danish Working Environment Authority (WEA). Executed under the Offshore Safety Act, the inspections' purpose is to assess compliance with the various applicable safety-related acts and regulations. The main types of visit are:
  - Those covering specific areas deemed by the WEA to require special attention.
  - Routine audits to assess selected elements of an owner's operation to ensure that Danish regulations have been complied with in task planning and execution.
  - Supervision visits intended to achieve the following objectives:
    - Assessment of general facility condition.
    - Follow up on focus areas.
    - Reassessment of corrective action applied to previously identified problems.
    - Meeting with the facility safety organization.

6.6.3 Consultancy Notices

The Danish authorities can issue consultancy notices which require owners to seek advice from a safety consultant in order to solve both specific safety problems and strengthen preventive safety measures. There are several types of consultancy notice:

- Related to either a serious and complex safety problem, or multiple safety problems.
- Associated with psychological working environment problems.
- Following repeated violations (at least 15 on two different facilities).
- Upon failure to draw up a satisfactory safety plan.

6.7 Norway

Safety in the Norwegian petroleum industry is regulated by the Petroleum Safety Authority (PSA) (psa.no, 2016). The PSA proposes to take a confident leadership role in improving safety, stating on its website, "The past 2 years have been characterized by serious incidents and safety challenges. This trend will be reversed in 2017—with us as the driving force and industry as implementer."

Similar to the transformation in the UK after the Piper Alpha incident, a philosophical change has taken place within the Norwegian regulatory regime, whereby the responsibility for safety has migrated from the regulator to the duty holder. This is emphasized by the characterization above of industry owners as "implementers."

6.7.1 Acts, Regulations, and Regulatory Framework

Safety-related acts under the PSA's authority are:

- Petroleum Activities Act.
- Working Environment Act.
- Fire and Explosion Prevention Act.
- Electrical Supervision Act.

Five sets of safety-related regulations are applicable, as below. Nonlegally binding guidelines associated with the regulations have also been issued.

Applicable both onshore and offshore

- Framework regulations (Regulations relating to health, safety, and the environment in the petroleum activities and at certain onshore facilities). Key framework elements include:
  - Assignment of responsibility (to the operator).
  - Principles for the reduction of risk.
  - Health and safety principles, including HSE culture.
  - Some rules extracted from the Working Environment Act.
  - The rights of all employees to be involved in the HSE process.
- Management regulations (Regulations relating to management and the duty to provide information in the petroleum activities and at certain onshore facilities). Key-specific requirements include:
  - Reduction of risk.
  - Barriers (referred to as SCEs in other regimes).
  - Management system elements.
  - Nonconformity and improvement processes.

Applicable offshore

- Activities regulations (Regulations relating to conducting petroleum activities). Key safety-related requirements governing the way activities are conducted include:
  - Operational startup preconditions.
  - Working environment factors.
  - Emergency preparedness.
  - Maintenance.
- Facilities regulations (Regulations relating to design and outfitting of facilities, etc, in the petroleum activities). Key safety-related requirements governing facility design and outfitting include:
  - Safety functions.
  - Materials.
  - Physical barriers.
  - Emergency preparedness.

Applicable onshore

Technical and operational regulations (Regulations relating to technical and operational matters at onshore facilities in the petroleum activities, etc.). Their requirements are equivalent to those of the offshore activities and facilities regulations.

6.7.2 Some Key Elements Included in the Regulations

Several fundamental safety elements and requirements of the regulations are discussed below, particularly those related to goal setting/safety case philosophies prevalent in many global regimes.

From the framework regulations

- *Assignment of responsibility*: The owner/operator is made clearly responsible for safety (and all other aspects) associated with the facility. This is a fundamental aspect of all safety case approaches.
- *Safety management system*: The owner is required to establish and effectively implement a safety management system designed to ensure compliance with all safety-related legislation.

From the management regulations

- *Risk reduction*: The operator is required to adopt technical, operational, and organizational solutions which reduce overall risk.
- *Risk analysis*: The operator is required to identify and analyze all credible risks associated with the facility, using a systematic process including:

- Hazard and accident situation identification (termed HAZID in many regimes).
- Identification of potential major accident causes.
- Accident sequence analysis.
- Consequence analysis.
- Identification and analysis of risk reduction measures (barriers).
- Identification of barrier performance requirements (termed "Performance Standards" in other regimes). Confirmation of compliance with these performance requirements comprises the primary scope of work of third party inspection agencies (usually classification societies) utilized on facility design, construction, and operation.

- *Barriers*: Similar to the terminology "safety critical elements" used in other regimes, barriers which prevent or limit the consequences of accidents are expected to:
  - Identify conditions which could lead to failures or accidents, for example, as provided by smoke/fire/gas detection equipment and high pressure indicators in pressure retaining equipment.
  - Reduce the possibility of failures and the occurrence and escalation of accidents.
  - Limit the extent of consequences, for example, as provided by deluge systems, ESD systems, and evacuation equipment.
- *Notification*: Operators are required to notify the PSA immediately upon occurrence of accidents, specifically including those resulting in death, serious, or acute life threatening injury or impairment of barriers.

From the facilities regulations

- *Design*: Facilities are required to be designed in the simplest and most robust manner possible, with specifically mentioned requirements as follows:
  - All loads must be able to be withstood without major consequence.
  - Major accident risk needs to be demonstrated to be as low as possible.
  - The principle of redundancy is to be applied—no single component failure can be allowed to result in serious consequences.
  - Main safety functions are to be in place and maintained.
  - A safe working environment is to be provided.
  - Barriers are to be provided which can detect abnormal conditions and prevent accident situations from occurring and escalating.
  - The design is to provide for the ability to conduct effective ongoing maintenance routines.

  Appraisal and confirmation of the suitability of design and its compliance with all regulations are key activity of the third party inspection agency/verifier/class society. For large construction projects a multidisciplinary team is often put in place, led by lead engineers for each of the main disciplines.

From the activities regulations

- *Simultaneous operations (SIMOPS)*: The operator is required to identify which simultaneous activities could contribute to an increase in risk and take the necessary mitigative precautions.
- *Installation and commissioning*: The owner is required to ensure that all loads experienced by the facility are within design limitations, and that all other technical requirements of the regulations have been complied with. This is usually part of the third party verifier/classification society scope of work.

6.7.3 Documentation Required to be Submitted

Safety-related documentation required to be submitted to the Ministry of Petroleum and Energy and the PSA, as stipulated by the framework regulations and elsewhere, includes the following:

- Health and safety objectives.
- Risk acceptance criteria.
- Organizational description.
- Management systems description.
- List of applicable codes, standards, and specifications.
- Maintenance plans and requirements.
- Safety zone description.
- Reports following accident occurrence.
- Reports of any damage to load bearing structures and pipelines.

6.8 Nigeria

Process safety regulation and compliance in Nigeria are primarily prescriptive and managed by the Ministry of Petroleum and its regulatory arm, the Department of Petroleum Resources. Areas of interest include any locations where petroleum is processed, stored, or sold, including producing wells, production platforms, and flow stations, crude oil export terminals, refineries, storage depots, pump stations, and pipelines.

Some safety-related activities of the government agencies include but are not limited to:

- Supervising operations being carried out under licenses and leases in the country.

- Monitoring flaring operations.
- Ensuring that health and safety regulations conform with international best oilfield practice.

6.8.1 Key Legislation

Safety-related legislation in the Nigerian petroleum industry includes the following:

- Petroleum Act.
- Oil Pipelines Act.
- Oil in Navigable Waters Act.
- Mineral Oils (Safety) Regulations.
- Petroleum Regulations.
- Petroleum (Amendment) Regulation.
- Petroleum (Drilling and Production) Regulations with Amendments.

6.8.2 Intent and Objectives of Key Legislation

The safety-related intent and objectives of some of the key legislation are as follows:

- *Petroleum Act*: Provides a framework whereby regulations can be made to ensure operations are carried out in a safe manner.
- *Petroleum (Drilling and Production) Regulations*: Identifies the documents required for an application for a lease. It imposes obligations on the operator to take necessary precautions to prevent and control pollution should an accident occur.
- *Minerals Oils (Safety) Regulations*: Intended to ensure safe handling of mineral oil. No specific requirements are included in this regulation. Instead, practices are expected to conform with good oil practice as prescribed by current Institute of Petroleum Safety Codes, the American Institute Code, or the American Society of Mechanical Engineers Codes.
- *Oil in Navigable Waters Act*: Specifies conformance with the International Convention for the Prevention of the Pollution of the Sea (and Nigerian navigable waters) by Oil.

6.9 China

Process safety in China is regulated within a prescriptive regime. The primary safety-related laws and regulations include:

- Safety Production Law.
- Offshore Oil Safety Production Regulations.

- Detailed Rules on Offshore Oil Safety Management.
- Provisional Regulations on the Supervision and Administration of Oil and Gas Pipelines Safety.
- Safety Rules for Coalbed Methane Surface Mining (for trial implementation).
- Health, Safety, and Environment Management System for Oil and Gas Industries.

Key state-owned oil company internal documents are also considered to be applicable to the wider industry, for example, China National Petroleum Corporation's (CNPC) Guidelines on Health, Safety and Environmental Management System for Oil and Gas Drilling, and China National Offshore Oil Corporation's (CNOOC) Requirements for Offshore Oil Operations Safety Training.

Operators are required to provide formal documentation to local governmental agencies prior to conducting hydrocarbon operations. The authorities review and approve the documentation and in some cases conduct a site visit. Some elements of the submissions include the following:

- Appropriate storage and management of life-saving and fire-fighting equipment and hazardous items.
- Appropriate maintenance of stand-by vessels and helicopters.
- Appropriate safety training of all personnel.
- Antihydrogen sulfide safeguarding measures.
- Establishment of an accident/emergency contingency plan.
- Establishment of an effective reporting program.
- The rights of workers to stop work and evacuate in the event of an emergency.
- The duty of workers to report any safety hazards to management.
- Suitable training and appropriate certification of management personnel, including full-time process safety managers.
- Documented confirmation that pipelines have been constructed, checked, and accepted in accordance with national standards.

6.10 Approach Where Safety Cases or Prescriptive Regulations Are Not Required by Legislation

In some cases, particularly regions where offshore oil and gas development is relatively new, prescriptive regulations or the development of a safety case and the associated requirement to comply with PSs are not mandated by legislation.

In most of these cases, however, as part of their due diligence process, multinational owners/operators nevertheless decide to establish safety cases and have them verified by an independent competent body, usually one of the classification societies. Compliance with industry best practice safety standards has come to be accepted as a matter of good business practice, particularly in light of the financial, reputational, environmental, and moral consequences of MAEs.

Duty holders usually choose the safety case process with which they are most familiar, often the one required by the country where the duty holder's headquarters is situated. In addition to safety case requirements, duty holders routinely request their contractors to comply with company-specific standards and practices, as well as any applicable requirements of the chosen classification society's rules. Achieving compliance with these three sets of requirements needs careful management by the duty holders, EPIC contractors, and the classification society, throughout the design, build, and operate phases. This is particularly true with respect to equipment package procurement.

6.11 Comparison of a Safety Result Indicator (TRIF) by Countries Considered

It is tempting to evaluate the approaches used in different countries by comparing the country-specific safety results achieved. There are many sources of statistics available in the literature and electronic media. For example, The International Association of Oil and Gas Producers (IOGP) generates an annual report showing accident statistics divided by many categories—by region and country, accident category, activity being undertaken, and time (International Association of Oil & Producers, 2015). The 2012–2014 average TRIR* in the eight countries considered in this section, as reported in the IOGP's 2015 report are shown in Fig. 7.

The reasons for the comparatively high TRIR value in Denmark and low values in Nigeria and China are unknown. However, despite differences in regulatory approaches, the TRIR results from five countries (Canada, the United States, the United Kingdom, Australia, and Norway) with similar levels of oil and gas experience are fairly consistent, ranging from 2.6 to 3.1. This might indicate that the differences in approaches in process safety regulation may not be as important as having some form of reasonably well defined and enacted form of regulation, regardless of, for example, whether the safety case is considered the primary tool for regulation, or its requirement is embedded in a defined prescriptive regulation.

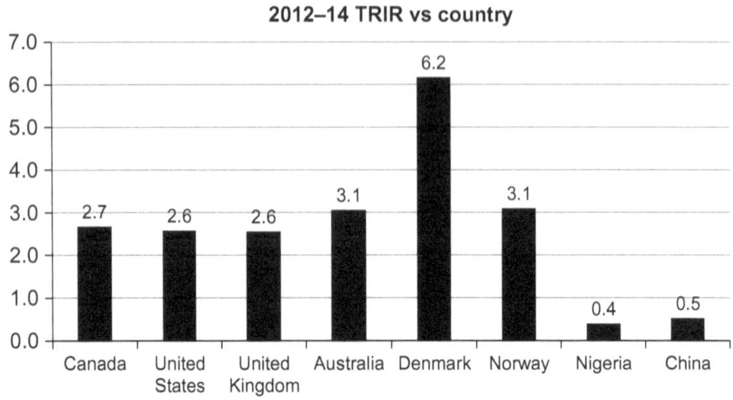

Fig. 7 Total recordable incident rate for eight countries. *International Association of Oil & Gas Producers. (2015). Safety performance indicators, 2014 data.*

**TRIR*: The number of recordable injuries (fatalities + lost work day cases + restricted work day cases + medical treatment cases) per million hours worked.

7. SOME HISTORICAL PROCESS RELATED ACCIDENTS: COULD EFFECTIVE REGULATION HAVE PREVENTED THEM?

The safe management of hydrocarbons within production facilities has challenged engineers for as long as oil and gas have provided energy for society. Unfortunately, shortcomings or errors in the management process have occasionally produced catastrophic disasters over the years. This section considers whether effective, or more effective, regulation could have prevented some of these well known accidents.

7.1 Piper Alpha: Offshore Oil and Gas Production Platform Explosion and Fire in the UK North Sea, 1988

The Piper Alpha platform exploded and burned in the North Sea on July 6, 1988 (Wikipedia, Piper Alpha, 2016), resulting in the deaths of 167 personnel and causing property damage of £1.7 billion. The platform, originally designed and constructed for oil only operation but subsequently modified to add gas production, accounted for about 10% of North Sea production. At the time, the accident was the worst ever offshore oil disaster, both in terms of lost lives and property damage.

The direct cause of the accident was overpressure in one of the two parallel condensate pumps, which could not be withstood by an improperly fitted temporary blind flange. The first pump, originally called upon, failed. The second, subsequently called upon, had its pressure safety valve removed and replaced by the loosely fitted blind flange by an earlier shift as part of partially completed routine maintenance. Gas then leaked out of the pump/flange and was ignited, causing an explosion which blew through a firewall designed to withstand fire, but not explosions. Some other issues associated with the accident were as follows, together with suggestions of how a rigorous regulatory program might have averted the disaster.

- Ineffective PTW process. A PTW covering the status of the pump undergoing maintenance was in place. It was generated by day shift personnel and stated that the pump was not ready and should not be switched on under any circumstances. Unfortunately, the situation was not discussed directly with the incoming night shift foreman. Instead, the permit, which could not subsequently be found, was simply left in the control center.

  PTW systems and procedures are key elements of safety cases and are also mandated by other regulatory requirements. PTW procedures are usually required to be part of the formal documentation submitted for review and approval to the regulator, classification society, and/or CA. Procedures generally require sign-off at all stages of the work, including handover from one group or shift to the next. The argument could be made that had the night shift foreman been required to formally sign-off all active work permits in place during his upcoming shift, the disaster may have been averted.

  Communication is also a key element of the safety case process. Usually a communications plan is required, which would include, for example, short tool-box style meetings between personnel ending and beginning their shifts. It is quite possible that had such a meeting taken place, the condition of the pump undergoing maintenance would have been known to the night shift staff.
- Unacceptable design. A contributory factor in the disaster was the inability of the firewall to withstand an explosion. In fact one panel from the firewall was violently dislodged and caused a rupture in another pipe and subsequently another fire. Design review and approval of all safety-related structures and equipment are primary elements of all regulatory programs, both prescriptive and those relying on safety cases.

It is quite likely that the unsuitability of the firewall for the facility's current configuration would have been picked up by the design appraisal process.

- Notification of changes to the facility. Regulatory programs require operators to advise the regulator, classification society, and/or CA when any significant changes are made to the facility. The Piper Alpha platform was originally intended to produce oil only, but was reconfigured to add gas production. This would certainly constitute a significant change.

  One of the significant changes associated with this would have been the upgrade of the firewall to a blast wall. Had there been a rigorous process in place to review all aspects of the changes it is quite possible that the effect of this contributory factor could have been minimized or eliminated altogether.

  A second and more fundamental result of the change from oil only to oil and gas production was the breaking of the basic safety concept that the most dangerous operations should be kept as far away as possible from personnel areas. The change resulted, for example, in gas compression equipment being located next to the control room. This change was also considered to be a contributory factor in the disaster.
- Ineffective interfield communications and levels of authority. The fire would have burned out more quickly except that oil from two adjacent platforms, Tartan and Claymore, continued to feed it. The managers of the adjacent platforms either believed they had no authority to stop production, or they were directly instructed by their superiors to continue production.

  Regulatory programs and safety cases have requirements for clearly defined, submitted, and approved communications plans, level of authority matrices and similar. It is quite likely that such plans would provide for communications protocols in emergency situations and give offshore installations managers the complete authority to take any appropriate steps they considered necessary to safeguard life and property in the event of an incident. This would have been an obvious mitigative step in dealing with the disaster.

7.2 Texas City Hydrocarbon Refinery Explosion, 2005

A refinery in Texas City, United States exploded on March 23, 2005 (Wikipedia, Texas City, 2016), resulting in the deaths of 15 workers and injuries to 170 others. The direct cause was the ignition of a flammable hydrocarbon vapor cloud, which was released from an overfilled and overheated pressure vessel. Secondary and root causes of the accident are listed below, together with some suggestions of how a rigorously enacted regulatory process may have eliminated the causes and prevented the disaster.

- Heavier than air hydrocarbon vapors coming into contact with an ignition source (suspected to have been a running vehicle's engine).

  Prevention of unintentional escapes to the external environment of hydrocarbon vapors is a key regulatory objective at several project/facility stages. The design approval process ensures that pressure vessels are sufficiently robust to withstand all applied forces for the life of the facility. Suitable corrosion allowances are required to be included, which specifically need to be approved by the authorities and/or classification society/CA. The effective and appropriate construction of vessels should be confirmed during mandatory new-construction surveys. The ongoing fitness for purpose of hydrocarbon containing equipment is to be confirmed during annual (or more frequent) poststartup surveys.

  A major component of modern safety cases is the identification and management of sources of ignition. Operators are required to execute systematic hazard identification exercises (HAZIDs and HAZOPs), usually in a workshop format with government representatives and the classification society present, in order to identify and manage all potential sources of ignition. It is possible that such a session would have led to the prohibition of vehicles in proximity to any hydrocarbon containing vessels having the possibility of leaking gaseous mixtures.

- Engagement of the overpressure protection system following overfilling and overheating of a production vessel.

  Regulatory regimes require that effective overpressure protection systems be designed, constructed, and installed according to industry recognized codes and standards. This includes the requirement that any release of hydrocarbons be only to a "safe" area, thereby minimizing the risk of ignition. In the present case, the protection system was effective in that it operated to relieve pressure from the vessel, but ineffective in terms of the location of gas release.

- Numerous technical and organizational failings at the refinery and within BP. Regulatory regimes, in combination with mandatory QA requirements, require that effective management systems be enacted. It is likely that the overarching system requirements required would address at least some of the organizational failings.

- Poor maintenance for several years (the refinery was built in 1934). All current regulatory regimes require that detailed maintenance plans be submitted and approved by the authorities, including the verifier/classification society/CA. Furthermore, annual inspections are required following startup of facilities, and these annual inspections always require a review of adherence to the maintenance program.

Postincident reviews of the conditions at the plant revealed the following situations, which, although they did not directly cause this accident, may have caused one having equal or even greater consequences.

- *Broken alarms.* Current regulatory regimes require that the design, installation, and subsequent effective operation of alarm equipment and systems is according to industry accepted codes and standards. Testing of alarm equipment and systems is a prime activity during annual (or more frequent) surveys. Such testing, done with sufficient rigor, will ensure that alarms remain in good operational efficiency throughout the operational life of the facility.
- *Thinned pipe.* This is addressed by regulatory regimes at both the design and operational phases of a facility's life cycle. At the design stage, a corrosion allowance for all equipment is included which should prevent dangerous pipe thinning. The annual survey acts as a backup to detect and manage corrosion before it gets to dangerous levels.
- *Chunks of concrete falling.* This is addressed by regulations covering the design, construction, and operational phases of the facility's life cycle. Concrete and any other main structural elements (e.g., steel for floating vessels or fixed offshore platforms) are required to be designed to effectively withstand (including agreed factors of safety) the effects of environmental and any other applied forces for the operational lifetime of the facility. The effective and code-compliant construction of the structure is required to be surveyed during new-construction surveys. Annual surveys confirm that the structure remains fit for purpose and that any unacceptable degradation is corrected.
- *Bolts dropping 60 feet.* This is addressed in both the design and ongoing survey stages of regulatory approval. At the design stage, a dropped object study is a common requirement, which addresses the types of objects that could fall, the consequences of potential events, and measures taken to prevent objects from falling. Other common design

requirements include, for example, kick boards, which prevent tools or other items from being accidentally kicked over the side of a deck to one below. At the annual survey, the ongoing fitness for purpose of the dropped object prevention measures are evaluated, as well as the observance of any areas which may have been missed at the design phase.

- *Staff being overcome with fumes.* Modern regulatory regimes require the implementation of an effective and rigorous notifications process. This would require, for example, the authorities (governmental, classification society, and CA) to be formally advised within specified time periods of any safety-related event, which would certainly include the observed presence of gas fumes at a facility. The reporting time and level of detail required to be submitted are dependent upon the seriousness of the event.
- *Five managers in 6 years.* Regulatory regimes require effective management systems, covering both safety and quality. It is likely that having so many managers over a short period of time would trigger questions, comments, or noncompliances within the regulatory process.

7.3 Bhopal: Toxic Gas Release From the Union Carbide Insecticide Production Facility, 1984

The Bhopal (India) disaster (Crowl & Louvar, 2002) occurred on December 3, 1984, when more than half a million people were exposed to 42 tons of toxic methyl isocyanate gas released from a Union Carbide pesticide plant. Thousands died immediately, and many more were trampled in the ensuing panic. More than 18,000 people are estimated to have died from the immediate, short-term, and long-term effects of the event. Cost cutting in response to poor selling of the plant's end product, Carbaryl, and the associated reduction in maintenance frequencies, was proposed as the fundamental root cause of the accident. It is noteworthy that at the time of the disaster, there was no production underway since there was a surplus of Carbaryl on the market.

The events leading to the incident have been proposed as follows:

- Unintentional water ingress into a pressure vessel during routine cleaning of adjoining pipe. This hypothesized initiating cause was later challenged, when in 1985 the scenario was tested in the presence of official investigators and demonstrated to be invalid. The plant owners proposed the alternative scenario of sabotage by a disgruntled worker who may have intentionally introduced water into the tank.

- A chemical reaction resulting from the presence of water, leading to a build-up of carbon monoxide. The rate and volume of the reaction were exacerbated by the rusty condition of the vessel.
- A resulting increase in internal temperature and pressure to a level higher than the vessel was designed to withstand.
- A resulting opening of pressure relief valves and release of gas to the environment for a period of approximately 2 h.

A 1985 report on the disaster suggested several causal factors, as below. It can be reasonably suggested that the effective and rigorous application of some of the basic fundamental safety case and management system requirements mandated by most regulatory regimes could have prevented the tragedy. The specific regulatory elements which may have eliminated each causal factor are described below.

- Failure to use, where possible, less dangerous chemicals than the ones actually used. One of the basic principles of risk management is to replace, where possible, a hazardous substance with a less dangerous one. The choice of chemical used may have been challenged as part of hazard analysis exercises, "what if" workshops, and similar, which are part of current safety case generation activities.
- Storage of chemicals in larger than necessary tanks. This would likely have been challenged within the design appraisal function at the early stages of plant design, possibly front end engineering and design (FEED), when overall configuration is considered. Regulatory or third party verifier review of the facility's safety case is another activity which may have discovered this problem.
- Possible piping and vessel corrosion. At the design stage, submission and approval of design plans, including a suitable corrosion allowance, are a requirement of most current regulatory regimes. Furthermore, the plans and procedures for routine maintenance, which would have included checking for excessive corrosion, are required to be submitted for review and approval to the authorities and classification society/CA. The problem would also have been apparent and likely found during mandatory annual third party surveys of the facility and the operator's routine maintenance activities.
- Poor maintenance after the plant stopped production. See the above related to submission of maintenance plans and annual surveys. More fundamentally, however, the mandatory safety cases in most of today's regulatory regimes require a "cradle to grave" philosophy, wherein the operator needs to demonstrate safety at all project phases, including

design, fabrication, procurement, installation, life extension, and decommissioning. The philosophy would require that appropriate, albeit probably reduced, maintenance was conducted even during periods of shutdown.

- Failure of safety systems. This is a general causal factor, but current safety case regimes require a facility wide rigorous and systematic approach which looks at all safety critical systems.
- Plant design modifications in response to economic pressures. One of the key requirements of all regulatory regimes is that the operator must advise the regulator, classification society, and/or CA when significant modifications affecting safety are made. In this case, weaknesses or errors associated with the modifications may well have been discovered and prevented, thereby averting the tragedy.

7.4 Flixborough, England: Explosion of the Nypro Limited Nylon Component Fabrication Facility, 1974

The Flixborough factory of Nypro Limited exploded in June 1974, completely destroying the facility, killing 28 people and injuring 36 others (Crowl & Louvar, 2002). The plant produced caprolactam, a material used in the production of nylon. The accident was believed to have been caused when inadequately supported and overflexing 20 inch feed stock, which was used to replace existing cracked and leaking 28 inch feed stock, ruptured under internal pressure. Thirty tons of cyclohexane were released, and the resulting vapor cloud was subsequently ignited by an unknown source. Several of the root causes mentioned below could reasonably be expected to have been prevented by an effective and rigorously enacted regulatory regime.

- Lack of a safety review following the replacement of the piping. Regulatory regimes require any significant repairs or modifications to be subjected to an effective internal safety review, including approval and sign-off by duly authorized personnel, some of whom are independent from the production and commercial departments of the company. Furthermore, the appointed classification society/verifier/CA/is required to be made aware of the details and approve the repair/modification. In this case, no safety review was conducted. A safety review and/or involvement of the third party agency may have discovered any safety weaknesses in the proposed modifications at the drawing stage and prevented them from becoming part of the physical changes, which may then have averted the disaster.

- Documentation of repairs and qualifications and background of supervisory and technical personnel. Regulatory regimes require repairs and modifications to be formally documented, approved, and signed off by experienced and qualified technical personnel. In the present case, the technical proposal was found to have been sketched on the machine shop floor using chalk.
- Inadequacy of design. Subsequent reviews of the accident showed that the design details of the repair were unacceptable. Design appraisal is a significant component of modern regulatory regimes. Approval is required both prestartup and before the physical implementation of any significant changes or repairs. An effective regulatory program would have required the submission of the repair details to the classification society/CA/verifier, and any deficiencies against either industry codes, standards, or good practice, may well have been discovered and rectified, thereby preventing the disaster.
- Excessive inventories (330,000 gallons of cyclohexane, 66,000 gallons of naptha, 11,000 gallons of toluene, 26,400 gallons of benzene, and 450 gallons of gasoline). The excessive inventories contributed to the duration and general destructiveness of the postexplosion fires, which burned for 10 days. Modern regulatory regimes require workshop-based exercises such as hazard identification (HAZID) and hazard and operability (HAZOP) studies. Potential consequences of all credible accident scenarios are considered, which contribute to the determination to each hazard's risk. It is probable that during the analysis of potential explosion consequences during HAZID/HAZOP/risk assessment exercises, the inventory level would have been challenged, and quite possible that alternatives would have been considered and evaluated.

7.5 Pasadena, Texas, Explosion of a High Density Polyethylene Plant, 1989

A high density polyethylene plant in Pasadena, Texas, exploded on October 23, 1989, resulting in 23 deaths, 314 injuries, and $715 million in property damage (Crowl & Louvar, 2002). The direct cause of the disaster was the ignition of an accidentally released 85,000 pound gas cloud comprised of ethylene, isobutene, hexane, and hydrogen. The subsequent accident investigation revealed that standard operating procedures had not been followed during routine maintenance procedures on some process valves. Some of the secondary and root causes are as below, together with suggestions of how an effective regulatory program may have prevented the disaster.

- *Failure to follow maintenance procedures*: Modern regulatory regimes require that procedures and other documentation describing maintenance plans and programs are reviewed and approved by the regulator, classification society, CA, and/or verifier prior to startup. Furthermore, the effective implementation of the approved maintenance procedures is the subject of regular (usually annual) inspections and surveys. It is probable that any weaknesses in the maintenance plan documentation would have been discovered in the initial review, and that any implementation nonconformances would have been discovered and rectified during the annual survey process. Either of these activities may have prevented the problems in compliance with documented maintenance procedures which contributed to the disaster.
- Failure to conduct ongoing routine process safety hazard analyses, resulting in the ongoing existence of serious safety deficiencies. The planning and execution of hazard analyses are key components of management systems required by current safety case/goal setting regimes. The ongoing inclusion of such activities may have identified safety concerns and prevented the disaster.
- Unacceptable design of process valves (not designed to fail to a "safe" condition). Modern regulatory regimes all include requirements for design appraisal. A main design principle is that, upon loss of power or other unexpected event, equipment fails to a safe position. An example is fuel supply line valves which fail to a closed position in the event of loss of power, so that a continuous fuel supply in an emergency (fire) situation is prevented. A rigorous design appraisal, which would include not just the valves themselves but the system in which they were installed, may well have picked up and corrected this deficiency.
- Absence of a permitting system covering plant activities. A PTW system is a key element of safety management systems required under all regimes, particularly those based on safety case philosophies. The permitting system generally requires sign-off by many levels of management as well as all participants in the activity. The requirement for sign-off generally ensures that all disciplines assume their appropriate share of the responsibility that a particular activity is conducted in both a safe and technically correct fashion.
- Absence of a combustible gas detection and alarm system. Modern regulatory regimes demand the presence of such systems, and their proper

design, construction, and operation are all checked. The overall design appraisal function would have noted its absence, or it would have been picked up during the mandatory safety risk assessment, hazard identification, or hazard and operability workshops. In the unlikely event that these functions failed to identify the absence of this key safety element, final prestartup surveys would almost certainly capture and rectify the omission.

- Proximity of high occupancy structures to hazardous operations. Current regulatory regimes encourage and in many cases mandate that a third party verifier become involved in plant design very early in the design process. The classification society, CA, or independent verification body is usually engaged at the FEED phase, when overall design aspects such as module and/or building occupancy and location are considered. This would have been the opportunity to identify and rectify a dangerous overall design flaw such as this.
- Inadequate building separation. Under a rigorously implemented regulatory regime, this type of problem would likely have been identified and corrected during the interaction between the designer and the third party agency during FEED review activities.
- Crowded process equipment. As in the previous root cause, this type of issue can be discovered at the FEED project stage, either by the designer or by the third party checking process mandated by the regulator.

7.6 ARCO Chemical Plant Explosion, Channelview, Texas, USA, 1990

A 900,000 gallon process wastewater tank at the ARCO chemical plant in Channelview, Texas, exploded on July 5, 1990, killing 17 people and causing $100 million in damage (Ness, 2016). The direct cause of the explosion was the development, during routine maintenance, of an explosive atmosphere in the tank, which was ignited when the plant was restarted. There were several secondary and root causes of the incident, which are listed below together with suggestions of how an effective regulatory regime might have averted the disaster.

- No MOC review was conducted. The technical root causes will be discussed below, but from an overarching management perspective, regulatory regimes require documented, and effectively implemented fundamental programs and procedures, such as MOC, to be in place. These programs require a series of formal steps to be taken by the appropriate group of duly certified and qualified management representatives, including those representing operations, design, and safety. The steps would include, for example, a review of the design of the equipment installed to ensure safety during normal operations and the suitability of steps taken to ensure ongoing safety during maintenance activities, when some of the main systems may be turned off. All management representatives are required to sign the resulting MOC report, which ensures that all are aware of the responsibility they are assuming. There is a reasonable chance that a rigorous MOC process would have averted this accident.

- No prestartup safety review was conducted. Modern regulatory regimes require safety reviews to be conducted at appropriate times. The most obvious of these is prior to the original startup of a facility, but legislation also requires that safety reviews be undertaken whenever significant changes are made, as well as following periods of shut down and/or maintenance. Such a safety review would most likely have identified and prevented the technical causes of the incident described below, thereby averting the accident.
- The wastewater tank was not considered to be part of the operating facility. Historical reviews of accident causation sometimes describe misjudgments about the degree to which apparently harmless equipment and processes can affect overall facility safety. Compared to the hydrocarbon containing tanks and equipment at the large and complicated chemical process facility, safety issues associated with the wastewater tank were not given the same consideration, with tragic consequences. A rigorous safety review, which would have taken an overview of facility wide safety considerations, may well have detected this misconception and ensured that the appropriate safety measures were taken.
- Nitrogen purge was significantly reduced during maintenance. The effectiveness of a nitrogen purge system, which was designed to maintain an inert condition in the tank, had been significantly reduced during the maintenance activity. Safety was expected to be maintained through the installation of a temporary oxygen analyzer (see below). It is not unusual for systems ensuring ongoing safety during regular operations to be turned off or reduced during maintenance activities, when less chemical is circulating through pipes. However, a safety review or MOC session might be expected to consider the consequences of reduced or eliminated safety systems and take the appropriate steps, thereby eliminating this accident root cause.

- A temporary gas analyzer was located incorrectly during the maintenance activity. The analyzer should theoretically have detected the dangerous build-up of gases. However, the analyzer was installed between roof beams in the headspace of the tank, which was effectively a stagnant dead zone, and therefore the dangerous condition was not detected. Unfortunately, this type of error, which can be well understood by most, is very common in historical accident accounts. There is no guarantee that safety reviews will pick up every safety concern, but their rigorous and systematic performance certainly increases the likelihood of accident prevention.

7.7 Summary of Accident Causation

A summary of the causes of the major accidents described in this section is shown in Table 3. Many, perhaps all, of the direct and secondary/root causes of the accidents could have been eliminated had an effective regulatory regime been rigorously implemented.

8. CLASSIFICATION SOCIETIES: AN INTRODUCTION AND THEIR ROLE IN PROCESS SAFETY

8.1 Introduction

Classification societies originated when insurance companies wished to assign appropriate premiums to goods being transported on ships about which they knew little or nothing. The first society surveyors were retired ships' captains and first mates to whom the insurance companies assigned the task of rating ships against an agreed series of categories, resulting in a "classification" of the ship. From those beginnings hundreds of years ago, class society activities have expanded to other industries, particularly the offshore oil and gas sector. Today's role in the energy sector is not philosophically dissimilar from that first assignment, except that now facilities are evaluated against the requirements of safety cases or prescriptive regulations.

Some details of four prominent classification societies are as follows:

- *Lloyd's Register (LR)*. LR was founded in 1760 at Edward Lloyds' coffee house in London, England (lr.org (2016)). Its corporate headquarters continue to be in London, but it has created centers of technical excellence in Southampton, UK, and Singapore. The world's oldest

Table 3 Summary of Major Accident Causes

Accident	Direct Cause	Secondary/Root Causes
Piper Alpha	Release (and subsequent ignition) of a flammable gas mixture from an unsecured condensate pump	• Lack of awareness of pump situation following maintenance by earlier shift • Failure to communicate contents of permit to work between day and night shifts • Inappropriate blast wall design • Poor interfield communications
Texas City	Release (and subsequent ignition) of a hydrocarbon vapor cloud from an overfilled pressure vessel	• Proximity to plant of ignition sources • Hydrocarbon flaring to unsafe areas • Poor maintenance processes
Bhopal	Uncontrolled release of toxic gas from an over pressured pressure vessel	• Cost cutting due to poor economic prospects for the main plant product • Release of water into storage tank • Chemical reaction producing high pressure gas mixture • Corrosion
Flixborough	Replacement piping failure producing a flammable gas cloud	• Lack of a safety review prior to pipe replacements • Failure to document repairs and obtain approval • Inadequacy of design
Pasadena	Ignition of gas cloud	• Failure to follow maintenance procedures • Failure to perform safety analyses • Absence of a permitting system
ARCO	Ignition of an explosive mixture inside a tank upon plant startup	• Lack of a management of change process • Failure to conduct a prestartup safety review. Lack of recognition of the tank as a potentially dangerous part of the facility

classification society is a "global engineering, technical, and business services organization." LR's more than 9000 employees operate in 78 countries across three regions: the Americas, Europe/Middle East/Africa, and Asia.

LR's areas of business are marine, oil and gas, low carbon power, QA, industrial manufacturing, utilities and buildings, rules and regulations, and research and innovation.

LR's mission statement is "We secure, for the benefit of the community, high technical standards of design, manufacture, construction, maintenance, operation, and performance for the purpose of enhancing the safety of life and property at sea, on land, and in the air ... because life matters. We advance public education including within the transportation industries and any other engineering and technological disciplines."

LR's historical strength in the oil and gas sector has been in the fixed offshore platform market and it has recently become involved in new-concept energy megaprojects in Kazakhstan and Australia.

- *Det Norske Veritas-Germanischer Lloyd* (DNV-GL, formed by the merger of DNV and GL). DNV was founded in 1864 and is headquartered in Oslo, Norway (Dnvgl.com, 2016). Germanischer Lloyd was founded in 1867 and was headquartered in Hamburg, Germany. The two organizations were considering joining forces from as early as 1868, but the combined organization finally became operational in September 2013. The combined group comprises about 15,000 employees operating in more than 100 countries.

  DNV-GL is organized functionally into the maritime, oil and gas, business assurance, energy, software, technology and innovation, and rules and standards sectors. It provides its clients with classification, technical assurance, software, and independent expert advisory services.

  DNV-GL's strength has been in the research, innovation, and rule development areas, although it competes effectively with the other societies in all traditional market areas.
- *American Bureau of Shipping (ABS)*. ABS was founded in 1862, and is headquartered in Houston, Texas, United States (Abs-group.com, 2016). It operates in the marine, offshore, oil, gas and chemical, power, and government sectors across three geographical regions—the Americas, Europe/Middle East/Africa, and Asia Pacific.

  ABS offers risk management services to industries that "power, fuel, and regulate our world." Its mission is "to be a leading global provider of technical services that better enables our clients to operate safely, reliably, efficiently, and in compliance with applicable regulations and standards."

  ABS has traditionally dominated the drilling rig classification and certification market and also participates in various other energy activities.

- *Bureau Veritas (BV)*. BV was founded in 1828 in Antwerp, Belgium (Bureauveritas.co.uk, 2016). BV is structured along eight global businesses: marine, industry, inspection/in-service verification, health, safety and environment, construction, certification, consumer product services, and government services/international trade. Its 66,000 employees operate from 1400 offices and laboratories, geographically divided into four regions: the Americas, Europe, Middle East/Africa/Eastern Europe, and Asia/Pacific.

  BV's mission is "To deliver economic value to customers through quality, health, safety, environment and social responsibility management of their assets, projects, products and systems, resulting in licenses to operate, risk reduction and performance improvement."

  BV has traditionally offered inspection services to all types of projects in many different disciplines. This strategy ensures a presence on a variety of energy projects, including ones where it is not the primary verifying agency.

8.2 General Role

Classification societies execute different roles within the overall initiative of maintaining and improving process safety. In some cases, specialist technical expertise is provided within a consultancy/subject matter expert role. More commonly, however, class societies act either in a third party verification capacity on behalf of flag administrations and/or national petroleum boards, or as second party inspection organizations on behalf of operators having no legal requirement to have a third party evaluation of their safety arrangements.

When acting as a second or third party verifier, the society evaluates a facility's safety arrangements from both the design and construction perspectives, with the objective of issuing certification attesting that the arrangements meet the requirements of the mandatory codes and standards and provide a level of safety which is ALARP.

The basic class society scope of work required to achieve the foregoing objective is as follows:

- For design appraisal, to identify the required drawings from a client generated master document register (MDR), review them against the agreed codes and standards (usually listed in the safety case/PS documentation), and issue comments and request changes until it can be confirmed that the design meets the requirements of the codes and standards.

- For construction, to attend at construction facilities to confirm that:
  - The materials and construction practices comply with the agreed codes and standards.
  - The finished product is in compliance with the requirements of the design appraisal process and the codes and standards.

Modern economics has driven multinational oil companies to favor megaprojects, with an associated requirement for the classification societies to provide much more detail and structure in their service offerings than previously. Where shipyards' assessments of the situation with the class society was at one time expressed simply by questions such as "is the Class man happy?," today's requirements are much more demanding. LR has adopted a structured project management approach to megaprojects, which includes:

- Internal and external review boards which oversee project teams' performance.
- Extensive project-dedicated organizations, including discipline specialists who remain on the project from FEED to abandonment. The organizations provide management, technical, commercial, and administrative functions.
- The specification of quality, documentation, and other project management requirements at all project phases from FEED to abandonment.
- Quality programs structured to cover each project component at all levels, from overarching quality and execution plans to detailed work instructions for site surveyors and design appraisal specialists.

LR's approach to megaproject classification, certification, and verification is described in the next sections (Attwood, Bates, & Price, 2013).

8.3 LR Project Management Framework

All projects, but especially large ones, benefit from a documented project management framework supported by robust tools and processes and with clear control points. Fundamental processes, for example, project risk management, are continuously applicable throughout all project stages.

The benefits, both to the specific project and the wider internal organization, of using a standard framework are:

- Consistent and repeatable approach to delivery of all projects within an organization.
- Increased likelihood of successfully achieving objectives.

- Continuous improvement of processes and tools.
- A scalable approach that can be adopted to both large and small projects.
- A common understanding of project roles, responsibilities, and terminology.
- Reduced risk from the loss/transfer of critical project staff from an in-progress project.
- Ability of management to quickly digest project reports, which will be in a consistent format across the project portfolio, and focus on principal issues.

The main aspects of a standard defined process are shown in Fig. 8. The processes, controls, and project management deliverables within the execution stage of a project lifecycle are highlighted. Each stage has inputs from the previous stage and expected outputs to be delivered to the next.

Some of the principles covered by the project framework are as follows. A rigorous implementation of the process will produce significant project benefits.

- A systematic approach to *risk management* is essential to successful project delivery. Risk identification, formal documentation in a project risk log, and subsequent management is an activity for all project team members.

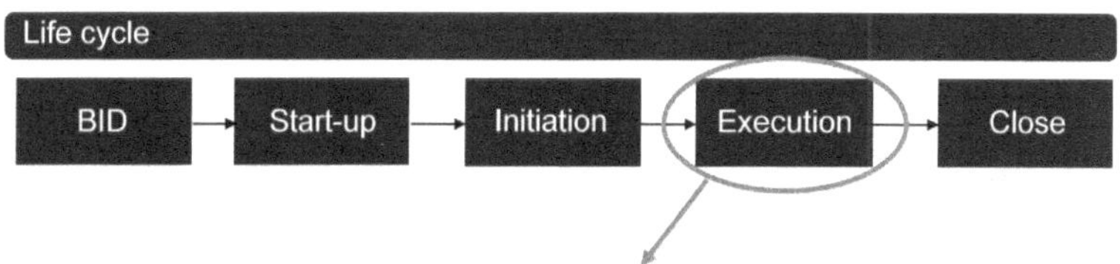

Execution subprocess	Primary document (s)	Supporting document (s)
Planning	Project Management Plan	Project Quality Plan(s) Project Safety Plan (s) Technical (e.g. Classification, Certification, Validation) Plan(s)
	Resource Planning	Human Resources Procedures
Control	Project Risk Management Strategy	Risk Log
	Project Control Logs	Issue Log Milestone Register Deliverables Register Project Action & Decision logs (internal, external)
Delivery	Project specific procedures and work instructions	
Communication and Reporting	Communication Management Plan	Periodical (e.g. weekly, monthly) reports Minutes of Internal Project Review Board Meetings Minutes of external meetings (e.g. Progress, Overarching, Sponsor).

Fig. 8 Classification project stages. *Attwood, D., Bates, E., & Price, S. (2013). Regulation and achieving compliance post Macondo. LR paper delivered at IMarEST conference, Houston.*

The review board also becomes involved in any intolerable or otherwise problematic risks.

- *Early engagement.* Most industry participants agree that the sooner the class society becomes involved in a project, the better. The purpose is to facilitate the documentation of a fully understood scope, broken down sufficiently to form detailed execution plans that have been agreed by the project team and client at an early stage.
- Maintaining open, regular, and accurate channels of *communication* with all levels of project stakeholders is vital to ensuring the smooth flow of instructions and sufficient warning of risks and changes to enable early assessment and appropriate management.

8.4 LR Approach to Megaproject Organization

A typical organizational chart for a large project is shown in Fig. 9. There are three main groups having the following roles:

- *Project team*—responsible for the effective execution of the project.
- *Governance group* (includes senior representatives up to the senior vice president/director level)—internal governance, guidance, support, and corporate accountability.

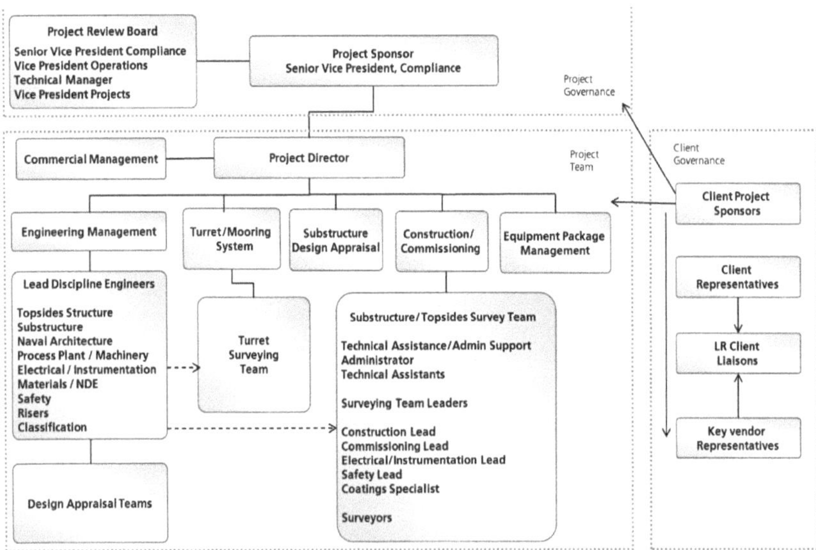

Fig. 9 Typical megaproject classification organizational chart. *Attwood, D., Bates, E., & Price, S. (2013). Regulation and achieving compliance post Macondo. LR paper delivered at IMarEST conference, Houston.*

- *Client group*—external governance, provision of information, and access to sites and vendors' works to allow the class society to execute its function.

A key governance element is regular project review board meetings, the objectives of which are:

- To give the project manager an opportunity to seek support from senior management.
- To give senior management an opportunity to assess project performance and plans.
- To give all internal stakeholders an opportunity to share ideas, ask questions, and assign and accept actions related to project activities.
- To share financial and technical project status and thereby enable informed decisions.

Some noteworthy points with respect to the organization are:

- The lead discipline engineers remain with the project throughout its lifecycle. The dotted lines between them and the site teams indicate that in addition to their design appraisal responsibilities they also provide support to the surveyors whenever necessary.
- The specialist team includes representatives for all critical disciplines, including a process, plant, and machinery (PPM) expert, whose responsibility specifically includes chemical process safety.
- Equipment package management can be very complicated on megaprojects, with in some cases more than 1500 individual certificates required. A separate project management group is required and is shown on the organizational chart.

8.5 LR Megaproject QA Program

The classification/verification process involves a series of activities described in quality documents such as project management execution plans, classification plans, survey work instructions, and similar. These documents have become more critical to the process as the geographical spread of megaproject activities widens and consistency expectations increase.

A typical map of quality documents for a megaproject is shown in Fig. 10. Some noteworthy points are as follows:

- The requirements extend from the very general corporate and local nonproject-specific requirements near the top of the diagram (e.g., corporate human resources procedures) to very project-specific instructions for site surveying and design appraisal near the bottom.

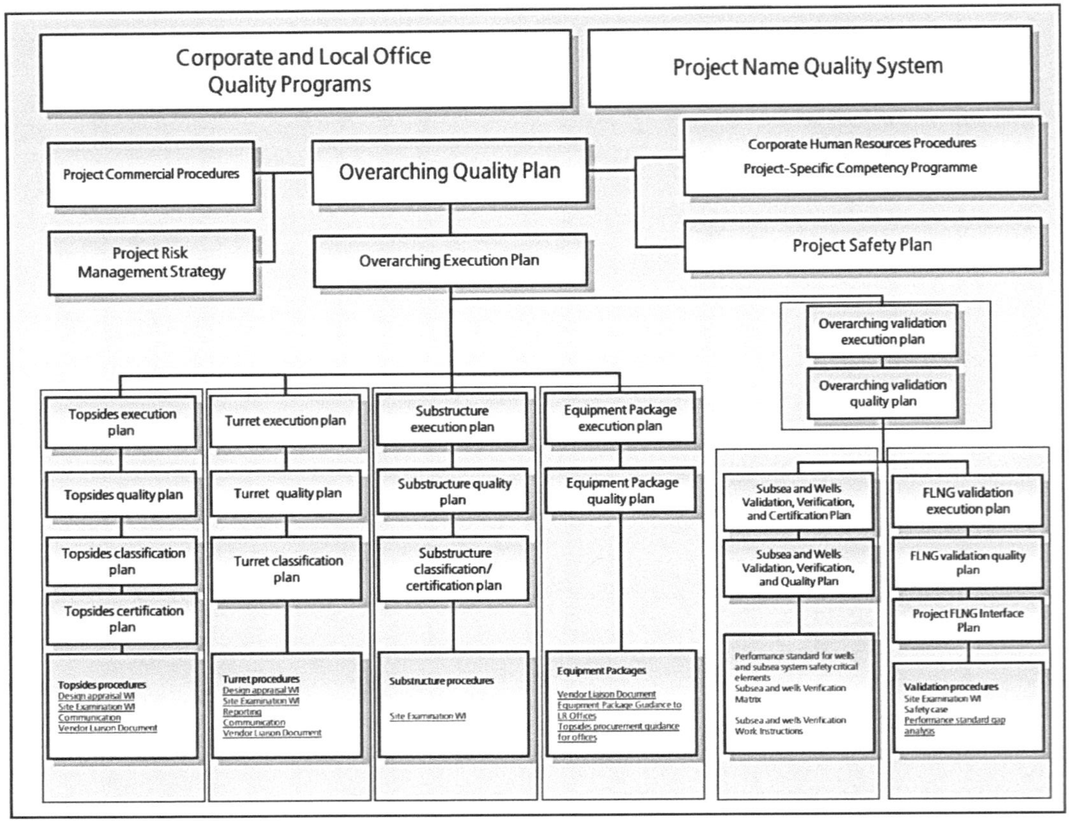

Fig. 10 Classification project quality system. *Attwood, D., Bates, E., & Price, S. (2013). Regulation and achieving compliance post Macondo. LR paper delivered at IMarEST conference, Houston.*

- Looking horizontally, the range of facility components is observable. For a megaproject, different modules are often designed and constructed by different EPIC contractors, resulting in separate contracts with the classification society. This provides challenges for the overall project managers and directors.
- In Australia, the regulator requires that Validation be kept independent from all other classification society activities. This concept is indicated by the lines surrounding Validation processes in the lower right corner of the diagram.
- Equipment package management is often the most complicated element of the megaproject classification process. Therefore separate dedicated execution and quality plans and procedures are required, as shown.

9. CONCLUDING REMARKS

Whenever a group of safety professionals gather to perform the unenviable and unpleasant task of investigating an accident, one of the first questions posed by the facilitator is often some version of "What went wrong?" At this point several hands are usually raised and the predictable responses include "poor communication," "inadequate training," "inexperienced staff," "too much cost cutting," "poor design," and others, depending on the particular backgrounds, preferences, training, and even prejudices of the attendees. The author's experience has been that most accidents have multiple causes. A previous section of this chapter describing historical process accidents included references to all of the causes mentioned earlier.

However, what is seldom, if ever, heard, is "the government should have enacted more regulation."

Although the principles and structures utilized differ, most countries have reasonable laws, acts, legislation, and regulation to ensure that, if effectively implemented, the risk of accidents is reasonably low, if not "ALARP". Furthermore, most process facilities, whether onshore or offshore, are operated under the umbrella responsibility of major international oil companies, which have libraries full of documented plans, procedures, and programs covering all aspects of safety management, including requirements for training, communication, safety review, MOC, and supply of personal protective equipment. Again, these programs, although differing from one company to the next, if effectively implemented, should maintain accident risk ALARP.

The situation is reminiscent of a recent analysis of the most successful half dozen teams in the major baseball leagues of the United States. The analyst wondered which approach to the game was best—attempt to assemble the best pitching staff? The best hitters? The best fielders? Go for a speed game? Recruit the best managers and coaching staff? The conclusion was that no single approach guaranteed the most championships. What was obvious, however, was that the most successful teams did, at least, have an approach and, importantly, stuck to it year after year. This lesson seems applicable to process safety regulation. Differences in approach are secondary to having a program and ensuring that it is effectively implemented.

So, to consider the title of the chapter and the various approaches and equipment utilized at chemical process facilities around the world, it is considered that government regulators have made a reasonable contribution to the generation of a framework to keep safety risk reasonably low. Accident history shows, however, that the system falls down when implementation fails to live up to both the spirit and the letter of the "laws" which have been put in place by the regulators.

Over time, hydrocarbon reserves have become more difficult to recover. Much of the remaining reserve base is located in areas that can be environmentally, technically, politically, and culturally challenging. Furthermore, the financial pressure to improve bottom line results using the latest, and sometimes untested, technology becomes ever greater. Under these circumstances, the challenge to keep all participants safe becomes ever more difficult. There is much at stake.

REFERENCES

Abs-group.com. (2016). www.abs-group.com. Accessed November 2016.

Attwood, D., Bates, E., & Price, S. (2013). Regulation and achieving compliance post Macondo. In *LR paper delivered at IMarEST conference, Houston*.

Bureauveritas.co.uk. (2016). www.bureauveritas.co.uk. Accessed November 2016.

Croston, J., Lloyd's Register Geoje. (2016). Personal discussions.

Crowl, D. A., & Louvar, F. (2002). Chemical process safety, fundamentals with applications. *Prentice Hall international series in the physical and chemical engineering sciences*.

Dnvgl.com. (2016). https://www.dnvgl.com. Accessed November 2016.

Ens.dk. (2016). https://ens.dk/en. Accessed November 2016.

International Association of Oil & Gas Producers. (2015). Safety performance indicators, 2014 data.

lr.org. (2016). www.lr.org/en/. Accessed November 2016.

Ness, A. (2016). *Lessons learned from recent process safety incidents*. New York: American Institute of Chemical Engineers, CEP, Aiche, IEG/CEP, Centre for Chemical Process Safety.

Nova Scotia Department of Energy. (2005). Code of practice, liquefied natural gas facilities.

Province of Nova Scotia. (2000). Pipeline act, Chapter 345 of the revised statutes.
Province of Nova Scotia. (2013). Gas plant facility regulations.
Psa.no. (2016). www.psa.no/?lang=en_US. Accessed November 2016.
Russell, D. (2016). *Lloyd's register Aberdeen. Personal discussions and email communication.*
Wikipedia.org/wiki/Piper_Alpha. (2016). https://en.m.wikipedia.org/wiki/Piper_Alpha. Accessed November 2016.
Wikipedia.org/wiki/Texas_City_Refinery_explosion. (2016). https://en.m.wikipedia.org/wiki/Texas_City_Refinery_explosion. Accessed November 2016.

Kolibri 360 GmbH
Ferdinand-Jühlke-Straße 7
99095 Erfurt, Deutschland
produktsicherheit@kolibri360.de